Extension Strategies for Agriculture and Rural Development

THE AUTHOR

Dr. Ashok K. Singh born in a village of Pratapgarh district (U.P.) in the year of 1958, he is presently working as Professor in the Department of Extension Education, Rajendra Agricultural University, Bihar. He received his Ph.D. degree from Magadh University and advanced training from Center for Psychological Study of Social Change and National Development, Department of Psychology, Allahabad University, Allahabad. He is teaching various specialized courses of social science and extension education at the undergraduate and post-graduate level and associated with numerous research projects in the area of agricultural and rural development of National and International importance. He supervised twelve M.Sc. and three Ph.D. Thesis of Extension Education and published more than sixty research papers/ articles in Indian and Foreign journals of repute, apart from various research reports/ monographs and two books (*i*) Technology Assessment and Refinement through Institution Village Linkage Programme, 2005, R.A.U, Pusa (*ii*) Pioneer Research in Extension Education, Scientific Publishers (India), Jodhpur, 2008. Dr. Singh also traveled many countries like Australia, Thailand, Sri Lanka, Philippines, Bangladesh and Nepal for his academic endeavour.

Extension Strategies for Agriculture and Rural Development

– Editor –
ASHOK K. SINGH
Professor, Extension Education
Rajendra Agricultural University
Pusa (Samastipur) – 848 125, Bihar, India

2014
Daya Publishing House®
A Division of
Astral International (P) Ltd.
New Delhi - 110 002

First Imression, 2010 Reprinted, 2014
ISBN 9789351240464

Publisher's note:
Every possible effort has been made to ensure that the information contained in this book is accurate at the time of going to press, and the publisher and author cannot accept responsibility for any errors or omissions, however caused. No responsibility for loss or damage occasioned to any person acting, or refraining from action, as a result of the material in this publication can be accepted by the editor, the publisher or the author. The Publisher is not associated with any product or vendor mentioned in the book. The contents of this work are intended to further general scientific research, understanding and discussion only. Readers should consult with a specialist where appropriate.
Every effort has been made to trace the owners of copyright material used in this book, if any. The author and the publisher will be grateful for any omission brought to their notice for acknowledgement in the future editions of the book.

Published by : **Daya Publishing House®**
A Division of
Astral International Pvt. Ltd.
– ISO 9001:2008 Certified Company –
4760-61/23, Ansari Road, Darya Ganj
New Delhi-110 002
Ph. 011-43549197, 23278134
E-mail: info@astralint.com
Website: www.astralint.com

Laser Typesetting : **Classic Computer Services**, Delhi - 110 035

Printed at : **Replika Press Pvt. Ltd.**

PRINTED IN INDIA

Foreword

Agriculture, which has been the key sector for Indian economy and rural prosperity, is going under serious threats nowadays in terms of dwindling land and water resources, rising the cost of cultivation, indebtedness, depleting nutrients base and production capacity of soils, ecological degradation and climate change. These issues required to be addressed by the policy makers and development planners for the welfare of burgeoning population. The desired inclusive growth rate in the field of agriculture is only possible if technology generation and dissemination system in agriculture would be accelerated by adopting suitable extension strategies. Therefore, there is need for appropriate National Policy for Extension in order to enhance the pace of agricultural and rural development process of the country.

The book "Extension Strategies for Agriculture and Rural Development" is a noble initiative taken by Dr. Ashok K. Singh, Professor of Extension Education who edited the important papers contributed by the esteemed personalities based on their rich experiences and successful endeavour, on the eve of Golden Jubilee Celebration of the extension discipline with the theme of Innovative Extension Strategies for Agricultural Development and Rural Prosperity held at historic campus of Pusa in Bihar.

I am fully confident that this book will provide valuable information pertaining to the several emerging issues related with crop productivity and quality management. Hope, the book will serve as source of reference for policy makers, development managers, scientist, teachers, students and extension workers who are involved in the area of agriculture and rural development.

G. Trivedi

G. Trivedi
Former Vice-Chancellor

Preface

During last fifty years, several alternative approaches of extension system are being attempted under various models *viz.*, KVKs, ATMA, IT enabled services, farmer's empowerment, contractual arrangements, entrepreneurs/NGO driven extension system for enhancing the crop productivity and ultimate raising the socio-economic condition of the farming community. But, the astronomical rise in cost of cultivation, indebtedness and causes of farmer's suicides compounded with dwindling land, labour and water resources, depleting nutrient base and soil fertility, lack of marketing avenues, escalating ecological degradation and ravaging climate change pose a serious threat to Indian Agriculture and Rural Prosperity. These emerging challenges call for introspection and revisit the whole concept, philosophy and methodologies of extension system under the changing global scenario. In this context, the discipline of extension education has crucial role to play in empowering and preparing the farmers and other stakeholders to overcome on these challenges by way of developing a synergistic and coherent pathways for enhancing crop productivity and providing livelihood opportunities to rural masses.

Since, the discipline has travelled a long way (more than fifty years of its existence) and it has always been involved in development of National Policy for agricultural technology development and its dissemination hence, as a part of Golden Jubilee Celebration of Extension Education, a National Seminar on the theme of "Innovative Extension Strategies for Agricultural Development and Rural Prosperity" was organized in the historic campus of Pusa (Samastipur), Bihar from December 18-20, 2008, in which more than two hundred eminent agricultural/social scientists, policy makers and stake holders were participated. The esteemed participants were also contributed their rich experiences in form of lead paper/views and research findings which were borne out from their systematic observations and interaction.

In order to brought these interactions to the glaxy of agricultural and social scientists involved in agricultural research and development a initiative has been taken to document the salient contribution of this event in the form of book. The book entitled "*Extension Strategies for Agriculture and Rural Development*" is a compendium of selected papers discussed on this occasion and it is divided into five sub- sections (*i*) Policy initiative in extension education (*ii*) Market- led extension and commercial farming (*iii*) ICT application and methodological advances in agriculture (iv) Entrepreneurship development and innovative approaches in extension and (v) Gender mainstreaming in farming system. The author wishes to express his sincere gratitude to all the esteemed participants and guest who contributed substantially on this occasion according to their own way. In this context, I have no words to express my deep sence of regards to Dr. Baldeo Singh, Jt.Director (Extension), IARI and Secretary, ISEE who kindly agreed and encouraged us to go ahead with this event at Pusa. I am also extremely grateful to Dr. G.Trivedi, Former Vice-Chancellor, RAU, Pusa for his encouragement and heartly support in organizing of this celebration. I am also thankful to other dignitaries like Dr. C. Prasad, Former DDG, ICAR and President VARDAN, Dr. R.P.Singh, Former Vice-Chancellor, MPUAT, Udaipur, Dr. R.P.Singh, Ex-Training Speciliatist, IARI and Dr. B.P. Sinha Ex-head (Agril.Ext.), IARI apart from our Hon'ble Vice-Chancellor Dr.M.L. Choudhary for their support and valuable contribution.

The management of the celebration required involvement of many stalwards and agencies of our university and outside of this institution; I am extremely indebted to all of them. My thanks are also due to my wife and both of sons Manas and Annu for their delicate affection. Last but not least, I admire all the learned delegates/ participants, scientist and workers who jointly made this National Seminar a grand success. At this juncture, I cannot forget the enthusiasm and support of Sri Anil Mittal, Daya Publishing House, New Delhi who kindly agreed to publish the selected deliberation of the seminar in form of this book.

Hope fully, this publication will prove useful for those researchers, social scientists, extension workers and policy makers who are engaged in advancement of agriculture and upliftment of farming community. The book will also serve as a guide to the policy planner and extension professionals who are interested in the area of agricultural development and rural prosperity.

Ashok K. Singh
Professor
Extension Education

Contents

Section V: Gender Mainstreaming in Farming System

Introduction

☆ ***Ashok K. Singh***[1]

The agriculture still continue to be known as the engine of the national prosperity and development. The rapid and balanced growth of agriculture is thus essential not only to achieve self-reliance at national level but also for household food security and to bring about equity in distribution of income and wealth resulting in rapid reduction in poverty among the rural poor. Although, the agriculture and its allied activities contribute nearly 25 per cent to the GDP of the country yet it faces a great challenge under the changing climate and economic scenario. The agricultural extension system, which was primarly, meant to transfer the new techniques, new knowledge and new technology from the research sector to the farmers has also undergone in considerable changes over several decades. Since, India has entered in the global market and signed WTO, hence farming community residing in rural villages need to be supported with information, knowledge and the skill to adopt improved technologies that would enhance productivity, employment opportunities and sustainability. Therefore, the ambit of agriculture extension has remarkably widened in the course of time and the onus of managing the complex affairs rests with agricultural scientist and extension specialist working with the farmers as per the requirement of the market. So, now it is imperative to design efficient, and location specific extension modules that would enable farmers and other stakeholders to reorient their action and thereby increase the profitability of their operation.

The state of Bihar, which is known as the motherland of agricultural research and education in the country. The state has also got the privilege to start Postgraduate

---

1 Professor and Organizing Secretary of the National Seminar–2008, Department of Extension Education, RAU, Pusa, Bihar.

Teaching and Research in Extension Education, for the first time in India at Bihar Agricultural College, Sabour (Bhagalpur district) now, a constituent unit of Rajendra Agricultural University, Pusa (Samastipur), Bihar in the year 1955. The discipline of extension education gradually recognized as one of the three main functions of an Agricultural University System, the other two being teaching and research, and it spread in other parts of the country. Since, inception of the programme leading to Master's Degree, submission of thesis based on systematic research pertaining to different facet of Agriculture and Rural Development was made an essential requirement for the award of degree.The research outcome in form of M.Sc. Thesis started to come out in the year 1957. Therefore, it is obvious that over a span of five decade Extension Education has played pioneer role for our agricultural and rural development process.

Although, half-century appraisal of extension efforts demonstrated that it paid rich dividends to the farmers in development, refinement and diffusion of technologies yet country is presently striving hard for inclusive growth rate in agricultural for the welfare of burgeoning population. In view of its historical importance, therefore it was thought appropriate and most proper to hold Golden Jubilee Celebration of the extension education discipline. Also, in a very humble way, a request was submitted to Indian Society of Extension Education, New Delhi to collaborate with its National Seminar on the occasion during the year 2007. Due to some unavoidable circumstances, the proposal was kept pending and in the year of 2008, the Indian Society of Extension Education, New Delhi is pleased to share the dais jointly with the Department of Extension Education, Rajendra Agricultural University, Pusa on the eve of Golden Jubilee Celebration of Extension Education discipline at the historic place of Pusa (Samastipur), in Bihar. Thus, it was really a opportune moment when ISEE and R.A.U, Pusa joined together in Golden Jubilee Celebration on this occasion and a National Seminar was hosted on the theme of "Innovative Extension Strategies for Agricultural Development and Rural Prosperity" during the December 18-20,2008 in order to provide suitable plateform for extension professionals of country to meet, interact together and share their rich experiences for setting new millennium goal of extension.

The theme of the National Seminar has specific relevance as the state of Bihar has declared the year of 2008 as "Agriculture Year" and has also initiated many innovative efforts such as, Kisan Samman Yojna, Kisan Maha Panchayat and Vagyanike Gaon Ki Ore etc.for the welfare and upliftment of the farming community.

The National Seminar was inaugurated by Dr. M.L. Choudhary, Hon'ble Vice-Chancellor of Rajendra Agricultural University, Pusa while Dr.C. Prasad, Former, DDG (Extension), ICAR presided over the function, Dr. R.P. Singh, Former Vice-Chancellor of MPUAT, Udaipur, Dr.G. Trivedi, Former Vice-Chancellor of RAU, Pusa, Dr. R.P. Singh, Ex. Training Specialist, IARI, New Delhi, Dr. Baldeo Singh, Jt. Director and Secretary, Indian Society of Extension Education, Dr. Madan Singh, DEE, RAU Pusa and other dignitaries were present on the dais. The inaugural function of seminar was formally started with the "Saraswati Bandana" followed by floral welcome to all the esteemed guests present on the occasion. The sacred lamp was lighted by the

Chief Guest of this occasion with the help of dignitaries of the dais. Dr. Madan Singh, Director Extension Education delivered the welcome address. The Hon'ble Vice-Chancellor Dr. M.L. Choudhary released the Souvenir of the National Seminar published on this occasion.

Dr. Baldeo Singh presented a brief outline about the Indian Society of Extension Education and relevance of the National Seminar on the occasion of Golden Jubilee Celebration. Dr. R.P. Singh, former Vice-Chancellor MPUAT, Udaipur has stressed for the need of market led extension and demand driven extension approach for enhancing the profitability of the farmers. Dr. G. Trivedi highlighted about the need of closer linkages and interface with the farmers and extension functionaries while Dr. R.P.Singh, Ex-Training Specialist emphasized the need for appropriate training methodology in the complex diverse situation of farming. In his Presidential Address Dr.C. Prasad, former DDG and President, VARDAN stressed the need for paradigm shift in extension approaches by way of incorporating more social science dimension. Hon'ble Dr. M.L. Choudhary, Vice-Chancellor, RAU, Pusa in his inaugural address emphasized the need for suitable extension strategies and tools for promotion of agri-business enterprises. He further said that extension should be viewed in a broader perspective to face the challenges and urged to the extension scientist for achieving the target of agricultural growth fixed by XIth Five Year Plan. The inaugural function was concluded with the vote of thanks proposed by Dr. K.R. Maurya, Dean, Agriculture, RAU, Pusa. The various technical sessions designed for the National Seminar were commenced after the lead paper presentation session, during the period of National Seminar.

Finally, a group of eminent extension scientists under the Chairmanship of Dr. G. Trivedi finalized various important recommendations on the basis of extensive discussion before the valedictory function of the seminar. The valedictory function of the Golden Jubilee Celebration and National Seminar was held on 20th December 2008. It started with the Welcome Song and floral welcome to Hon'ble Minister of Agriculture, Government of Bihar, Sri Nagmani Ji and Dr. V.V. Sadamate, Advisor (Agriculture), Planning Commission, Govt. of India who was the Guest of Honour on this auspicious occasion. After welcome address, the Hon'ble Vice- Chancellor Dr. M.L. Choudhary briefly discussed about the priority areas of extension in the context of the salient recommendations that were borne out from the deliberation of the National Seminar. The other important event took place during the valedictory session was the Award Ceremony which has been convened by Dr. Baldeo Singh, Secretary ISEE. The different awards constituted for the ISEE were distributed to the eligible extension scientist/professionals by the hands of Hon'ble Minister of Agriculture, Govt. of Bihar. On the occasion Dr. V.V. Sadamate in his keynote address spoken about the policy issues and framework under which extension activities has to be carried out for agricultural and rural development. He stressed for the need of appropriate extension reforms for revamping the present extension system. Finally, Hon'ble Minister of Agriculture, Govt. of Bihar Sri Nagmani Ji concluded the National Seminar with the urge for greater participation of farm community in the process of agricultural development. Dr. Ashok K. Singh, Professor and Organizing Secretary of the National Seminar–2008 convened the entire proceeding of inaugural and valedictory functions of this occasion.

Recommendation

The salient recommendations which were emerged out of with the systematic deliberation during the National Seminar are given here as:

1. There is need for establishment of a Council for Transfer of Technology in Agriculture (COTTA) at the Centre by integrating the Directorate of Extension of the MOA and the Division of Extension of the ICAR.
2. There is need to develop a "Cadre" of Extension Education Services comprising both extension personnel as well as the Subject Matter Specialist.
3. The extension agencies, both public and private, should assess the information needs of agribusiness-oriented farmers and organize training for them.
4. The public extension service needs to be strengthened as per the demand driven extension. It should be reoriented as per the requirements of agri-business enterprises in order to maintain the quality and standard for export potential.
5. The private extension should be geared up to supplement the efforts of the public extension system towards the promotion of agri-business enterprises.
6. The different farmer's organization and self help groups, specially the women's SHGs should be encouraged to take up agri-business enterprises and its members should be trained with the desired knowledge and skills required for successful functioning of such enterprises.
7. The communication strategies have to be translated into practice at the field level for creating impact visibility and comprehensive field extension programmes.
8. Agri-business enterprise should be promoted by exploiting the agro-ecological diversity of the country. In this context, efforts should be made to promote the eco-friendly inputs of agriculture.
9. As several fruits and vegetables are available with us which has the export potential hence, there is need to develop a close linkages with farmers-agro-industries-consumer network. Also, it is required to produce export quality in order to capture the world market in WTO era.
10. As the high quality of agriculture produces fetch attractive prices in the metropolitan cities, the extension agencies should promote the Hi-Tech cultivation for sending the produce to the needy market in order to have maximum profitability out of them.
11. The small and marginal farming groups need to be encouraged by the development agencies to undertake farming system approach in order to make the farming more viable and generate additional self-employment. In this context, integrated farming system approach can become more attractive and sustainable for their livelihood.

12. The cooperative societies should be strengthened. They can work for promotion of different entrepreneurship in the area of soil testing, vermi-composting, dairy, and other agro-based allied activities.
13. The self-employment and entrepreneurship development are necessary to face the current changing agricultural crisis in rural areas; therefore area-region and location based planning may be designed for integrated development.
14. The village based market led agriculture extension approach for transfer of technology being carried out by IARI may be promoted and validated by other organizations and agencies.
15. The agricultural resource banks (seed, manure, fertilizer, energy, machineries etc.) may be created at the block and panchayat level for easy and timely availability of resource to the end users.
16. The farmer's participatory technology development approach should be encouraged in research and extension programme for developing need based and demand driven technology in order to enhance productivity and improve livelihood.
17. The FEIPC (Farmer's Entrepreneurship Institute and Public Chain) may be established for entrepreneurship development in the rural areas.
18. For holistic and sustainable development of agriculture, there is need for more operational research programmes in the rainfed/watershed, submerged, pond and draughts areas. Under these programmes farmers of the local areas should be involved in planning, implementation, monitoring and evaluation of the programmes.
19. There is need for strengthening of Kisan Call Center (KCC) and regular updating of agricultural database is needed for effective delivery mechanism.
20. Strengthening of Village Resource Center (VRC) is required as it serves single window delivery system for providing technology inputs and services to the rural population at the village level.
21. There is need for clear-cut application of the terminologies used in extension as; extension education; extension services, extension work and transfer of technology. The different organizations and institutions may work for which they are meant.
22. The SMS (Extension) in each KVK must take a micro-level empirical document about the institutional linkage. They must capture the process, outcome and bottlenecks of each technology, which is available in their location.
23. The successful experiences of contract farming by Pepsico in case of tomato growers of Punjab gives sufficient reason to replicate it for other crops like spices and potato.

24. The community radios may be popularized in rural areas whereby extension organizations can facilitate information sharing, capacity building and developmental communication. The extension policy may incorporate community radios as a viable extension method.
25. The knowledge and skill of multimedia tools and software's should be imparted to the entrepreneurs. The increasing use of ICT may be useful for providing Internet connectivity in the village/clusters of villages to the Agri-business entrepreneurs.

List of Contributors

1. **Dr. C. Prasad**

   Former Deputy Director General, Agricultural Extension (ICAR) and President, VARDAN, EB-106, Maya Enclave, New Delhi-64

2. **Dr. P.N. Jha**

   Former Vice-Chancellor and Member of Bihar Kishan Ayog, Pant Bhavan, Bailey Road, Patna-1

3. **Dr. B.P. Sinha**

   Former Head, Division of AgriculturalExtension (IARI) and Consultant, TIFAC, Paliganj Project, Paliganj, Patna (Bihar)

4. **Dr. Baldeo Singh**

   Joint director (Agricultural Extension), IARI, Pusa, New Delhi-12

5. **Dr. Dipak De**

   Professor of Agricultural Extension, Department of Agricultural Extension, IAS, BHU, Varanasi-221005

6. **Dr. Basavaprabhu Jirli**

   Lecture, Agricultural Extension, Department of Agricultural Extension, IAS, BHU, Varanasi-221005

7. **Dr. R.N. Padaria**

   Senior Scientist, Agricultural Extension, Division of Agricultural Extension, IARI, Pusa, New Delhi-12

8. **Dr. A.K. Singh**
Zonal Coordinator, Zonal Coordinating Unit (ZCU), ICAR, I.I.P.R., Kanpur-24 (U.P.)

9. **Dr. Shantanu Kumar**
Senior Scientist, Indian Institute of Pulses Research (ICAR), Kanpur-24 (U.P.)

10. **Dr. Uma Shah**
Scientist, Indian Institute of Pulses Research (ICAR), Kanpur-24 (U.P.)

11. **Dr. S.K. Singh**
Scientist, Indian Institute of Pulses Research (ICAR), Kanpur-24 (U.P.)

12. **Dr. D.K. Sujan**
Professor of Agricultural Extension, Department of Agricultural Extension, IAS, BHU, Varanasi-221005

13. **Dr. K. Ghadei**
Lecture, Agricultural Extension, Department of Agricultural Extension, IAS, BHU, Varanasi-221005

14. **Dr. S. Panigrahi**
Scholar in Department of Agricultural Extension, IAS, BHU Varanasi-221005

15. **Dr. P. Adhiguru**
Senior Scientist, National Centre for Agricultural Eco. and Policy Research (INCAP), D.P.S. Marg, Pusa, New Delhi-110012

16. **Prof. Ram Pratap Singh**
Former Vice-Chancellor, MPUAT, Udaipur, B 5/5 Vinay Khand, Gomti Nagar, Lucknow-226010 (U.P)

17. **Dr. C.S. Arneja**
Professor of Extension Education, Department of Extension Education, PAU, Ludhiana (Punjab)

18. **Ms. Shabana Yasmin**
Research Scholar, Department of Extension Education PAU, Ludhiana (Punjab)

19. **Dr. Virendra Kumar Yadav**
Scientist, Directorate of Maize Research, Pusa Campus, IARI, New Delhi-110012

20. **Sri C.Y. Manikanhaiya**
Ph.D Scholar, Division of Agricultural Extension, IARI, Pusa Campus, New Delhi-110012

21. **Dr. Samir Ranjan Dash**
KVK, Kandhamal, G. Udayagiri, O.U.A.&T., Bhubaneshwar, Orissa

22. **Prof. S. Satapathy**
Director, Amity Business School, Bhubaneshwar, Orissa

23. **Dr. Sabita Mishra**
Senior Scientist (A.E.), N.R.C. W.S., Bhubaneshwar, Orrisa

24. **Dr. S.N. Nigam**
Scientist, ICRISAT, Hyderabad (A.P)

25. **Prof. P.G. Khalache**
Head, Department of Extension Education, MPKV, Rahuri, Ahmednagar-413702

26. **Dr. H.P Sonawane**
Assistant Professor, Department of Extension Education, MPKV, Rahuri, Ahmednagar-413702

27. **Dr. J.H. Gaikwad**
Assistant Professor, Department of Extension Education, MPKV, Rahuri, Ahmednagar-413702

28. **Prof. M.C Varshneya**
Vice-Chancellor, Anand Agricultural University, Anand (Gujrat)

29. **Dr. R.S. Parmar**
Principal, Extension Education Institute, Anand Agricultural University, Anand (Gujrat)

30. **Dr. Arun Patel**
Project Coordinator (I.T), Anand Agricultural University, Anand (Gujrat)

31. **Dr. Sarita Anand**
Head, Department of Developmental Communication and Extension, Lady Irwin College, Sikandra Road, New-Delhi-1

32. **Ms. Suchi Gaur**
C/O Dr. Sarita Anand, Head, Department of Developmental Communication and Extension, Lady Irwin College, Sikandra Road, New-Delhi-1

33. **Dr. Prabhjot Kaur**
Professor of Extension Education, Department of Extension Education, PAU, Ludhiana-141004 (Punjab)

34. **Sri Vijay Bharat**
Founder Cum Managing Director, Mobile Agricultural School and Services, Ranchi (Jharkhand)

35. **Dr. Dharminder Singh**
Assistant Professor (Extension Education), KVK Moga, Moga, Punjab

36. Prof. D.S. Dhillon

Professor and head, Department of Extension Education, PAU, Ludhiana – 141004 (Punjab)

37. Ms. Sonika Gupta

Research Scholar, Department of Agricultural Extension, R.B.S. College, Bichpuri, Agra (U.P)

38. Dr. J.P. Singh

Head, Department of Agricultural Extension, R.B.S. College, Bichpuri, Agra (U.P)

39. Dr. R.P. Singh

Associate Professor, Department of Agricultural Communication and Rural Development, SVBP University of Agricultural and Technology, Modipuram, Meerut (U.P)

40. Dr. Pradeep Kumar

Department of Extension Education, CCS, H.A.U., Hisar (Haryana)

41. Dr. Nikulsinh M. Chouhan

Programme Coordinator, KVK, Mangalbharti – 391 125, Ta-Sankheda, Vadodara (Gujrat)

42. Dr. A. Saha

Lecturer, Department of Agricultural Extension, Agricultural Ecoonomics and Agricultural Statistics, Palli Sikcha Bhawan, Institute of Agriculture, Visva Bharti, Srineketan-731236 (West Bengal)

43. Dr. Swapan B. Mukhopodhya

Professor and Head, Department of Agricultural Extension, Agricultural Ecoonomics and Agricultural Statistics, Palli Sikcha Bhawan, Institute of Agriculture, Visva Bharti, Srineketan-731236 (West Bengal)

44. Prof D. Das Gupta

Professor, Department of Agricultural Extension, Agricultural Ecoonomics and Agricultural Statistics, Palli Sikcha Bhawan, Institute of Agriculture, Visva Bharti, Srineketan-731236 (West Bengal)

45. Dr. N.B. Jadava

SMS, Krishi Vigyan Kendra, JAU, Jamnagar (Gujrat)

46. Dr. P.R. Kamani

Professor, Department of Extension Education, JAU, Junagadh (Gujrat)

47. Dr. P.S. Gorfad

Agricultural Officer, KVK, Millet Research Stations, JAU, Jamnagar (Gujrat)

48. Dr. G.L. Bagdi

Senior Scientist (Agricultural Extension), CSWCR and T.R.C. VALSAD, Anand (Gujrat)

49. Dr. Anupama Shah

Professor and Dean, Department of Extension and Communication, Faculty of family and Community Science, Mahraja Shivaji Rao University of Baroda, Vadodara

50. Dr. R.P. Singh

Associate Director Extension, KVK, Hastinapur, Meerut, U.P

51. Dr. Amit Kumar

SMS, Extension, KVK Hastinapur, Meerut, U.P

52. Dr. V.S. Katiyar

Scientist, Central Soil and Water Conservation Research and Training institute Research Centre, Datia, M.P.

53. Dr. S.K. Meti

Professor, Department of Agricultural Extension Education, College of Agricultural, Raichur, UAS, Dharwad, Karnataka

54. Dr. Arvind Saxena

SMS, KVK Ujjain, RVS, KVK Ujjain (M.P)

55. Dr. A.K. Dixit

Programme Coordinator, SMS, KVK Ujjain, RVS, KVK Ujjain (M.P)

56. Dr. R.B. Prasad

Head, (Agricultural Extension), ARI., Mithapur, Lohianagar, Patna

57. Dr. Meera Singh

Dean, College of Home Science, Rajendra Agricultural University, Bihar, Pusa (Samastipur)-848 125

58. Dr. Sudhir Das

Programme Coordinator, KVK, Sheohar, RAU, Pusa, Seohar, Bihar

59. Dr. Barsati Lal

Scientist (S.S) Agricultural Extension, CPRS, Near Phulwari Sarif Jail, BVC, Patna, Bihar

60. Dr. N.K. Pandey

Senior Scientist and Head, Division of Social Sciences, CPRI, Simla (H.P.)

61. Dr. Om Prakash

Sr. Scientist, CSWCRTI, Research Centre, Datia, Datia - 475 661 (M.P.)

62. Dr. Y.V.R. Reddy

Central Research Institute for Dry land Agriculture, Santosh Nagar, Hyderabad, A.P.

63. Dr. R.N. Sharma

Scientist, Krishi Vigyan Kendra, Janjger Champa, IGAU, Raipur-492006

64. Dr. G. Trivedi

Former Vice-Chancellor, Vill+PO-Mutlupur, Via-Pear Muzaffarpur, Bihar

65. Dr. A.P. Dwivedi

SMS (Agronomy), KVK, P.G. College, Ghazipur, U.P

66. Dr. B.P. Singh

Reader, Department of Agronomy, P.G. College, Ghazipur, Ghazipur, U.P

67. Sri B.K. Mandal

Ph.D Scholar, Department of Agricultural Extension, IAS, BHU, Varanasi-221005

68. Dr. U.S. Gautam

Zonal Coordinator, Zonal Coordinating Unit VII, ICAR, Jabalpur-482004, M.P.

69. Dr. D.K. Paliwal

SMS (Agronomy), KVK Kasturbagram, Indore-452020, M.P.

70. Dr. R.S. Suman

Scientist, IARI, Regional Station, Katrain-125129, Kullu Valley, H.P.

71. Dr. M.M. Patel

Senior Scientist, Department of Extension Education, College of Agricultural, Indore, M.P

72. Dr. Seema Naberia

Assistant Professor (Agricultural Extension), College of Agriculture, Ganjbasoda, Vidisha-M.P

73. Dr. R.F. Thakor

Programme Coordinator, Gujrat Vidyapith, Krishi Vigyan Kendra, Valsad, Gujrat

74. Dr. N.B. Jadav

SMS (Extension Education), KVK Jamnagar, JAU, Jamnagar-6, Gujarat

75. Dr. S.M. Kenchanagoudra

Department of Agricultural Extension, College of Agricultural, UAS, Dharwad, Karnataka

76. Dr. Dayaram

Associate Professor (Microbiology), FBS and H, RAU, Pusa (Samastipur)-848125

77. Dr. Ajay Kr. Singh

Zonal Coordinator, Zonal Coordinating Unit Zone-II, ICAR, Salt lake, Kolkata

78. Dr. Reeta Singh

SMS (Home Science Extension), KVK, Bikramganj, Rohtas, Bihar

79. Dr. Madan Singh

Director, Extension Education, Rajendra Agricultural University, Pusa (Samastipur)-848125

80. Dr. Shivnath Das

Jr. Scientist cum-Asstt. Professor, IRS, Bikramganj, Rohtas, Bihar

81. Dr. Maya Kumari

SMS (Home Science), KVK, Pakur, Maheshpur, Jharkhand

Section I:
POLICY INITIATIVE IN EXTENSION EDUCATION

Chapter 1

Major Reforms Needed for a Dynamic Agricultural Extension System

☆ ***C. Prasad***[1]

Introduction

After Training and Visit extension Project of the World Bank, since 1985 the extension education system in agriculture has come to a cross-road–the agricultural extension system of the Ministry of Agriculture, has been entrusted with major responsibility of Transfer of Technology (TOT) in the country. The ICAR/SAUs extension system plays the first-line extension role: (*i*) bringing latest farm technologies promptly to the notice of (*a*) limited progressive farmers of the states/districts; and (*b*) all the state extension functionaries. While the ICAR/SAUs extension system has continued mainly through its KVK project, but the state extension system was weakened due to resources of T&V system being withdrawn in fact, a vacuum has been created which needs immediate attention and reforms.

As a matter of fact, our extension system since 1952–the era of planned development, continued with the foreign supports–C.D., AES, IADP, NYVP and T&V, so much so that even the name of national extension system could not be stabilized. Lately, it was proposed to the government at the highest level as to why we do not have our own extension system–the name does not cost anything, but gives stability and sustainability of the system. If not anything else, it may be conveniently called,

---

1 Former Deputy Director General (Agriculture Extension), ICAR & President, VARDAN, EB-106, Maya Enclave, New Delhi–64.

the "National Agricultural Extension System (NAES)." Perhaps, this has been accepted by the Ministry of Agriculture, Govt. of India in a subtile manner. The USA extension system was named as the "Cooperative Extension Work" since 1914 and still it is the same. We adopted foreign projects with their names and resources and forgot about our own.

The vacuum created by the T&V system, could not, perhaps, be perceived by the government which functions on its administrative strength, not on technical and intricate educational system. Extension is an informal and out of school education– a more difficult concept; Ministries, by and larges, continued with the agricultural extension more offering agricultural extension services than education.

Now the ATMA (Agricultural Technology Management Agency) at the district level and SAMETI (State Agricultural Management and Extension Training Institute) at the state level are being prompted only in half the districts in the country (252 out of 600).

The functional linkages between the ICAR/SAUs system and the state extension mechanism remained always a new subject, for linkages and convergence occur when more than one system have (*i*) common goals and interests; and (*ii*) they are equally committed to the goals. Do such conditions exist anywhere in the country?

Then there are gaps between a system and its functioning. How much are we committed to that? Almost we are reaching a point of no return; God is helping us as the Russian Expert said while leaving the country. He never believed in God; now he does?

Conceptual Clarity

Defining Extension

The basic discipline for TOT is Extension Education, which is applied to agriculture (Agriculture Extension, Veterinary Extension or Industrial Extension etc.). Education is the keyword ? all TOT efforts, no matter who does it, are educational in nature/character–education, training, demonstrations, fairs, exhibitions, use of mass-media, information technology, all are ment for education of the farmers which may lead to acceptance of recommended technologies and then finally the adoption by the farmers and diffusion in the farming communities. Therefore, the saying that the ICAR and SAUs do the extension education work and the State Government does the extension service is simply misnomer a misunderstanding. Both systems are only expected to do extension education work-imparting knowledge and skills to the farmers through various extension programmes, activities, methods, designs and strategies.

Agricultural extension services are inputs support (seeds, fertilizers, chemicals), pricing policy, crop insurance, credit policy, land reforms, state farms management, soil and water testing, etc, which are the responsibilities of the Governments. In reality, the extension services they are doing; they are hardly doing any extension education work. The extension projects, training institutes, etc. have mostly diverted to the SAUs, ICAR Institutes, and KVKs. One can very well examine the negligible state extension presence in the farmers' fields.

This has serious implications for the Extension Education System in the country. The question being raised here is should the State Governments take up the Extension Education and if they do, what approach and changes are necessary?

Agriculture is defined as all the land-based activities including crops, horticulture, animal husbandry, fisheries, and other allied areas. But in actual practice, agriculture is taken for all the crops; Agriculture, Horticulture, Animal husbandry and fisheries are separate departments and by and large have separate extension systems/sub-systems and programmes. Thus, we have to take it as they are-four extension systems or sub-systems in practice? The problem is how they can be functionally linked and integrated, for the farmer is the only one who does all these-crops, animal husbandry, horticulture or fisheries. The system has to be visualized which can integrate all these sub-components in the best interest of integrated and well coordinated extension activities leading the effectiveness and efficiency.

Resources Distribution

As per policy in the ICAR/SAU System, 10 per cent of the total budget and 20 per cent time of the scientists are expected to be spent on TOT. This is a reasonable proposition but are they happening? If we add to this, the extension education role of the Ministry of Agriculture/State Governments (exclusive of services role), the budget, staff etc. should not be more than one-third of the total agricultural budget? Are we investing even that in the extension education programmes?

Extension Functionaries

There are two categories of staff who are directly involved in Extension Education Programmes like two sides of the coin: (*i*) Extension Education Professionals, and extension workers and (*ii*) the Subject Matter Specialists SMSs (Agronomists, Horticulturists, Entomologists, Animal Scientists etc.). This dimension of TOT has to be understood by both the professionals' ? one cannot do without the other. At present, the understanding is that TOT means the job and responsibilities of the extension professionals only or if at all, the SMSs are the light support to extension programmes without owning it as their own programmes or responsibilities. This dichotomy has to be cleared so that both feel responsible to the extension programmes and they share the credits accruing out of these activities. Staffing of the TOT System has to plan accordingly.

Existing Extension System

There are two sub-ordinate extension systems: (*i*) the main extension system represented by the Directorate of Extension, Ministry of Agriculture (MOA), Govt. of India; and (*ii*) the Division of Extension of the ICAR representing the first-line extension system, mostly through KVKs and IVLPs. The MOA, by and large, worked with foreign Projects and Agencies from 1952 till 1985–the termination of T&V extension Project. After that they experimented with ATMA under the NATP of the World Bank implemented by the ICAR. Finding them in order, they are gradually promoting it with SAMETI at the state level. They have covered, as stated earlier, half the districts in the country–252 out of nearly 600 districts. The ICAR continued with their system with KVKs now nearly 575.

The organizational arrangements and the programmes and resources are one thing and how they are governed and managed is another thing. We are poor managers said late Hon'ble Prime Minister of India as explained earlier. The Ministry always worked as the agricultural extension service providers rather than stress on extension education–the heart of TOT. As is reported nearly half the KVKs of the North India are suffering because of paucity of funds–the KVKs' funds have been diverted for other purposes by the SAUs where most of the KVKs have been allocated. The ICAR has not been able to control that.

As a policy of the ICAR, the ICAR Research Institutes and the SAUs are to invest at least 10 per cent of their budgets and 20 per cent time of the farm scientists for the TOT. Have they been happening? The scarcities of funds have affected TOT badly, for that is not the priority of the government or the ICAR, or are they?

The Zonal Research Stations (ZRSs) were launched in the country (120) with the international support, which had a component of extension staff and a farmers' hostel. This never worked.

This is why; we are at the crossroad. The annual growth rate of agriculture is not moving up from around 2.5 per cent vis-à-vis 4 per cent being the target. We imported Wheat from Australia, a blow to our Green Revolution. In future this is going to be aggravated unless governments get alive to these challenges. On every critical occasion, the Planning Commission has increased the budget of the ICAR/Ministry but where they have been spent–mostly in research and education or services sectors? You know as extension professionals where we are heading to–the bridge of extension has to be much improved and strengthened. When this big issue was brought to the notice of Hon'ble Prime Minister of India, he took it in the right spirit and at his instance; the governments have been advised to strengthen Agricultural Extension System in the 11th Five Year Plan. God willing let us hope for the best.

Skewed Development

After over 60 years of Independence, we are still struggling on poverty and hunger for at least 25 per cent of the rural population; they are deprived, vulnerable and handicapped. The salvation has to come from agriculture and cottage industries including agro-industries. So far there has been skewed development in the country in terms of economic gains and consequently the social side of development has not only been ignored, but also almost forgotten. This has led to "mental poverty" among well-to-do people–greed for wealth–more wealth; power–more power and absolute power- leading to corruptions and chaos in the society at large. Leadership, which should drive people to prosperity, peace and salvation, is promoting criminalism to remain king among the poor, dalits and minorities–a brutally negative approach. Such trend has to be arrested by bringing the positive social values in people in general and development professionals and social workers in particular. A balancing act has to be done between economic and social development very seriously. Agriculture permeates all sections of the rural population as farmers, labourers or agri-traders. In this sector, social dimensions at all levels have to be promoted and enriched. Agricultural scientists as well as social scientists allied to agriculture must realize this reality to balance the social and economic sides of the inputs both in research and development. A conducive environment for agricultural R&D has to be created through appropriate technology generation as well as their accelerated transfer to the end-users–the farmers.

Agriculture remained the main occupation of the rural people in developing countries, which could hardly prove beneficial poorer sections of the society. Manual farming converted majority of rural masses into labourers especially women. Infrastructural development like roads, electricity, water and irrigation, transport system, housing, etc. were not given adequate attention. Less amount was spent on health and education. Consequently, the skewed development created a wide gap between urban and rural societies. Urban people availed better facilities, whereas rural people rolled in ignorance, poverty, malnutrition, ill-health, and unemployment.

Challenges for agricultural R&D today remain primarily in food insecured countries. They include growing population, the protracted debate on biotech crops, markets and non-tariff barriers, environmental issues and several policy issues related to institutional and organizational management.

Attempts to increase agricultural yields are no longer enough because the benefits for the poor are getting weaker with globalization and agriculture is becoming less rewarding to the livelihood of the poor; the gap between the rich and the poor has widened even more. With climatic adversities, poor rural infrastructure, risks of finance instability, conflicts, urban migration, lack of pro-poor policies, health and nutritional crisis, natural resource degradation, social inequality, and changing roles

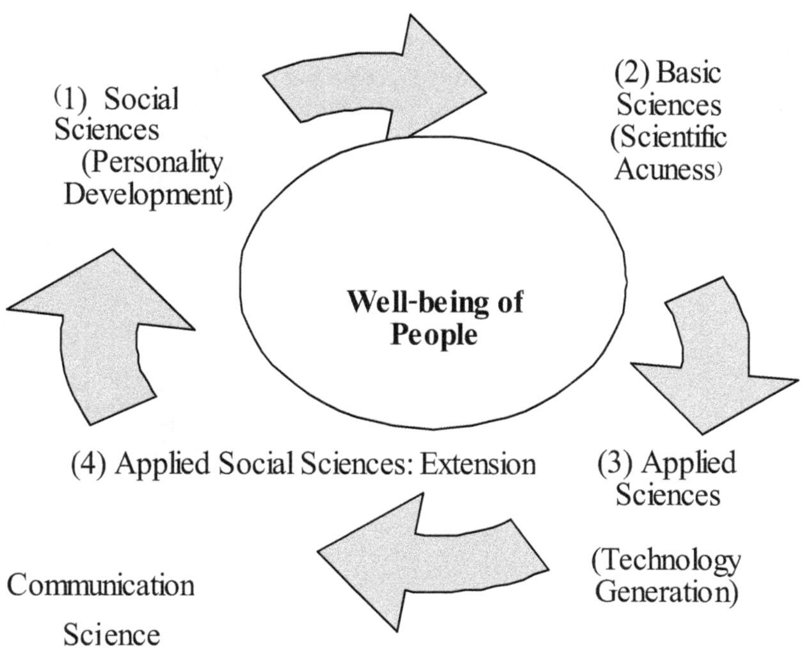

Figure 1.1: Theorizing Social Sciences–Social Sciences First

of governments, the challenges of agricultural R&D is getting bigger and harder, which needs technical backup of social scientists for analyzing these complexities.

Nobel laureate Amartya Sen has so definitively demonstrated that empowerment leads to entitlements and entitlements lead to enrichment. The skewed distribution of the fruits of the impressive progress the country has made, requires immediate course correction because economic progress in itself cannot be our national goal.

Major Policy Issues

In spite of our struggling agriculture, deeper thoughts are not going into the crucial and critical policy issues and serious concerns have not been shown to the alarm of importing Wheat from Australia. And on the critical analysis of the facts, agricultural extension could have emerged as the most critical factor to be reckoned with for the livelihood security of the people. Agricultural extension is the bridge between the Farm Science and Technology on the one hand and the scientific farming and the farmers on the other. This bridge has been weakened in absence of the foreign projects as the Ministry of Agriculture was used to it. This vacuum has not yet been perceived by the Planners, Managers, and the Farm Scientists. Agricultural extension has been, somehow, a neglected discipline, but without strong extension there is no salvation for food and nutritional security for the country. These basic and burning facts must be noted by our decision-makers in agriculture, but as I have expressed earlier, the present extension system has no 'teeth and voice'. Therefore, we are bound to suffer, unless some miracles happen.

The Planning Commission, GOI, normally have a long exercise before the next Five Years Plan. The extension subject is always chaired by one who has been an Administrator/Manager, but not an extension professional. Extension people are involved at the lower levels as Members or Chairman of the sub-groups. Then highest level Committee of the Planning Commission on extension is always headed by some one who has the highest stature, but without much background of field work and extension; there not even a Member comes from the extension discipline. And therefore, look at their recommendations Plan after Plan–nothing substantial emerges out in spite of our critical agriculture. What to do except crying horses?

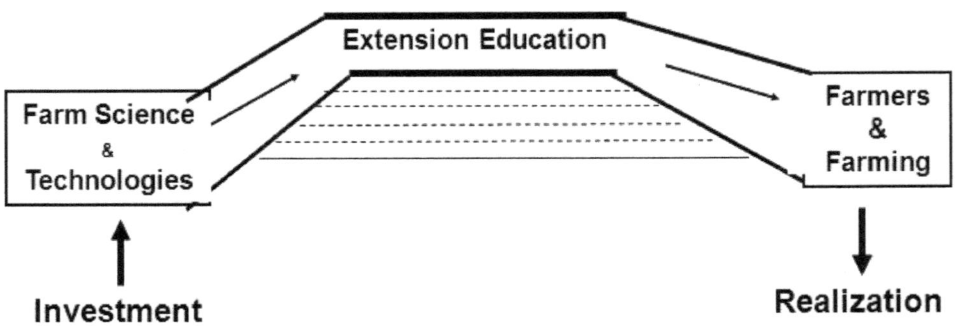

Figure 1.2: Bridge of Extension Education

Have we ever studied thoroughly, how much we are investing in agricultural R and D, and agriculture services vis-à-vis agricultural extension education? No doubt, the results will be startling.

Reforms Needed in Extension System

- ☆ The basic and drastic change in the system demands a 'Council for Transfer of Technology in Agriculture' (COTTA) at the Centre–an autonomous body for aggressive extension leadership under the Ministry of Agriculture. This could be easily done by integrating the Directorate of Extension of the Ministry of Agriculture, Government of India, and the Division of Extension of the ICAR. These two subordinate Offices do not have their 'teeth and voice'. The Council would have virtually no financial implication when the resources and staff of both the system/sub-systems are brought together, integrated and rationalized.
- ☆ The two arms of this Council–State Extension Education and the ICAR/SAUs Extension System may continue with some modifications: the 'extension education' component (non-existent at present) and the 'extension service' component of the States must be separated; they can supplement each other but as independent functional units with their own objectives and methodologies. Extension education professionals must not indulge be in regulatory–routine functions.
- ☆ The time has come to develop a 'Cadre' of extension education services comprising both extension personnel as well as the subject matter specialists (SMSs). Today it is extremely loose–anybody could become anything in extension system any time irrespective of his or her training, field aptitude and outlook.
- ☆ The demand-driven extension has been the talk of the town: bottom-up vis-à-vis top down. To bring about this change, we may start with the 'progressive farmers' further trained in extension and their farms to be developed as the 'model farms'. These model farms and the farmers should, then, become the hub of all extension activities for neighboring farmers. The government agencies and staff should support this system, no Village Level Workers would be needed, rather educated unemployed rural youth can serve this purpose with adequate training on reasonable honorarium basis as Para Extension Workers.
- ☆ The "Knowledge Centres' may be established at the Block level with ATIC like approach (Agricultural Technology and Information Centre of the ICAR/SAUs) and the e-Chaupal at the Panchayat level. The old '*Charcha Mandal*' may be re-vitalized for information dissemination–added adoption and diffusion.
- ☆ The 'Self-Help Group' model of organizing the rural women (now open for men as well) may be the foundation of all agricultural and rural development activities including crops, horticulture, animal husbandry, fisheries and agro-industries as well as cottage household level industries.

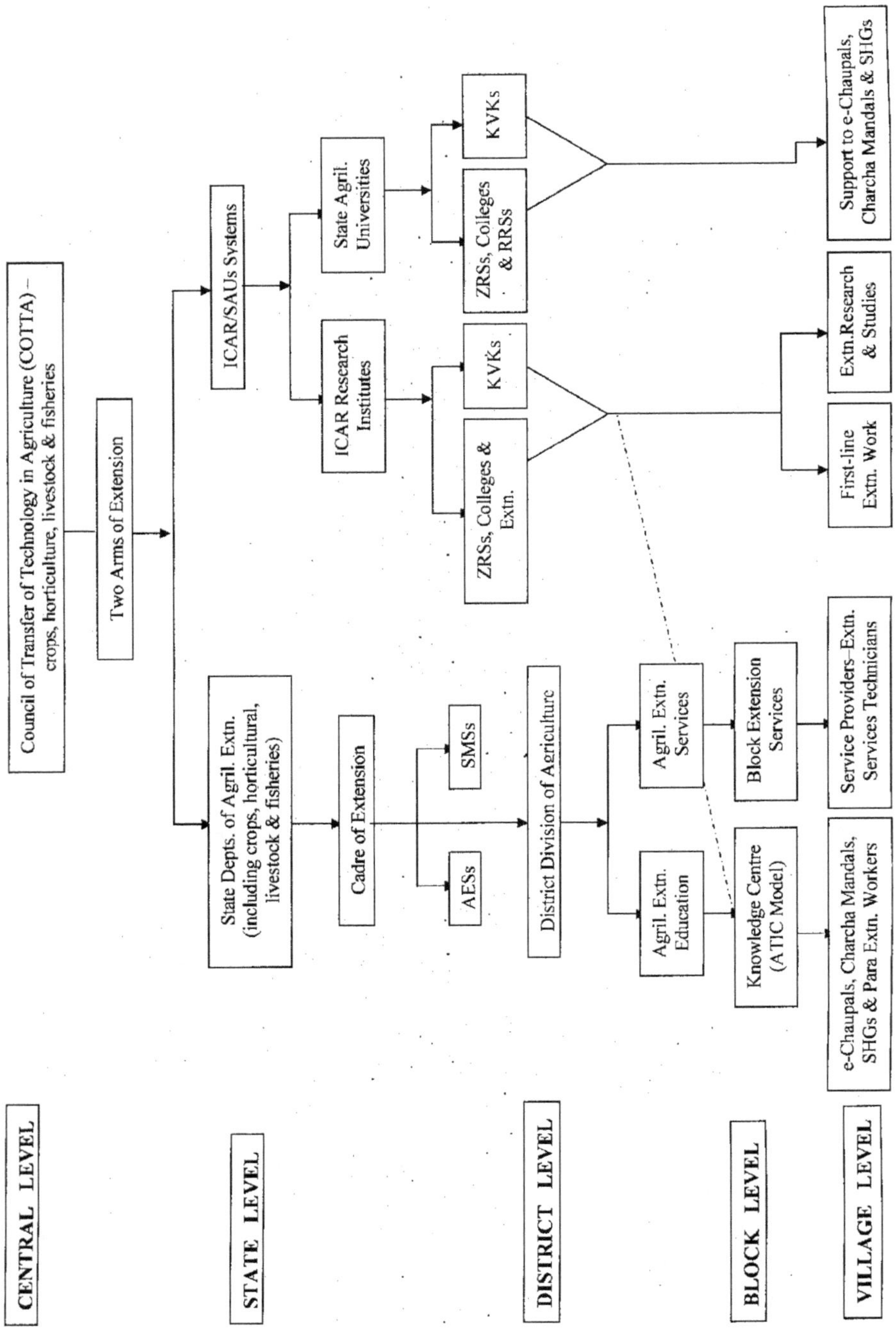

Figure 1.3: Broad Organizational Extension Structure Proposed for the Country (Second Alternative)

- ☆ 'Management' of programmes and resources both men and material will be the key to the success of the agricultural and rural development efforts stresses the Planning Commission, Govt. of India. So far we have done very badly in this respect and, this is why, we are nowhere near the eradication of poverty, which should not have taken more than 10-15 years, if the resources ear-marked for them were utilized sincerely and devotedly.
- ☆ Privatization of extension education services has been advocated since long with partial success, but private-public partnership (PPP) can work. This has to be strengthened by the functional linkages and convergences. But this occurs when the interacting systems/sub-systems have common goals and similar purpose and commitment. This is why, 'linkages and convergences of programmes at the field level are a far cry'–an impossible task in the present socio-economic scenario where everything can be sold and purchased–a self-centered approach unfortunately.
- ☆ We adopted the Land Grant College/University model of the USA where teaching, research and extension are fully integrated; government gives the resources and management support. This system has functioned successfully in the U.S.A. since 1914–the Cooperative Extension Work. However, in India we accepted the truncated model of the U.S. System: (*i*) the ICAR/SAU Extension System was to play the limited first-line extension role to be followed by (*ii*) the main extension role of the Ministry of Agriculture and the State Departments of Agriculture. This approach was conceived when there was to be only one Agricultural University in each State so that their research and academic programmes might not suffer due to the heavy load of extension work and responsibilities. But the situation has changed drastically–there are more than one Agricultural University in most of the States irrespective of the Model Act of the ICAR–the Maharashtra State has five SAUs? Under this condition, the area jurisdiction being reduced, the fully integrated model of the USA comprising teaching, research and extension is fairly possible.
- ☆ 'Training' is another important component of development–so far an integrated part of the extension system. But in view of its scope and importance, especially for vocational areas and entrepreneurship development–the potential areas for job-creation and self-employment, demands an independent treatment. While being part of the SAUs and the ICAR Research Institutes, a Directorate of Training or an Advanced Centre for Vocational Training and Entrepreneurship Development with limited core staff and resources can help to promote this approach. The laboratories, farms, livestock and fisheries facilities of the respective institutes/SAUs fully utilize for such training programmes/courses.
- ☆ There could be two alternatives for the broad and basic organizational reform for agricultural extension education:

  First: as has been explained earlier, the U.S.A. Land Grand College/University Model may now be adopted fully for two reasons: (a) there are more SAUs now with limited area jurisdictions in the states; and (b) the agricultural extension projects/programmes of MOA/SDOA even now, are mostly being implemented by the ICAR/SAUs system.

OR

Second: With an autonomous Council (COTTA) at the Centre (explained earlier), the two arms of extension education may be continued and strengthened. The Cadre of extension educators should come up in the departments of Agriculture in larger interest of the extension education services and the systems.

☆ The proposed extension organization may be seen in Figure 1.3.

References

Prasad, C. and Babu, S., 2008. *Social Science Perspectives in Agriculture–A Thrust for Integration*, VARDAN, New Delhi; IFPRI, USA and ISEE, New Delhi.

Prasad, C., Choudhary, B.N. and Nayar, B.B., 1987. *First-line Transfer of Technology.* Projects of the ICAR, Krishi Bhavan, New Delhi.

Report on National Commission on Agriculture, Part-XI, Research, Education and Extension, Ministry of Agriculture and Cooperation, Govt. of India, New Delhi, 1976.

Chapter 2

The Needed Shift in Agricultural Development Domains and the Commensurate Extension Strategy for Ushering into the Era of Ever-Green Revolution and Commercial Agriculture

☆ *P.N. Jha*[1]

Introduction

India achieved self- sufficiency in food grain production mainly on account of intensive agriculture and high-yielding varieties programme which led to green revolution. The main emphasis was on advocating high-yielding varieties of crops along with their package of practices in areas having assured irrigation facilities.

During the last few years, it is being observed that the momentum of green revolution is virtually standstill and there is virtually a plateau like situation. Further, vast rainfed areas, problematic soils, marginal land, deep-water area etc. are yet to be exploited. The green revolution which is fatigued now is to pursued vigorously by

---

1 Former Vice-Chancellor, RAU, Pusa, Samastipur (Bihar).

attempting at second rather ever green revolution and making farming sustainable, rewarding, profitable and commercial so that the overall socio-economic status of farmers/villages rises and is comparable to their urban counterparts. We have to ensure food security, nutritional security and also environmental security and check the massive exodus/migration of rural people to urban areas in search of jobs and employment but for all this, Agricultural Extension System has to be given new dimension and orientation.

Let us first see what are the shifts needed now in agricultural development domains so that the extension approach and strategy could also be dovetailed accordingly.

The Needed Shift in Agricultural Development Domains for which the Extension Strategy is to be Modified and Recast

- ☆ Shifting to diversified and Commercial farming
- ☆ Emphasis on rain-fed, dry land and problematic soils.
- ☆ Tackling the problem of waterlogging water management, deforestation, drought and flood management.
- ☆ Shift from commodity- centered approach to farming system approach.
- ☆ Examining various land use alternatives.
- ☆ Farming to be viewed not only as a source of livelihood and subsistence living but also as a commercial and profitable venture for decent living.
- ☆ Modern technology to be wedded to I.T.K. and ancient wisdom.
- ☆ Eco-friendly environment-prone technology generation and dissemination.
- ☆ Market decision to precede production-decisions.
- ☆ Demand-driven and Market-driven approach rather than the hitherto supply driven and produce-driven approach.
- ☆ Emphasis on value-addition, organized market, processing and Agri-Exports.
- ☆ Entrepreneurship development both at micro and macro level.
- ☆ Agro-based Small Scale Processing unit at block/Taluka level; Medium Scale unit at Regional level: and Large Scale unit at state level.
- ☆ Compatibility of Technology with the Public Policy in Vogue.

The Gradual Trend and the Current Thinking to be Kept in View while Re-orienting Extension System

Agriculture is going through metamorphosis in India during the last 55 years from mere single/mono cropping to double cropping to relay and multiple cropping to cropping system and Green Revolution Agriculture. Now it is to shift to farming System, commercial Agriculture, complex diverse and risk-prone Agriculture moving towards Ever-green Agriculture bringing stability, sustainability equitability and gender neutrality in agriculture.

The Needed Paradigm Shift in Extension Mechanism for Ushering into the Era of Evergreen Revolution

The transfer of technology programme and the overall Extension efforts has to be directed not only at routine transfer of earlier technology but have to include the second-generation technology as priority areas. The technology transfer pertaining to Organic Farming, Integrated Nutrient Management, Integrated Pest Management and Scientific Water Management has to receive priority. Bridging the miss-match between production and post harvest technology, identifying the components of the farming System based on the farmers needs and perception and the principles of economic and ecology of the respective area popularizing minimum tillage, soil improvement measures, technology for rain-fed agriculture and agro-meteorology-based recommendations of technology are also very important. Strategy for bridging the gap between knowledge and practice *i.e.,* Extension has to shift from merely communicating techniques of know-how to do-how and from technology transfer to technology application. Empowerment of farmers, farm women and farm youth as well as school and college drop-outs and unemployed graduates with knowledge, information, skill and managerial empowerment for gainful self-employment and earning livelihood are the needs of the hour. Extension has now to concentrate on disseminating the technology and monitoring the adoption pertaining to precision farming so that no scare resources of the farmers are wasted and every rupee spent by the farmers gives proportionate divided. In selecting the technology for dissemination, the cost has to be reduced without reducing the yield. The earlier notion of farmers being at the receiving end has to be substituted by farmers first not the last and, therefore, farmer participatory research/extension and the national participatory research network with farmers has to be given a fair trial for the involvement of farmers right from selecting of a problem for research up to the adoption and saturation of the development technology in the farmer's fields. Farmers will have a say at different stages of technology use and also ultimately the feed-back to the research and organized extension system for bringing improvement and perfection in the technology as well as in the communication channel. The major required shifts are as under:

- ✰ Group Approach to Agriculture Extension
- ✰ Transfer of technology mode to technology application mode
- ✰ Broad-based extension system
- ✰ Involvement of community-based organization
- ✰ Associating stake-holders in extension
- ✰ Formation of self-help groups and extension through them
- ✰ Selecting Para Extension workers and assigning them the roles
- ✰ Partnership mode between public sector extension, participatory extension and private extension
- ✰ Multi-dimensional and multi-agency pluralism approach to extension
- ✰ Co-existence of all the systems and sub-system and agencies *i.e.,* Government agencies, Stake-holders, Private agencies, Input agencies, Credit Agencies, NGOs/VOs

- Rural Location Specific Village Knowledge Centre
- Implementation of Bio-village concept for mass movement and Mass involvement at Grass-root level and for job-led growth not job-less growth.
- Formation of field management committees for day-to-day monitoring of the extension programme and activities/demonstrations/farmers participatory research and resource mobilization.
- Emphasis on and drawing programmes for *Antyodaya*-the all round progress of down-trodden hitherto un- reached and the resource-poor farmers/rural poor.
- Need-based, problem solving, skill-based training for self-employment
- Public- sponsored private training and training sponsored by stake holders
- Increasing use of mass-media umbrella and newspapers and magazines for farm extension and agricultural communication.
- Agricultural technology management agency in each district along with KVKs as the Nodal Agency (ATMA)
- State agricultural extension management and training institute (SAMETI) in each state.
- More and more use of electronic media and the digital revolution in extension system also.

Improving the availability and transfer of knowledge based inputs at Grass root Level for achieving Green Revolution.

- Information shop/village knowledge centre as well as input sale centre with single door delivery system in every Panchayat along with Farm advisory system.
- Agricultural technology information centre at Block level.
- Formation of Self-Help Hroups and Raitu Bazaars in order.

Chapter 3

An Innovative Approach to Field-Extension–Paliganj Experience

☆ *B.P. Sinha*[1]

Introduction

Field extension may mean many things to many people, but what I mean by this is the production of desirable change in farmers' (human) behavior, which is reflected in their fields (visibility of change). For this to happen effectively and efficiently, our communication must not be 'prescriptive' (which is so common), but explanatory, comprehensive and scientific in nature with explicit or intelligible logic and reasoning.

How this approach in communication achieved was would be explained with live examples drawn from the field experiences. Prescriptive approach is quite common with which almost everybody is familiar. In this approach, the client is given the package of practices with the advice to adopt them in the same sequence/order without raising any question with assurance given that if this package is adopted, the crop yield will substantially increase, profitability will go up and economic condition will improve (or some times economic condition will leap frog.) The prescriptive approach derives its energy or drive from the assumptions that the farmers are ignorant of their development needs and potentials due to their being uneducated, tradition oriented, incapable of understanding science and its logics and they are accustomed to have blind faiths. Thus they are told 'to do this and don't question

---

1 Former Head, Division of Agricultural Extension, IARI, New Delhi–110 012

this'; 'you can't understand beyond this and so follow this blindly'. This is a serious underestimation of the farmers and their abilities and thus this approach sometimes becomes dangerously counterproductive. I would like to give an example of this in the following anecdote.

Once we visited a village in Gurgaon district of Haryana. Having heard that scientists from IARI had come, a good number of farmers gathered quickly and started making strong request that you should visit our village regularly and help us in solving our farm problems, I suggested them that 'you have an agriculture university at Hissar and its K.G.K. at Gurgaon itself. Why don't you visit them to get your farm problems solved'? On this one of the farmers got quite excited and almost shouted, "Sahab Unki baten mat karo we to ghane bebkoof hain" (Don't talk of them, they are great fools). I was quite stunned and asked how do you say this? In his excitement, the man ran to a nearby farm and returned in a few minutes with a white grub in his hand. He said (showing the white grub), "This is a great menace to our crops. I once went to them with this and asked them the remedy; they suggested 'light trap' as its control measure. Now you tell me can this fly? If not then how will it go to the lamp flame and die"? I got the clue. The scientist simply prescribed the control measure without talking about the life cycle of the insect and without explaining that the months would get attracted to the light flame and would get dropped in the water mixed with kerosene oil kept below the flame. This would reduce their number resulting in reduced number of eggs laid and subsequently reduced number of white grub and hence reduction in damage to the crops. See the danger of prescriptive communication and implied effectiveness of explanatory and comprehensive commutation.

In our field extension work we resorted to the explanatory or comprehensive approach in communicating technology to them. For this, we followed the following steps:

Clear Identification of Issue to be Addressed to

One of the issues identified in our project was 'very poor yields of crops despite lands being very fertile with assured irrigation facilities through a network of sone canal system in Paligunj area of Patna district. Take for example the most major crop–rice which occupies over 96 per cent of cultivable land, had an average yield of 20.13 quintal/hectare as against the potential of over 50 quintal/hectare.

Analysis of the existing package of practices to identify the practices which are counter productive, or which inhibits the production potential of the plants: it is quite obvious that for such a low yield the cultivation practices in vogue must be responsible: This analysis clearly brought to light a number of practices being followed by the farmers which were counterproductive or production inhibitive. Some such practices, for example, are:

Seed Rate

In nursery 25 to 30 kg of paddy seeds used to be sown in 1kattha (1/32th acre) of land for raising seedling to transplant 1 acre of land. Obviously the plant populations in the nursery are exorbitantly high. Seedlings are too close to each–other. The lands are available to each plant for nutrient up take was too small and too competitive, resulting in grossly inadequate nutrients supply to them. Hence, the seedlings were unhealthy, slow growing and fragile. The consequence was that the seedlings used to take 50 to 60 days to become transplantable, they lost their vigour, many got broken while uprooting for transplantation and after transplanting, they remained poor in growth, tillering and earhead bearing giving lower yield per unit of land.

No. of Plants Per Hill and Overall Plant Population

Seeding being very thin, weak and some broken during uprooting, 10 to 15 plants per hill used to be transplanted.This practice became counterproductive in variety of ways

The seedlings were too competitive in to deriving nutrition from soil.

- ☆ Because of large number of seedling per hill planting distances were kept wider to cover the entire field resulting in less dense plants population.
- ☆ Because the seedlings were weak and tall (50 to 60 days old), they used to be placed deeper in the mud so that they could remain standing, Hence first node of the plant (considered the best for producing healthy tillers) were inside the soil and tillering used to take place from subsequent nodes 15 to 20 days after planting. Consequently the tillers used to hear small ear heads and the later ones remain without bearing any ear head.
- ☆ Because of poor health the seedlings used to fade way and top portion dry up during the period they establish their roots in the soil after transplantation and start taking nutrition from the soil (about 6-7 days after transplantation). After they establish their roots in the soil, they produce tillers; hence initial tillers become the main plants after drying up of the tops. This reduces the life span of the plants since maturity dates remain unaltered resulting in poor productivity.

There were many other counter productive practices like home grown poor quality seeds with high mixture and low vigour, improper selection of variety, late sowing, defective fertilization of land, poor weed and disease pest control etc.

Finding Out Substitutes to Replace the Counter Productive Practices

After identification of counterproductive exiting cultivation practices, the substitutive practices were identified from the recommended package of practices with full scientific rationale and then a communication package was developed.

Communication Strategy

The package of practices selected to constitute the local practices were not communicated to them in a prescriptive manner. An approach of participative communication was adopted. For this, training camps were organized in selected villages in which group interactive process was started with discussion on the local practices being followed by them, and how they were counterproductive taking individual processes one by one. The scientific reasons behind the counterproductive impact on plant growth processes were discussed in simple language with smiles or resembling processes drawn from day to day life. For example, comparing weak and unhealthy seedlings with weak and unhealthy newly born child and his growth related problems. Similarly with high seed rate, in small space many seedlings germinate which cannot have enough food for them, like if many children are made to sit around a plate full of food for eating what will happen? No child will get enough food; each one of them would remain hungry and grow weak. Like these scientific facts and processes were made so intelligible. That they not only learnt, them but they started seeing reason and became logical in their thinking. In other words, they became science oriented seeing meaning to every act in the farming process. This also generated conviction in them about scientific package of practices. The farmers have resistance to the blind faith demanded by the prescriptive approach, but in the participative approach of this type they go by reasoning and understanding of scientific logics, hence no resistance in acceptance of the substitutive practices.

After analysis of existing practices, substitutive or desirable practices were explained to them again with emphasis on the scientific logic or reasoning behind them. For example, instead of 25-30 kg of paddy seeds, sow only 5 kg of seeds in the same area. There will be less number of seedlings, each at a reasonable distance from the other. Your initial perception would be that there would be inadequate number of seedlings for transplanting one acre of land, but only after a week of germination, you would see that tillers had started sprouting from every seedling, which may be 3 to 4 per plant after a fortnight, and your perception will then change. Since every seedling has enough space to grow, they would get adequate nutrients, grow healthy and you will ultimately get 4 to 5 healthy seedlings from every grain of seed sown in the seedbed.

The farmers in this sort of interactive process used to raise their doubts and relevant questions to seek clarifications. For example, if the seedlings are sparce, there would be more weed growth–and roots of the seedlings would go deeper in the soil due to which uprooting would be difficult and these would be more breakage. After solutions to these problems were discussed, they used to feel delighted. This is how the process became interactive and there was no trace of any prescription in the whole process of communication.

Under this process the entire package of practices got completely transferred with great amount of acceptability and many more farmers turning to be innovators. Hence adoption became faster.

Translating Communication Theories into Practice

In order to make the communication more effective not only to help them acquire scientific knowledge, but become more adoption prone, the communication theories of 'source credibility', arousal of achievement motivation etc. were put to practice during the village training camps. 'How' is being explained below.

Source Credibility

We know that if credibility of the communication source is high, acceptance of communication contents will also be high which will result in higher knowledge gain. We also know that credibility of the source has two components 'expertise' and 'trust'.

Now in every training camp, a subject matter specialist–a senior scientist from Agril. University or IARI used to be called to impart them training. While introducing the scientist or the trainer to the farmers, his expertise in terms of research work done, number of years spent in scientific pursuit, some special achievements made, were highlighted which gave a clear impression about his expertise. Along with this his concern and desire to see his scientific endeavour and achievement to bear fruits on farmers' fields and the satisfaction derived form this has mobilized him to come to you and share his experiences with you, was highlighted which generated trust upon him. The farmers felt that he had not come merely to perform his official assignment but has come with a genuine desire to help farmers make their crop more productive. The official assignment used to be skillfully denied and converted into willing participation. These details certainly highlight the credibility of the trainer in the farmers' eyes, which made the knowledge transfer process more effective. Similarly, a few words were used to convey homophillic perception to the farmers (sense of homophilli with the trainer).

Motivation Arousal

We know that "motive is an emotionally charged thought" which generates energy in an individual to make him move and take action or in other words 'sets the motivational process'. This used to be achieves by making a comparison with known achievers like Punjab and Haryana farmers and stressing that they are in no way inferior to them, rather they have the potential and nature's endowment (like good soil and water etc.) to go ahead of them, they only need to accept this challenge. This is how by using simple tricks of the trade, based on our theoretical understanding of communication process and managerial skills, the training camps were made very lively and effective.

The Impact Visibility

The impact of these efforts were quite clearly visible on the farmers' fields which used to give encouraging feedback, sense of success and desire to achieve further excellence in the extension team carrying out the programme. Some of such visible impacts were:

- ☆ Reduction in seed rate of paddy from 25-30 kg to 2 to 5 kg (per kattha of soil bed) by an overwhelming majority of rice growers.
- ☆ Plantation of 1 to 2 seedlings per hill instead of 10 to 15 seedlings by the same majority of farmers.
- ☆ Transplantation of 3 weeks old seedlings against 50 to 60 days old seedlings as before.
- ☆ Balanced use of fertilizers in place of an over dose of nitrogenous fertilizers alone being used earlier.
- ☆ Discontinuing of home grown grains being used as seeds and tendency to procure improve seeds from outside and many farmers started scientific seed production–certified as well as truthful seeds on their own.

- ✰ Chemical seed treatment became a common practice.
- ✰ Proper plant protection measures became quite common.
- ✰ Raised seedbed technique and strip planting methods of paddy were adopted by a large number of farmers.
- ✰ System of Rice Intensification (SRI) method of rice cultivation was adopted by good number of farmers.
- ✰ As a result of all these measures taken up by farmer's paddy yield in the area increased by 150 per cent. Similar changes in other major crops like wheat and pulses were also seen.

Comprehensiveness of the Field Extension Programme

The field extension programme was not limited merely to promotion of improved farming technology and increased productivity. It did take care of the other pressing needs for a system of progressive farming to set in. For example, availability of good quality seeds was a big problem since, the state seeds corporation became defunct and the private seed dealers in the market used to sell all kinds of seeds, authenticity of which were repeatedly questioned by the farmers after having used them. This obviously generated a need of establishing a seed producing institution in the area with trust and reliability. Thus a farmers' institution called "Patliputra Beej Utpadak Swablambi Sahkari Samiti" was established in the project area and got registered under the Bihar Cooperative Society Act. The members of the samiti were given training in seed production technology at institutions of repute like Regional Research Station of IARI at Karnal, which has long experience of seed production, Seed Science and Technology Division of IARI and Rajendra Agricultural University, Pusa, Bihar. Then certified seed production activities were started on the registered seed grower's fields that were trained for the purpose. For this, foundation seeds used to be procured from ICAR institutions, Agril. Universities of Andhra Pradesh and Bihar. The seed farms got registered under the Bihar Seed Certification Agency and the seed production programme was initiated with strict supervision and monitoring by the project staff. The seeds produced were named "Patliputra Beez" and were given certification tag by the Bihar Seed certification Agency. Soon 'Patliputra Seed' became very popular for its purity, germination and seedling vigour, not only in the project area, but also in the neighboring districts like Bhojpur,Kaimur, Nawada,Gaya,etc. Another problem, which the farmers faced, was availability of pure (mixture free) fertilizers, and plant protection chemicals. In order to mitigate this problem, another institution called 'Kisan Seva Kendra' was established again with farmers' participation, which undertook the services of input supply. This not only improved the inputs availability but also impacted the local market. Good quality chemicals and fertilizers found their places in the local dealer's, stores also.

Field Extension is a Continuous Process

As stated above, the crops productivity substantially increased in the project area, yet the farmers were not as well off as they should have been. Obviously then, something more needed to be done to bring the desired prosperity in the farming community. One of the idea which cropped up was that crop diversification was needed to include a cash crop in the cropping system. Perhaps this would improve their financial capability. Hence a search began, and we finally settled with introduction of aromatic plants cultivation in the cropping system since aromatic oils had good market and have reasonably high price. Two aromatic plants were selected for cultivation.

Japanese mint

This fits very well in rotation with rice crop. Since majority of rice fields are low lands, high soil moisture persists till mid December, making wheat showing much delayed which adversely affected wheat yield. But sowing time for mint is January (direct sowing) and even February for transplanted crop. Hence this fits better than wheat and even with one cutting by March–April, the oil yields of mint generate much more money than the sale proceeds of wheat crop. After April the land is again available for paddy plantation. With experience, a more intensive cropping system was discovered–Rice–Wheat–Japanese Mint. After rice early maturing wheat is planted and in January seedbeds of mint are sown to produce more than 1 month old seedlings by March when wheat is harvested and then the field is given flood irrigation and mint seedlings are transplanted without giving any cultivation to the wheat field. The mint crop gives one good cutting by April or early May and leaves the field for paddy cultivation. This substantially increased cash flow in the farming system and provided desired liquidity to the farmers.

Lemon Grass

There is some upland in the area, which are not very productive. Generally only one crop in a year is taken. Such lands are ideally suited for cultivation of lemon grass, which is a five-year crop and gives regular income at three months intervals. These two aromatic plants cultivation was introduced in the area and a distillation plant was installed at the project head quarters for extraction of oils. This caught up very well and now five more distillation plants have been installed at various locations and the cultivation is spreading quite fast. This is going to create a lot of economic activities and employment generation in the near future. With this project in action for which financial and technical support have been provided by TIFAC, a Council under Department of Science and Technology, Govt. of India, the Paliganj area of Bihar is on move. Transformation not only in agricultural system but also in the rural economy as a whole has started taking place. His Excellency Dr. A.P.J. Abdul Kalam, Former President of India visited this project three times, and Shri Nitish Kumar,Hon'ble Chief Minister of Bihar visited it once and both of them desired that this effort must be multiplies in other parts of the state and the nation to bring prosperity to the rural areas.

Chapter 4

Preparing Extension for Emerging Opportunities and Challenges in Agriculture

☆ ***Baldeo Singh*[1]**

Introduction

The agriculture sector continues to provide employment to about 58 per cent of the country's labour force and sustains over two-thirds of our population. A recent survey by the NSSO revealed that nearly 40 per cent of the farmers wanted to give up farming if option was available. Smaller land holding combined with large family size, shrinking natural resources, increasing input costs, diminishing net returns from conventional crops, unreliable weather conditions etc. are gradually proving agriculture an uneconomical proposition leading to grower's inclination towards urban migration. With WTO and increasing role of foreign players, the Indian farmers finds himself further marginalized. The shift to cash cropping has hurt more than it has helped the resource poor farmers. Close to 60 per cent of rural households do not have bank account and only 21 per cent have access to credit from a formal source. In spite of increase in production and productivity, and in export potential, the poor farmers do not get any benefit except minimum support price. Even existing services provided by state do not reach rural poor and so, rural people suffer from ill health, illiteracy and malnutrition as well.

Improvements in agricultural education, communication, technology and markets have made the country a global village. While remarkable progress has been

---

1 Joint Director (Extension), IARI, New Delhi–110 012.

made in a number of directions, the fruits of development have not benefitted the country's growing number of rural poor. In order to ensure the fruits of development reach the most needy, we need to introspect and analyze our strengths and weaknesses in agricultural extension and develop our capabilities to exploit the opportunities and address the emerging challenges for better socio-economic order. We need a strong and stable National Extension system with adequate manpower and modern ICTs networks to tackle these issues and harness the advantages.

Emerging Directions and Challenges in Agriculture

The emerging extension directions and challenges that require due attention are discussed here:

1. Today, a new paradigm of agricultural development is fast emerging. The overall development of rural areas is expanding in new directions, old ways of delivering important services to the people are being challenged and traditional societies are being transformed into knowledge societies. Developing extension capacity to generate, absorb, disseminate and protect knowledge and exploit it, as a powerful tool is necessary to derive societal transformation.
2. With expanding frontiers of science and technology and growing pace of technology generation, tremendous opportunities are being created in all walks of life and agriculture is no exception. This signifies the paramount importance of research and extension integration, speedy and effective transfers of technology and ICTs use to realize full potential and advantages of technologies.
3. We are more likely heading towards a polarization of work force into a high technology user elites and a group who are technologically illiterate and economically poor. Specific extension strategy and treatment is required to address the concerns of each group. Each work force must be seen as institution or system and accordingly approached, organized and developed using appropriate training and education strategy and labour market requirement.
4. The ensuing number of rural-urban migration is creating serious problem in the cities. Self-employment and entrepreneurship development are the necessary demands to face this in more challenging way along with creating necessary infrastructure for small-scale entrepreneurship in rural sector. We have to make investment on skill development and organize the people better at local level to make their requirement not only known but satisfied also.
5. The policy environment for agricultural research and extension has changed giving rise to increasing involvement of private players and increasing emphasis on ICTs uses. We need to look into these developments and identify the competitive advantages in terms of facilities, roles and responsibilities of different players/stakeholders. The structural adjustment needs to be more attuned to sectoral needs and demands. Hence, partnership

between public and private extension with ICT networking for production, value addition and market advantages on area basis has become more relevant to some of the sectoral demands to derive advantages of emerging opportunities. The technology on information communication front and its increasing space is posing its own challenge.

6. We are living in paradox. On one hand there is revolution in agricultural technology and ICT and on the other the farmers are in utter desperation and committing suicide. The poverty, credit indebtedness, family division, problem of availability of quality inputs, high cost of production etc. are the serious issues influencing farmers' profits. A critical appraisal and redressal of these are essential to bring smile on farmers' faces. This supports the demand of a stable and responsive extension system.

Appraisal of Extension System

Organized attempts for agricultural development predate the planning era in India. After the independence, the first well conceived multi-disciplinary extension system for holistic development of rural community was *Community Development (CD)* programme and then, over the years many extension and development strategies/ approaches with varied focuses were tried along the CD programme that helped increase and enhance farm production, productivity and overall quality of human life. Superimposed on CD programme was the T&V system funded by World Bank with emphasis on the role of extension in technology transfer to encourage utilization of research results. After its withdrawal, the extension system collapsed and gave rise to the birth of many private players in the scene.

The ICAR also took initiative to implement various extension programmes like ND, KVK, Lab to Land, ORP, Special programmes on tribal and hilly areas, IVLP, ATIC etc. to accelerate technology transfer. The SAUs are also implementing extension programmes in their areas. In addition, the Call Center, Help Lines, Agri-clinic, Agri-business and National Watershed Development Programme for Rainfed areas etc. by Govt. of India were also initiated. Under National Agriculture Technology Project (NATP), effort was made to integrate the activities of pubic and private agencies in some selected districts under Agriculture Technology and Management Agency (ATMA) through Directorate of Extension, Department of Agriculture and Cooperation, Govt. of India. The ATMA programme is now expanded in almost all the districts, but with old development staff to coordinate activities of all Govt. and non-government functionaries engaged in agriculture and rural development in the district. The results ATMA are yet to be seen and realized. Currently under NAIP (ICAR) funded by World Bank, livelihood security programme under consortia approach involving a number of Govt. and Non-Govt. organizations has been started, besides communication awareness networks for the empowerment and development of gender for sustainable development in agriculture.

All these programmes/approaches considered farming problems essentially as 'individual' concern and never as a 'collective' issue. The approach relied heavily on diffusion of innovation model and less on group or social action model. Contrary to the expectation, the development efforts or modern technology did not trickle down to most of the farmers at the bottom of development. At present, the agricultural extension scenario is more fluid and confused. The demand for a strong and effective extension system is now a real challenge for following reasons:

Information Now has Real, Measurable Value

Private information providers- crop advisors, business consultants, private education and training providers- are cropping up everywhere. The attraction is the information to meet specific audience needs.

Essential Shift from "Provider Mentality" to "User Mentality"

A 'provider mentality' focuses on process of distribution, meaning it is driven by what we want to distribute. A 'user mentality' focuses on *access* and customer, meaning we now need to anticipate customer needs and provide them *access* to our knowledge base.

Public Extension Service is No More Sole Source of Information

Technology has created the opportunity to provide and promote access that wasn't realistic before. Because of this, the traditional educational market is now more easily accessible for competitors–Agri-business and commercial crop market advisors. Technology combined with open access to the rapidly expanding knowledge marketplace entices private information providers to step up the competition for clients, who previously were dependent on public sector extension as a sole-source education provider.

Required Shift from Broadcasting to Narrow Casting

Database is being created for almost every subject matter area be it pesticides, chemicals or any thing else. People want to choose where they get specific information and not lock themselves into choices made by one provider. Specific information is available more efficiently than public extension system.

Paid Services Used More Effectively

It is argued that paid services are more effectively used than the services provided free of cost and clients regard this as more reliable.

Increasing Transformation

The innovative development in information technology and increasing commercialization in agriculture has provided new sets of problems, opportunities and challenges, which require client friendly, market friendly and export/trade friendly interventions. A market is developing for skilled and specific agricultural advice.

Instance Performance

The market place demands customized, immediate response to problems and questions. Therefore, the challenge is to synthesize the data and make effective need based prompt decisions using the call centers and information kiosks. Here, partnership to link the facility and scientific knowledge is advantageous.

Reorienting Agricultural Extension

Agriculture is the most complex commercial system that requires inputs from a large number of sources like soil, water, environment, society and trade. To reach 104 millions farm families spread over more than 590 rural districts and 6 lakhs villages is an up-hill task. Farmers' needs are now much more diversified and the knowledge required addressing calls for improving the capacity of existing research and extension system. Studies on total factor productivity of crop sector in the Indo-Gangetic Plains of India conducted at IARI indicate that extension accounts for around 45 per cent of the growth in total factor productivity, followed by research (36 per cent), literacy (10 per cent), infrastructure (8 per cent) and urbanization (1.5 per cent). Research and extension together accounted for nearly 80 per cent of the growth in total factor productivity. In view of this, we need to introspect about the strengths and weaknesses related to extension research, education, training and field extension functions of the discipline and prepare a roadmap to reorient these with modern facilities and adequate manpower to meet the future challenges. The following aspects require due consideration:

Reorienting Extension System of SAUs/ICAR Institutes

The extension education role of SAUs/ICAR institutes encompasses four functions *i.e.* extension research, education, training and field extension (TOT) and these are carried out through extension directorate/extension divisions/sections including PG/Ph. D. and others researches. We need to assess the outcomes of these functions and identify critical gaps/areas requiring reorientation and infrastructure support. The extension system requires adequate strengthening in terms of investment, manpower support, physical facilities, modern ICTs networks, transport etc. We need to focus research efforts on frontier and burning areas; and develop capabilities to undertake necessary interventions in coordinated manner. The researches already conducted need to be compiled to project important findings and implications thereon affecting the extension policy. The present scenario demands more knowledge and skill based development. A planned programmes for research, education, training and extension interventions is needed for impact analysis and developing training plan for the human resources development in agriculture and allied areas at various levels. The frontier areas of research and training interventions are farming system approach; project planning, monitoring and evaluation; participatory action research; entrepreneurship development, gender empowerment and local capacity building (SHG mobilization, farmers' institution/cooperatives building etc.), information technology and media for development communication etc.

We need to give sufficient focus on professional growth of students by providing specialization and restructuring course curricula. Here, we need to provide answers to the following questions:

- ✰ What changes should be made in current ways of education?
- ✰ What are the impediments to change?
- ✰ How can these impediments be addressed?
- ✰ What do the students need to prepare for global world?

- How do we determine these needs?
- What contents and specific courses should be added or modified?
- How can these be delivered?

Training is an important function of SAUs/ICAR institutions. We need to develop the professionals in different areas and upgrade them at frequent intervals; besides imparting training to entrepreneurship development and technological empowerment among farming community. The 'Advanced Training Centers (ATCs)' may be created in selected SAUs/ICAR Institutes.

The extension function needs to be more aggressive in its TOT approach and strategy not only to accelerate the transfer process for integrated agricultural development, but also to provide knowledge support to post-harvest practices, value addition and marketing. For catering the multiple needs of farming community and ensuring participatory development, the requirements for affecting the change are different kind of extension support, knowledge based extension, need for community action, gender sensitization and capacity building (SHGs/Associations), linkages and partnership development among stakeholders, participatory technology development and transfer, marketing extension–ITKs/IPR issues etc. and entrepreneurship development.

Reorienting KVK Functioning

The KVKs should be equipped with manpower, facilities and resources to play pivotal role for agricultural transformation in their respective areas. They should work as role model and lighthouse for agricultural development with new technologies and need based trainings rather than just a center for technology testing. It is imperative to have KVK in-charges with agricultural extension qualification. It is also felt that all KVKs should be brought under the umbrella of ICAR/SAUs. At present, there are plenty of KVKs, which are non-performing due to various reasons. One of the serious limitations is technological back up among private KVKs.

Reorienting Public Extension System

We need a strong and responsive extension system, which can cope up the challenges and meet the needs and aspirations of rural poor. The CD extension system is the most vital and befitting system. The number and role of private players need to be restricted and their involvement is limited to some sectoral needs and demands at different levels, but under the coordination of CD extension system to ensure accountability. The involvement of *Village Panchayat* in planning and implementation of development initiatives should be mandatory. The extension system is regarded as a part of government bureaucracy and internal articulation on its strengthening in terms of proper directions, human resource requirements, transport, ICTs facilities etc. has never been encouraged. Lack of vibrant internal review and planning process is the major factor behind early erosion of human resource and institutional productivity. There is a need to revamp CD extension network with leading role and involve selectively the private sector particularly in the post harvest processing, value addition and marketing. The Community Development System at block level and below is required to be revamped for catering multiple needs of farming community and ensuring participatory development. Thus, the CD extension system should be reoriented with respect to required human power (gender-wise) and adequate logistic support of modern information and communication technologies, adequate fund, transport facilities etc. to cater emerging farming needs.

Public-Public Partnership

The country is bestowed with varied agro-climatic conditions that can be used advantageously to take up technology transfer in different parts of the country. There is a need for effective methodologies to transfer them at a much faster rate so that the varieties or breeds, products, techniques developed with the required traits can be transferred to the fields without much loss of precious time. The SAUs/ICAR institutions can join hands to participatory technology development and transfer and farming systems research and development for developing location-specific models/strategy for different areas.

Public-Private Partnership

The integration of research, extension and client systems for sustainable development have assumed greater importance today. The private service providers approach by corporate sector and the NGOs in agriculture are increasing in number. The private extension works on the specific areas and can meet the demand for the specific information more efficiently than public extension system because of its network. Therefore, the private services need to be properly channelised and utilized to support the main CD extension system, particularly for post-harvest management, entrepreneurship development through small-scale cottage infrastructure and export promotion. In this context, the public-private partnership has been expressed explicitly as follows:

Promotion of Agri-business through Development of Agri-business Center

The agribusiness through development of agri-business centers at Block and Panchayat levels should be promoted. This objective cannot be achieved without the partnership with the private players. For example ITC tobacco programme in Andhra Pradesh is complemented by the varietal development efforts of Central Tobacco Research Institute, Rajamundri. Similarly, Punjab Agricultural University is supporting the Pepsi contract-farming programme by providing hybrid variety seeds.

Establishment of Agro-clinics

Because of commercialization and market orientation, the demand for the specific farm information is on the rise on one hand and on the other hand a large number of agricultural graduates are in dire need of employment. Under government initiative through MANAGE with involvement of recognized agricultural institutions/ universities, agricultural graduates are being trained to open information shops as an entrepreneurs. These shops can render advisory services on both crops and animal husbandry and act like single window for farm products, input information, inputs sale etc. at the block/district levels. Private agencies can play active role in supporting this initiative for its successful implementation.

Marketing and Retailing

Yield, price and market risks remain major constraining factors for smallholders' participation in retail marketing. Smallholders are risk averse and their effective participation in regional, national and global markets requires a mechanism that could absorb such risks. Organized retail involving farmers/their associations can help reduce wastage that currently characterizes farm production. The studies indicate that about 50 per cent of the population, below 30yrs of age, is aspiring for higher production, consumption and innovation. There is a need to promote entrepreneurship in agricultural and allied areas with financial and infrastructure support by corporate on benefit sharing basis.

ICTs Infusion (Info-tech and Agro-biotech)

ICT diffusion and infusion in agriculture provide necessary digital opportunities or advantages for productivity increase, income generation, decrease in national disparity and improving linkages with the market. Net working of people and networking of information are required for reaching the unreached through public extension system. Preparation for future development and communication among masses require an extensive thinking regarding information technologies and convergence with mass media including the Internet. An important development is the emergence of e-Commerce. More and more companies are setting virtual showrooms, which can be assessed by any users.

Area Coverage by Extension Agents

Several research studies conducted on extension organizations have revealed that the delivery of goods is effective when the grassroots extension worker covers a small area of jurisdiction, with multiple purposes. The existing system of large jurisdictions, each with a narrow range of activities, is less effective. The broad basing requires grassroots extension workers to be at the cutting edge of extension and master of many trades. IT strengthening of public extension system can help enable extension staff to gather, store, retrieve and disseminate a broad range of information needed by farmers, thus transforming them from extension workers into knowledge workers to realize much talked about bottom-up development. Human-ware is not the substitute of techno-ware. At least two extension workers (male and female) are necessary to meet the demands of one *Village-Panchayat*.

ICAR Front-line Extension Projects

ICAR led previous front-line extension projects like Lab to Land Programmes (LLP), National Demonstration (ND) and Operational Research Project (ORP) should be reintroduced for socio-economic betterment of small and marginal farmers.

The Socio-cultural Inputs

Socio-cultural inputs are hardly utilized in development programmes, although it has considerable influence on programme sustenance. These inputs work better in development initiatives for local capacity building at individual and collective levels. Self-help and participation have always been seen as integral part of extension strategy. Rural society is a group of people who have lived together and become organized to be considered as a unit and social structure. The social structure meets the needs of the people and society and facilitates its functioning.

Area Approach

A systematic area planning in concentric zones with vertical and horizontal links keeping in view the important commodities in production cum food value chains is an appropriate strategy. Linking small farmers to processing industries and fast food chains would be a good proposition. The priority need is to thoroughly study the existing retailing of agricultural products and promote it on three tiers basis: First tier–high-tech diversified agriculture, processing, value addition and trading at District level; Second tier–integrated agriculture, processing, value addition and retailing at Block level and Third tier–commodity based integrated agriculture, pre-processing and petty retailing at Village *Panchayat* level. While retailing at first tier should be promoted in partnership of corporate sector, but the retailing at second and third tiers should be restricted to rural entrepreneurs to create employment market and utilize local market avenues. The financial support at these two lower levels is essential. The key areas of food retailing are storage, transportation, segmentation based on class of buyers, packaging etc.

IARI Initiative 'Village-Based Integrated Models for Market-Led Agriculture'

The extension initiative of IARI is to develop 'Village based integrated models for market-led agriculture' in different parts of the country by involving institutions. The programme aims at creating location-specific models to utilize properly prime land and rural resources for agriculture; realize the untapped production potential by bridging yield gaps; ensure household food, income and employment security; develop system approach with all attention to the links in production to food value chain; develop local capacity for participation, leadership, group action and entrepreneurship; sensitize, upgrade and strengthen the extension support and linkages; strengthen input-output chain from farm to market and sustain retail boom and promote quick spread of technologies through farmer-to-farmer extension in wider areas.

The IARI has established project field sites in north, south, east and west parts of the country by involving institutions to carry out TOT interventions in participatory mode in the fields of diversified/commercial farming. The attempt is to understand the problems faced by the farmers related to their livelihoods and adoption and diffusion of farm enterprises, and address them to strengthen the technologies/ enterprises movements through farmer-to-farmer extension. The local partners being involved include development departments, SAUs/ICAR institutes, NGOs or corporate sector. Some selected nearby sites are being connected with IARI for information communication and knowledge sharing. The project is being carried out in phased manner in concentric mode in collaboration with selected SAUs/ICAR institutes across the country.

1. Peri-urban agriculture (up to 75 kms. from IARI): Three (3) sites in NCR of Delhi + Gurgaon (Haryana)

2. Integrated agriculture (above 75 to 500 kms. from IARI): Eight (8) sites (Sonepat in Haryana, Western districts and Mathura in UP, Patiala in Punjab, Churu, Jhunjhunu and Bharatpur in Rajasthan)
3. Integrated farming systems (beyond 500 kms. from IARI): Nine (9) sites (Gulbarga and Banglore in Karanataka, Parbhani and Rahuri in Maharashtra, Varanasi and Mirzapur in UP, Ranchi in Jharkhand, Udaipur in Rajasthan and Navsari in Gujarat).

A cluster of 2-4 villages/a village *Panchayat* has been selected at each project site. The innovative farmers in different project locations are also being linked and utilized for technology transfer. A mutual agreement for collaborative efforts between the IARI and SAUs/ICAR Institutes has been made. The SAUs/ICAR institutes involved are MPUAT, Udaipur (Rajasthan), MPKVV Rahuri (Maharashtra), MAU Parbhani (Maharashtra), UAS Dharwad and UAS Banglore (Karnataka), BAU, Ranchi (Jharkhand) and BHU (UP), IIVR, Varanasi and CIRG, Mathura (UP), NAU, Navasari (Gujarat) and NRC Rapeseed-Mustard, Bharatpur (Rajasthan) from Kharif 2007. The programme encompasses village-based natural resources management and lays emphasis on quality seeds production, protected horticulture, diversification with high value crops/vegetables/flowers, processing, value addition and marketing, capacity building for entrepreneurship development by integrating indigenous knowledge and practices. The exchange of technologies among partners for trial and adoption in suitable locations and avoiding duplication of research efforts are also the concern of IARI initiative. It involves a farming systems approach with following emphases:

1. The concept of Seed Bank by producing enough good quality seeds and planting materials of important crop/vegetable varieties and hybrids through appropriate training and education to meet requirements of village cluster and nearby areas.
2. The concept of Water Bank by harvesting rain water and judiciously using it by adopting efficient water management techniques and choosing the crops and varieties according to available water quantity.
3. Development of Fodder Bank by introducing quality fodders and their conservation techniques using Feed Block Making Machine through farmers' association/SHG/*Panchayat* for successful rearing of animals for dairy, meat, wool, draught purposes.
4. Development of Energy Grid/Energy Bank by harnessing renewable energy from different sources such as solar, wind, water, biomass, bio-diesel etc. to meet major demand of energy.
5. Development of Manure Bank by utilizing cattle dung and other biomass for improving soil health and physical properties, providing renewable energy and reducing pollution.
6. The concept of Knowledge Bank by developing knowledge society through, information support in production, processing, value addition and marketing.

7. Development of Villagers Cooperative Bank completely owned and managed by the rural people with appropriate directions and risk mitigation fund for helping the resource poor under unforeseen adverse situations.

The IARI extension is to create 'Village-based Models of Market-led Agriculture' in various locations so as to improve livelihood security and facilitate adoption and diffusion of technologies through farmer-to-farmer extension. The model envisages building up knowledge base of biodiversity experiments of agro- enterprises; promoting rural knowledge systems; empowering people and their institutions for sustainable management of rural natural resources; developing local capacity for effective implementation of agro-enterprises through field demonstrations/research and information networking and establishing partnerships among various institutions involved in agro-enterprises development.

Conclusion

There is a need for extensive and intensive increase in investment in extension. This is a priority area and needs adequate funding support for extension works.

Urgent need is to revive national extension system-Community Development (CD) system- adequately equipped and revamped in respect of additional human power, physical facility including transport, ICT-computer Internet facilities etc. to cater the multiple needs of farming community and agri-preneurs.

The public-private-client partnership should be selectively built and strengthened according to location specific needs and aspirations.

The IARI models of 'Village-based Integrated Development for Market-led Agriculture' need to be emulated and promoted for sustainability, profitability and partnership development. A comprehensive area planning is required to develop agriculture on farming system perspective with vertical and horizontal links of concentric zones.

Farmers' needs are now much more diversified and hence, integration of research and extension for farming system research and development is essential. The infusion of Agribiotech (agricultural bio-technology) and InfoTech (information technology) in farming system is needed to catalyze progressive changes in more sustainable ways and help to attack the problem of rural livelihoods.

References

Chamber, R., 1994. *The Origins and Practice of PRA*. IDS, University of Sussex, Brighton, UK: Challenges, Potentials and Paradigm.

Jha, D.N., 2006. Out-of the-Box framework for public R&D in Agriculture. In: *International Conference on 'Social Science Perspectives in Agricultural Research and Development* held at IARI, New Delhi, February 15-18.

Moni, M., 2006. Digital inclusion for fostering rural prosperity and grassroots developments: A much needed domestic strategy for sustainable livelihoods in rural India. *Ibid.*

Rao, T.V., 1996. *Human Resources Development.* Sage Publication, New Delhi.

Report of Third Deans' Committee on Agricultural Education in India, 1995. Indian Council of Agricultural Research, New Delhi, published in 1996.

Schultz, T.W., 1990. *Restoring Economic Equilibrium: Human Capital in the Modernising Economy*. Oxford, Basil Blackwell.

Social Sciences Education in Agriculture: Perspectives for Future, 1996. (Eds.) Rasheed Sulaiman V. and S. Selvarajan. National Centre for Agricultural Economics and Policy Research, New Delhi.

Chapter 5

Staratiform Information and Participatory Extension Management: A Paradigm for In-Tension of Ex-Tension

☆ *Dipak De*[1] *and Basavaprabhu Jirli*[1]

Introduction

The communication science has to be viewed from different perspective due to the growing challenges to be met out. It has passed through simple interaction to highly complex process of information network resulting to growth of information society. Today's world can be viewed on information continuum from least to highly informed. There are nations, which are highly informed, and there are nations, which are moderately informed, and striving to join the highly informed category. The third group consists of those nations, which are living in a least informed state. The 2nd category can be classified as developing nations and 3rd category can be grouped as underdeveloped nations. Similarly, communities and individuals can also be categorized in terms of information accessibility and use.

In this perspective, Corea's (1990) viewpoint on role of knowledge (information) in human resource development deserves attention.

But no less important in the area of human resource development is a new factor; the crucial and strategic role that knowledge will play in the world economy of the

---

1 Department of Extension Education, Institute of Agricultural Sciences, BHU, Varanasi–221 005.

1990s and of the century to come. Already, it is being said that the gap between countries is no longer a gap between rich and poor counties, but rather a knowledge gap, a gap between those with scientific and technological knowledge and those without. The future progress and prosperity of nations will depend less on their geographical position, less on their natural resources endowments, less on their reservoirs of unskilled manpower. It will depend on the extent to which those nations acquire command over modern scientific and technological knowledge.

This clearly indicates the importance of information as an input and power in Nation/Community/Individual development. In a broad sense, those who are powerful and dominant have accumulated wealth and possess ability to control, create and flow of information from their point of view. All sorts of clubs, associations and structures, which we witness at International level, are in fact mechanisms by which Developed World is constantly controlling and manipulating information flow.

The topic has been dealt in the following sub-heads: (*i*) Information System; (*ii*) Information Stratum; (*iii*) Characteristics of people in Information Stratum; (*iv*) Participatory approaches; and (*v*) In-tension of extension

Information System

Webster's dictionary defines information as "Knowledge Communicated by others or obtained by personal study and investigation", or alternatively as "Knowledge of special events, situation or the like". Hence, all information is knowledge. Legans (1971) defined information as a message, a communicator wishes his audience to receive, understand, accept and act upon. He further stated that information might consist of statements of scientific facts, descriptions of actions being taken, reasons of why certain action should be taken, or steps necessary in taking action. Thus, information is generalized idea, opinion, fact, feeling, piece of knowledge that precedes action.

Ackoff (1961), defined system is an entity, conceptual or physical, which consists of interdependent parts. More or less similar views have been expressed by Arundale (1968) about the system. To him system refers to a boundary maintaining set of interdependent particles or components, while a system's components can be separated for the purpose of study, but such a study must also consider the interactions between the components. Any collection of elements that interact in some way with one another can be considered a system (Dahl, 1975). A system to Mathew (1964) is collection of independent parts devoted to the accomplishment of some goal or goals with the parts maintained in a state in relation to each others and the environment by means of (*i*) standard modes of operation, and (*ii*) feedback from the environment about the consequences of the system-action. Rapport (1968) in his foreword to the book "Modern Systems Research for the Behavioural Scientists", maintained that a whole, which functions as whole by virtue of the interdependence of its parts, is called a system. To be even more specific for the purpose of research, a system is defined as a set of components interacting with each other and a boundary, which possess the property of filtering both the kind and rate of the flow of inputs and outputs to and from the system (Berrien 1968). Briefly Hall and Fagen (1968) defined a system as a set of objects together with relationship between the objects and between their attributes. To Cassata and Asante (1979) system is a group of constituents plus the relationships between them plus their attributes.

Therefore, information system refers to a group of ideas having a set of attributes and their relationship.

Information Stratum

We can classify a social system in distinct three strata in terms of information. The individuals who find themselves within the boundaries of an information system are seldom equally concerned with information use. Some are indifferent and others more deeply involved. Even among those who are seriously involved, only some actively hold and seek information. These groups can be classified as: (*i*) Aphonic persons (AP); (*ii*) Information Seekers (IS); and (*iii*) Information Holders (IH).

Aphonic Persons (AP)

These are the person(s) who are indifferent to seek and use information. They are the majority members of the society.

Why do people remain indifferent to information? Some of the reasons could be:

1. If they place less value on the particular information in terms of rewards to be gained. To many people information relating to Scientific and Technological (S and T), political, etc. may be less gratifying than working on one's job.
2. If they think that alternative information is not going to fulfill their immediate need(s).
3. If they think that the new information is beyond their comprehension level; and
4. If they think that the new information use requires resources more than their capacity.

Information Seekers (IS)

These are the person(s) or groups who are constantly engaged in seeking information. Generally people have desired to know the unknown. They employ multiple means to achieve their goals. Not all people use their resources to gain information, yet some. Why do some people seek information? The answer to this question can be grouped into following categories as opined by Dahl (1975):

1. Men seek information in order to achieve collective good;
2. Men seek information in self interest; and
3. Men seek information from unconscious motivation.

Depending on culture, society and political system, information can be used to acquire fame, security, respect, wealth and many other values. It is a compound of conscious and unconscious motives. They may be progressive members of the society.

Information Holders (IH)

These are the person(s) or group (s) of persons who create and hold the information. They may be scientists, technical experts, change agents and elite of the society. Why do some people hold information? This question can be answered from several angles:

1. Person(s) or group (s) who generate the information, have tendency to hold due to inherent tendency of fulfilling ego;
2. To exercise authority over others; and
3. To derive benefits in favour of self.

At a global level it means control on scientific and technological knowledge (information) by the developed nations. There could be several examples; one of them is sharing of atomic information with Third World countries. It is they who decide what to give and what not to give; where to give; whom to give; why to give; how much; and at what speed to give. This is how they hold and play with information. At the individual level it is the fulfillment of ego and interest, which governs the tendency to hold information. This is at the point of information generation.

At the level of information dissemination, the person(s) or group(s) also holds information. As for example Extension Agencies do not pass on the total information received from generation sources to the ultimate users. Similarly at the community level the elite holds the information and uses in its own interest.

Our examination of AP, IS and IH reveals endless variety of human motivations and personalities at work in information use. A particular person's orientation toward information can be explained in some degree in the light of his personality or character and cultural milieu in which he lives and his life situation.

An attempt has been made to present an orderly view of their characters in the Table 5.1.

Information Types

- ☆ *Scientific information*: refers to ideas, thoughts, concepts etc. Which are the product of systematic endeavour
- ☆ *Technical information*: refers to the application of scientific information in day today life situation
- ☆ *Policy information*: refers to ideas, thoughts, opinion etc. For making choices of alternate strategies
- ☆ *Managerial information/operational information*: refers to ideas enabling production, maintenance and services

Information Use

- ☆ Know–why–scientifically oriented
- ☆ Know–how–technically oriented
- ☆ Show–how–operationally oriented

Table 5.1: Characteristics of the People in the Different Strata of Information System

Types of People in Terms of Information System		*Characteristics*	*Key Concepts*
Information Holders Scientists, Change Agents and Elites	(*i*)	Originator/Generator/ Multiplier	They are the persons who invent/discover/ create/accumulate information.
	(*ii*)	Information owner/ ownership (Cassata and Asante 1979)	It usually involves control of information production.
	(*iii*)	Apathetic sharing	They have least desire to share information.
	(*iv*)	Brahminical Attitude	Feeling of superiority causing non-sharing of information with other members of social system.
	(*v*)	Gate keeper (Westly and Maclean 1957, Fett 1974)	Controlling of information flow.
	(*vi*)	Highly receptive	Well developed capabilities of operating simultaneously on different wavelengths.
Information Seekers Progressive members	(*i*)	Multiplier	They are the persons who accumulate process and store information.
	(*ii*)	Inquisitor	They are the persons who decide the desirability of information.
	(*iii*)	Receptive	Capabilities of operating on different wavelengths.
	(*iv*)	Gate keeper (Westly and Maclean, 1957; Fett, 1974)	Controlling of information flow.
	(*v*)	Non-insulated	They are constantly in interaction with happenings in the surroundings.
	(*vi*)	Space maker	They are having coping capabilities with the recent happenings in the social system.
Aphonic Persons	(*i*)	Apathetic attitude	They are inactive in seeking information.
Indifferent persons	(*ii*)	Meaninglessness (Seeman 1959)	The individual is not clear as what he ought to believe. He cannot predict with confidence the consequences of acting on a given belief. The problem of meaninglessness, therefore, is a problem of uncertainty. In this he finds that his existence is irrational, absurd and meaninglessness.
	(*iii*)	Self estrangement (Fromm 1956, Marcuse 1964, Laing 12973)	It is the state of mind where individual does not experience himself/herself as an active agent for bearer of human power.
	(*iv*)	Non-receptive	Lack of capabilities of operating on different wavelengths.
	(*v*)	Powerlessness (Pandey 1984, Marx 1972, Chambers 1983)	It is the state of the individual where he does not have the access to the resources available.
	(*vi*)	Insulated (Chambers 1983)	They are not in touch with the happenings in the surroundings.

Participatory Approaches

Participation Means Different Things to Different People

Passive Participation

People participate by being told what is going to happen or has already happened. It is an announcement by administration or project management without listening to people's response.

Participation in Information Giving

People participate by answering questions posed by researchers by using questionnaire surveys or similar approaches. People do not have the opportunity to influence proceedings.

Participation by Consultation

People participate by being consulted, and external agents listen to views. This does not concede any share in decision making, and professionals are under no obligation to take on broad peoples' views.

Participation for Material Incentives

People participate by providing resources, for example labour, in return for food, cash or other material incentives. Though this is called participation, people have no stake in prolonging activities when the incentives end.

Functional Participation

People participate by forming groups to meet predetermined objectives related to the project. These institutions tend to be dependent on external initiators and facilitators, but many become self-reliant.

Interactive Participation

People participate in joint analysis, which leads to action plans and the formation of new local institutions or the strengthening of existing ones. These groups take control over local decisions, and so people have a stake in maintaining structures and practices.

Self-Mobilisation

People participate by taking initiatives independent of external institutions to change systems. They develop contact with external institutions for resources and technical advice they need, but retain control over how resources are used (*Extension Digest*, 1995, Vol. 2).

Common Principles of Participatory Approaches

- ✰ *A Defined Methodology and Systematic Learning Process:* The focus is on cumulative learning and their use has to be participative
- ✰ *Multiple Perspectives*: A central objective is to seek diversity, the assumption being that different individuals and groups evaluate situations in different ways leading to different actions

- ☆ *Group Inquiry Process*: All emphasize the process of group enquiry. This implies multidisciplinary, multisectoral mix and mixes of outsiders (professionals) and insiders (local people, farmers etc.)
- ☆ *Context Specific*: Each application is flexible enough to be adopted to suit new conditions and actors
- ☆ *Facilitating Experts*: The role of the expert is best thought of as helping the people in their situation carry out their own study
- ☆ *Leading to Action*: The enquiry process leads to debate about change. Action is agreed, and implement able changes represent an accommodation between the different conflicting views. The debate defines changes which would bring about improvement and seeks to motivate people to take action to implement the defined changes (*Extension Digest*, 1995, Vols. 3 and 4).

Five Phases of Participatory Process

1. *Entry Phase*: Establishing rapport and forming groups
2. *Preparatory Phase*: Preparing groups for making action plans
3. *Planning Phase*: Preparation of action plans in suitable formats
4. *Implementation Phase*: Execution of action plans
5. *Follow Up and Evaluation Phase*

The process of participatory management in each phase is broken down into a series of successive operations by conceptualizing participatory process into a series of major undertakings–each broken into discrete tasks–each task broken down to a series of steps. The basic structure of this process is based on the principle of learning by doing and incremental progression in small steps.

Monitoring and process documentation are an essential ingredients of the participatory process, in terms of the need of participants to assess what has been achieved, identify problems and find solutions for further improvement. (*Extension Digest, 1995*)

Process and Methods of Participatory Extension Management

1. Farmer to farmer extension should be encouraged and capacity of farmers for experimentation and extension should be strengthened. Since each farm is different and needs change over time, external institutions must be flexible and responsive, and ready to learn with farmers.
2. The pace of programmes and projects must be slow to build motivation, confidence and rapport all most all the groups involved.
3. Village based management of programme funds and planning are important prerequisites (*Extension Digest*, 1995, Vol. 2).

Inter Institutional Arrangements for Participatory Extension Management

1. Widespread training and competence building is needed to encourage and sustain participatory approach. There is need for expanding training in participatory methods that focus on joint learning for action.

2. Joint approaches increase contacts and linkages between farmers and external institutions, increasing the likelihood of policies and practices emerging that satisfy all actors.
3. Roles of different institutions involved (*e.g.* In developing joint watershed management need to be clarified)
4. Development of strong inter-village institutions can effectively manage activities like credit, marketing, and protection of common lands with limited extension support.

In-Tension of Ex-Tension: A Paradigm

The concept of extension/extension science has a history of about 160 years, but in Indian context the concept of extension is as old as mankind. Yes, it was not termed as extension but it was realized as extension. At the same time we do not find any text where the term extension is used for the activities that were going on, or being taken up by the agencies responsible for development but they were meeting all the qualities of extension service/extension education. But we depend heavily on the available history of the term extension suffixed with work, service, education, science etc. Also we find so many prefixes for the term as agricultural extension, dairy, fishery, home science, forestry, veterinary, university etc.

By attaching the meaning to the term extension, "extending" relevant and useful information to the adult population at large, extensionists has assumed their responsibility on their own. Also they have tried their level best to differentiate education of extension science with formal education and they tried to find their own place in the ocean of education.

The term Extension Education as defined by Leagans (1961) Extension education is an applied science consisting of content derived from research, accumulated field experiences and relevant principles derived from behavioural sciences synthesized with useful technology into a body of philosophy, principles, content and methods focused on the problems of out of school education for adults and youth. In all these efforts it is emphasized enough on 'Ex-Tension', the term 'In-Tension of extension' remained unexplored. What we mean to say is that, now it is opportune time to look into the term 'in-tension of extension'. *The term 'in-tension of extension' is conceptualized as "the degree to which the concept, philosophy, objectives, principles, theory and models of 'ex-tension' have been internalized by the extensionists to make the farmers/people responsible to get the work done".*

Since long experiments are going on with the extension system in India. Extension system at the national level was first introduced in the year 1952, but before that extension efforts were going on for the upliftment of rural life. A brief list of extension efforts will make a bench mark for development of extension system in India. There were two distinct types of efforts made to do the welfare of the rural people. First some benevolent persons and private agencies made some efforts to improve rural life, second government efforts were made to initiate some projects to solve the emerging problems.

A bird's eye view of the various projects/prgrammes taken up throughout the nation may be with internal resources or with the help of external resources. The outcome of the programmes/projects can be summarized in the perspective of relationship of the project/programme with the farmers. According to Rivera (1989) there were four types of relationship that existed with the farmers. The relationships that existed were,

1. Take it or leave it
2. Take it or demand different packages or programmes
3. Take it Or else
4. Take it or run away

The earlier extension efforts taught the field extension functionaries as well as extension educators how the farmers perceive the extension initiatives. All efforts made so far have been emphasizing on participation, may be directly or indirectly. But all these efforts put together could achieve little as far as 'participation' of community is concerned. The study conducted in this regard reveals the farmers perception of different approaches (Hagmann *et.al.* 1996) as under: Forceful methods were followed; only a few people could benefit; we were told to do things without questioning; there was no dialogue between farmers and extensionists; there was little cooperation among farmers; Extension agents treated our fields as if they owned them

The National Commission on farmers has drawn attention to the knowledge deficit, which constrains agricultural productivity. To over come this, farmers need to have an effective linkages with Universities and best practices. A good extension system is the means for achieving this linkage, which for the present has virtually collapsed in most states, partly as a result of constraints on non-plan expenditure. As a result farming practices in large parts of the country are sub-optimal.

Emerging Challenges

Recommendations of working group on Agricultural extension for formulation of Eleventh five-year plan, Constituted by Planning Commission, Govt. of India (2007-12) which submitted its report on 12th January, 2007 emphasized on significant growth in agriculture as pre-requisite for engineering country's economy on fast tract. Support to agri-based and expanding export market besides ensuring national food security needs focus in agricultural growth. The combined effect of an unabated population growth, improved family income, increasing industrial needs and rising export requirement has been causing drastic changes in the demand pattern for agricultural commodities in the liberalized era.

The problem is compounded by the fact that the farm holdings in the country are shrinking in size, production costs are rising, and the resource drain from the farm sector is mounting in recent decades.

By the year 2010 and 2020, we need to reach a production level of about 247.8 MT and 296.6 MT of food grains respectively from the present production level of 206.39 MT (2004-05). The projected production must emanate from improved resource productivity. Since, the output of new technology gradually slowed down and rather, for more than two decades no new break-through technologies have been offered. Substantial yield gaps still exist in major crops with the available technology. Research indicates that the available technologies have not been adopted effectively and efficiently, further more notable unevenness and distortions were observed wherever adopted. Additionally, the main global forces of change that are affecting or are likely to affect the existing structure, mandate and practices of national agricultural extension systems are globalisation and market liberalization; privatization; commercialization and Agribusiness; democratization and participation; environment concerns; information technology break through; sustainable development; Eco-technology and genetic engineering; market led extension; diversification and value addition integrated, multidisciplinary and holistic development.

It is true that the growth in the agriculture sector is not the sole responsibility of the extension service. Many other factors including policy interventions, price incentives, institutional support and market facilities also play a significant role. Since agricultural production is entirely in the hands of the farmers, the extension service has the central role in informing, motivating and educating the farmers about the available technological, managerial and market opportunities and enable them to improve their farm productivity and income.

In order to address the key constraints faced by extension system and to meet the emerging challenges the Innovations in Technology Dissemination component of National Agricultural Technology Project (NATP) was implemented in seven states in the country through four project districts in each state. Agricultural Technology Management Agency (ATMA) approach and the impact were documented by an independent agency, the Indian Institute of Management (IIM), Lucknow. It revealed that there is a remarkable achievement of institutional and operational reforms in addition to the significant project impact. Thus, with the backdrop of changes and keeping in view the futuristic needs, direction and vision, Extension Reforms was initiated in 252 districts all over the country during the Xth five-year plan based on the NATP experiences.

Extension reform was a major intervention in overhauling the extension system for making it farmer driven and farmer accountable through process and institutional reforms mechanism in the form of Agricultural Technology Management Agency (ATMA) at district level. It operationalises the extension reforms with focus on reforming public sector extension, Decentralized decision making, Farming systems approach, Bottom-up planning, Group approach in Extension, Promoting private sector, Augmenting media and information technology, Mainstreaming gender and Capacity building of various stakeholders. Commitment to promote public-private partnership in agricultural extension management is demonstrated by reserving minimum 10 per cent of the funds. Similarly, considering the need for gender concerns to be mainstreamed in agricultural extension, minimum 30 per cent of resources on programs and activities are earmarked for women farmers. ATMA promotes an active participation of farmers/farmer groups, NGOs, Krishi Vigyan Kendras, Panchayati Raj Institutions and other stakeholders operating at district level and below. In addition to support to State Extension Programmes for Extension Reforms, the Department of Agriculture and Cooperation (DAC) has initiated number of schemes to revitalize the agricultural extension system in the country, duly incorporating the elements needed for reforms. These Schemes are Mass Media Support to Agricultural Extension–utilizing infrastructure of Doordarshan and All India Radio; Kisan Call Centres–for providing agricultural information through toll free telephone lines; Establishment of Agri-Clinics and Agri-Business Centres–by providing self-employment opportunities for professionally qualified agricultural graduates facilitating delivery of value added extension services and finally Extension support to central institutions.

Salient Recommendations on Policy Reforms in Agricultural Extension in XIth Five Year Plan

(As mentioned in Recommendations of working group on Agricultural extension for formulation of Eleventh five-year plan, constituted by Planning Commission, Govt. of India (2007-12):

- ☆ Extension is the most important instrument for agricultural development; hence, extension Reforms Scheme may be up-scaled in all the 588 development districts in the country for strengthening extension system by establishing Agricultural Technology Management Agencies (ATMAs).
- ☆ There is a need to establish a well equipped State Agricultural Management and Extension Training Institute (SAMETI) at state level, full complement of staff in ATMA, and strengthening a training institution at the district level for effective delivery of extension system.
- ☆ Since all the districts have Krishi Vigyan Kendras (KVKs), generally one per district, the mandate and functioning of KVKs should be clearly defined, and KVKs and ATMAs must complement activities of each other and should avoid duplication. Synergy between these institutions should be clearly promoted.
- ☆ Progressive farmers from different commodity groups may be identified, trained at District, State and National level. They may be recognized as resource persons in extension activities, further promoting the farmer-to-farmer extension.
- ☆ Wide publicity is required for popularizing Kisan Call Centres among farming community and other stakeholders.
- ☆ Diploma in Agricultural Extension Services for Input Dealers (DAESI) programme may be up scaled to entire country so that the services of input dealers can be utilized as Para extension workers at village level.
- ☆ Create mechanism at district and block levels for integration of the efforts of multiple agencies involved in research and extension activities both in public and private sectors.
- ☆ To monitor effectively the Operationalisation of Research–Extension linkages, a separate Monitoring and Evaluation Cell (Research-Extension-Farmer and Market Coordination Committee) to be operationalised with the Agricultural Production Commissioner (APC) in each state. Similarly, at district level ATMA should ensure R-E-F-M linkages. Farm schools and

farmers organizations would be promoted on a massive scale in extension work.

- ☆ National Institute of Agricultural Extension Management (MANAGE) may be promoted as International Centre of excellence in Agricultural Extension Management laying greater emphasis on capacity building activity at National level and SAMETI at State level.
- ☆ Establishment of a national forum for research and extension as a clearinghouse for various developmental issues in research and extension.
- ☆ Adequate provisions are made under Extension Reforms for involvement of PRI's, which need to be effectively implemented.
- ☆ Agri-preneurs may be identified, trained under Agri-Clinics and Agri-Business Centres Scheme and established with the help of ATMA's. Such Agri-preneurs may be linked to commodity specific groups. AC-ABCs may also be associated in implementation of ATMA activities.
- ☆ Agri-Clinics and Agri-Business Centers Scheme may be continued with strengthened handholding mechanism involving state governments, ATMA's, Agri-Business companies and Banks.
- ☆ Agro-Polyclinics may be established at block level by private sector or in Public Private Partnership (PPP) mode to provide multiple services to farmers.
- ☆ Awards may be instituted at different levels in recognition of services/ achievement of farmers and extension service providers.
- ☆ Sustained Capacity Building of Public Extension Functionaries should be taken up for providing quality extension services.
- ☆ Extension intervention by the private sector should preferably be in project mode with quantifiable results and accountability aiming at maximizing profit of farmers. Private initiatives may be supported by government schemes, fiscal incentives.
- ☆ There is a need to strengthen Extension Education Institutes (EEIs) as Zonal level Training Institutions liberating from State Agricultural Universities (SAUs) control.
- ☆ In all the trainings programmes at all levels, women participation must be ensured. Specialized training for women trainers; women farmers have to be organized. Across all the programmes, allocation of minimum 30 per cent of the resources may be ensured for training of women functionaries/ women farmers.
- ☆ Involvement of private training/Human Resource Development/ consultancy institutions in training programmes in specialized areas.
- ☆ Placement of core faculty in management areas is needed at SAMETIs to inculcate managerial competencies in extension functionaries.
- ☆ To organize the Farmers field schools on massive scale, adequate trained manpower may be ensured by State Agricultural Universities (SAUs), KVKs, SAMETIs and State Agriculture and allied sector departments.

- ☆ Training Need Assessment (TNA) at grass root level to be conducted by ATMAs/SAMETIs.
- ☆ Specific schemes like Women in Agriculture with modifications should be continued till gender mainstreaming is made operational at various levels.
- ☆ All the stakeholders in agricultural extension may be sensitized in gender issues.
- ☆ Self Help Groups may be supported with better credit, marketing facilities and support services.
- ☆ Gender budgeting may be introduced meaningfully in agricultural extension programmes. The programme content, service delivery and the business models of Mass Media and Information and Communication Technology (MM&ICT) should specially address the needs of small and marginal farmers.
- ☆ KVKs should play a key role in support of the Mass Media and ICT initiatives in each district. The Community Radio Centres may be established at each KVK so as to provide location specific information to the farming community.
- ☆ A dedicated TV channel for agriculture may be launched, if feasible.
- ☆ Mass Media and ICT attempts by public, private and NGOs in the field of agriculture and related sectors need to be converged and shared among them.
- ☆ The gaps identified in Strategic Research and Extension Plans (SREPs) of each ATMA district and Frequently Asked Questions (FAQs) of Kisan Call Centers may be used for production of the messages/programmes for Radio, TV, print and content for training Programmes.
- ☆ There is a need to improve the awareness among farmers on Kisan Call Centres (KCC)- particularly its cost free services through toll free telephone so as benefit needy farmers.
- ☆ Expertise of level-I of KCC needs to be improved.
- ☆ FM Radio should have a provision for advertisement while broadcasting of agriculture programmes as a business model to generate revenue under Public-Private-Partnership mode.
- ☆ For decentralized decision-making, effective coordination, monitoring and concurrent evaluation of centrally sponsored Programmes, Zonal level coordination units similar to Zonal coordination unit of KVKs need to be established.

Conclusion

A vast majority of the social system nominclated as Aphonic Persons (in this case), have not been able to participate actively in the process of democratic development. Participatory method is an attempt of harnessing their potentialities for making them active partners in the process of change. This will make them aware of their rights, arouse initiations, realize their inner strength, absorb useful information and utilize them for achievement of meaningful goals. This will be a regenerative process for empowering them. The public extension system needs change for better services by structural reforms and improving function through transferring the ownership to and sharing responsibility of its management with community and by building synergic relationship between community and the public extension system. Understanding the in-tension of ex-tension would strengthen the role of public sector and it would facilitate introduction of common regulations against the particular interests of individual members. Appreciation of the in-tension of ex-tension by public extension system can be achieved through participation of the community. Support to State Extension Programmes for Extension Reforms, the Department of Agriculture and Cooperation (DAC) has initiated number of schemes to revitalize the agricultural extension system in the country, duly incorporating the elements needed for reforms. These Schemes are Mass Media Support to Agricultural Extension–utilizing infrastructure of Doordarshan and All India Radio; Kisan Call Centres–for providing agricultural information through toll free telephone lines; Establishment of Agri-Clinics and Agri-Business Centres–by providing self-employment opportunities for professionally qualified agricultural graduates facilitating delivery of value added extension services and finally Extension support to central institutions. It can be concluded with saying of Lao Tsu, China "Go to the people, live with them, learn from them, love them, start with what they know, build with what they have, but with the best leaders, when the work is done, the task is accomplished, the people will say we have done this ourselves.

References

Ackoff, C., 1961. Quoted by Ambasta, C.K. Unpublished *Ph.D. Thesis.* "Communication Patterns in Farm Innovation Development, Extension and Client Systems in Bihar: A Systems Approach". Division of Agril. Extension, IARI, New Delhi 1974.

Agrawal, A. and Narain, S., 1989. *Towards Green Village: A Strategy for Environmentally–Sound and Participatory Rural Development*. Centre for Science and Environment, New Delhi.

Arundale, A.C., 1968. A System Theory: A Sociological View. In: *Walter Buckley. Sociology and Modern Systems Theory*. Eglewood Cliffs, New Jersey, Prentice Hall, Inc.

_____1990. *The Manila Declaration on People's Participation and Sustainable Development*. Ifda dossier 75/76 January/April.

_____1989. *World Commission on Environment and Development Our Common Future*. Oxford University Press.

Beltrans, L.R., 1974. Rural Development and Social Communication: Relationship and Strategies. In: *Cornell CIAT International Symposium on Communication Strategies for Rural Development*, March 17-22. CIAT Headquarters, Cali, Colombia.

Berrien, F.K., 1968. *General and Social Systems*. Rutgers University Press, New Brunsurck.

Byrnlo, F.C. and B.K.J., 1971. Agricultural extension and education in developing countries. In: R.Weltz (Ed.) *Rural Development in Changing World*, Cambridge Press.

Cassata, Mary and Asante Molefi, K., 1979. *Mass Communication Principles and Practices*. Macmillan Co. Inc., New York.

Chambers, R., 1983. *Rural Development: Putting the Last First*. Longman, London, Lagoes, New York.

Corea, G., 1990. *An International Development Strategy for the 90s Global Stakes Require a New Consensus*. Ifda dossier, July/Septemebr 78.

Dahl, Robert A., 1975. *Modern Political Analysis*, 2nd edition. Prentice Hall of India Private Ltd., New Delhi.

Extension Digest, 1995. *Participatory Watershed Management*. MANAGE, Hyderabad. Vol. 2.

Extension Digest, 1995. *Handing Over Stics to Farmers*. MANAGE, Hyderabad, Vols. 3&4.

Felt, J.H., 1974. Gatekeepers in agricultural information dissemination. In: *Cornell-CIAT International Symposium on Communication Strategies for Rural Development*, March 17-22, CIAT Headquarters, Cali, Colombia.

Fromm, E., 1956. *The Same Society*. Routledge and Kegan Paul Ltd., London.

Hall, A.D. and Fagen, R.E., 1968. Definition of systems. In: Walter Buckley (Ed.) *Modern systems Research for Behavioural Scientists*. Aldin Publishing Company, Chicago, U.S.A.

Jacobs, P. and Munro, D.A., 1978. *Conservation with Equity Strategies for Sustainable Development*. IUCN Publication, U.K.

Khoshoo, T.N., 1990. *Towards a Sustainable Society*. The Times of India, New Delhi, 2nd March.

Laflin, M., 1980. *Environmental Communication: New Roles Hard Choices*. Development Communcation Report, No. 65, Arlington, U.S.A

Laing, R.D., 1973. *The Politics of Experience*. Ballomtine Books, Inc. 101 Fifth Avenue, New York.

Leagans, Paul J., 1971. Extension education and modernization. In: Leagans, Paul J. and Loomis, C.P. (Eds.) *Behavioural Change in Agriculture Concepts and Strategies for Influencing Transition*, Cornell University Press, Ithaca and London.

Marcuse, H., 1964. *One Dimensional Man*. Routledge and Kegan Paul Ltd., London.

Marx, K., 1972. Alienated Labour. In: W.A. Finifter (Ed.) *Alienation and the Social System*. John Wiley and Sons, Inc., New York.

Miles, Mathew B., 1964. *One Temporary Approach*. Dell Publishing Company Inc., New York.

Pandey, S.R., 1984. *Alternative Development Communication Efforts for Science Education in Rural India*. Ifda dossier, 39 January/February.

Pietila, H., 1990. *Environment and Sustainable Development*. India dossier 77 May/ June.

Rapport, A., 1968. *Modern Systems Research for Behavioural Scientists* (Ed.) Walter Buckley. Aldin Publishing Comany, Chicago, U.S.A.

Recommendations of Working Group, 2007. Agricultural extension for formulation of Eleventh Five-year Plan (2007-12) 12th January, 2007 by Working Group on Agricultural Extension Constituted by Planning Commission, Govt. of India.

Seeman, M., 1959. On the meaning of alienation. *American Sociological Review*, 24.

Shiva, V., 1989. *Staying Alive Women, Ecology and Survival in India*. Kali for Women, New Delhi.

Stephen, T.S., 2003. *Basic Principles of Project Formulation for Voluntary Organizations*. PDC Network, Bhubneshwar.

Westley, Bruce H. and Maclean, Malcolm S., 1957. A conceptual model for communication research. *Journalism Quarterly*, 34(31-38).

Chapter 6

Transgenic Crops, Biosafety and Sustainable Development: Policy Reflection

☆ *R.N. Padaria*[1], *Baldeo Singh*[2], *Leela Ram*[3], *Surya and Pankaj Kumar*[4]

Introduction

Amidst several concerns and controversies about transgenic crops, India embarked upon deployment of *Bt.* cotton for commercial cultivation to address the disquitening agrarian issues of escalating cost of cultivation and indebtedness as well as pesticide-led ecological disturbances and health hazards in cotton belts. There is no denying the fact that crop biotechnology has emerged as a potential tool for addressing the problems of agricultural production and food security. Successful development, deployment and adoption of transgenic crops across the world have drawn the attention of policy makers to provide impetus to biotechnological interventions to harness potential edge and benefits. Beginning with acreage of 1.7 m/ha across six countries in 1996, the global acreage under approved biotech crops increased to 114.3 million hectares in 2007 as per ISAAA Brief, 2007. Development of Bt.cotton

---

1 Senior Scientist

2 Joint Director (Extension)

3 PG Research Scholars

4 SRF, Division of Agricultural Extension, IARI, New Delhi – 110 012.

hybrids and their use among 12 million farmers across 23 countries amply demonstrate the supremacy and the cutting edge of biotechnological tools and techniques in crop improvement and progress of mankind.

In India, research programmes have been taken up for development of various transgenic crops (Table 6.1). As far as deployment is concerned, India remained watchful for some years and after some successful trials, gave consent to commercial cultivation of Bt. cotton. The first approval for commercial cultivation of *Bt.* cotton in India was granted to three cotton hybrids MECH 12 Bt, MECH 162 Bt, MECH 184 *Bt.* developed by Mahyco. Following the approval of GEAC, commercial cultivation of *Bt.* cotton was undertaken during 2002 in 6 states in India *i.e.* Andhra Pradesh, Gujarat, Madhya Pradesh, Karnataka, Maharashtra and Tamil Nadu and the total area under cultivation was 0.038 m/ha.

By 2007, its acreage soared to 6.2 million hectares. It speaks of growing conviction, acceptance and demand of *Bt.* cotton hybrids among the farmers.

Table 6.1: Transgenic Crops Under Field Trial in India

Sl.No.	*Crop*	*Trait*	*Organization*
1.	Brinjal	Insect resistance	Indian Agricultural Research Institute (IARI), MAHYCO, Sungro Seeds Ltd.
2.	Cabbage	Insect resistance	Sungro Seeds Ltd.
3.	Cauliflower	Insect resistance	Sungro Seeds Ltd.
4.	Corn	Insect resistance	Monsanto (India), Metahelix life sciences
5.	Cotton	Insect resistance	University of Agricultural Sciences (UAS, Dharwad), Mahyco and other leading seed companies
6.	Ground nut	Virus resistance	ICRISAT
7.	Mustard	Cytoplasmic male sterility	University of Delhi South Campus (UDSC)
8.	Okra	Insect resistance	MAHYCO
9.	Pigeon pea	Resistance to fungal pathogens	ICRISAT
10.	Rice	Resistance to pests: bacterial blight, fungal pathogens, salt tolerance	IARI, MAHYCO, Methelix Life Sciences
11.	Tomato	Insect resistance, Virus resistance	IARI, MAHYCO

Source: *Bio safety News Letter*, 2005, Vol. 1(2).

Benefits of Bt. Cotton

The cultivation of *Bt.* cotton offered many advantages and benefits to the farmers. American bollworm has been the most dreaded insect pest of cotton and of the total pesticide application in cotton it appropriated the maximum for its management. In India, an estimated about 28 billion rupees worth of pesticides is used in agriculture, of which rupees 16 billion is used on cotton. The most devastating insect pest, the bollworm, alone takes a heavy toll, costing the farmers an annual rupee 11 billion. This accounts for more than a third of pesticides sales in India (Datt, 2001).

However, with advent of Bt. cotton hybrids, wherein the incorporated *Bt.* gene (*cry 1AC*) provides security to the hybrids against the American bollworm, there is no need of pesticide application for its management. As the insect feeds upon the leaves or any part of plant of *Bt.* cotton, the crystal protein in the gene *cry 1AC* binds and kills by destroying the intestinal gut of the insect. In a study the farmers of Punjab and MP reported that with adoption of *Bt.* cotton hybrids in the initial years, the average number of total pesticide spray has reduced by about 60 per cent and as a result the average cost on total pesticide spray reduced by about 70 per cent. Though with emergence of mealy bug insect pest, farmers are baffled about euphoria of *Bt.* cotton. However, in comparison to pre-*Bt.* scenario when cotton area witnessed rampant and frequent sprays of lethal pesticides, the average number of health hazards due to pesticide spray (number of persons affected by pesticide induced health problems per season per village) had reduced significantly, even by over 89 per cent. Besides these benefits, farmers secured 38-67 per cent higher average yield and over 142 per cent higher monetary return from *Bt.* cotton hybrids than the non-Bt cotton hybrids/ varieties. Farmers of Rajasthan too reported that the number of pesticide sprays had reduced with introduction of Bt. cotton. While majority of Bt. cotton farmers (63.7 per cent) practised 1-3 number of sprays, the corresponding percentage in case of non-Bt. cotton was 15 per cent. Fifty five per cent of non-Bt. cotton farmers applied more than 5 sprays. Highly significant inverse relation of plant protection with profit showed by the chow test suggests over use of pesticides by farmers. They could earn profit even with low level of pesticide use. The higher Benefit Cost Ratio (2.6:1) for Bt. cotton in comparison to non-Bt. cotton (1.6:1) amply showed that Bt. cotton was profitable.

Experiences of Other Countries

The use of genetically modified cotton in China led to several benefits like reduction in pesticide use by 15,000 tonnes, higher yield and lower cost of cultivation. At present China is the biggest producer of cotton in the world.

Similarly, *Bt.* Cotton has proved to be highly advantageous to the US farmers as well. Total acreage of cotton production increased from 12 per cent of total land in 1996 to 36 percent in 2000, estimated increase in the yield was about 260 million pounds and the revenue increased by approximately $ 99 million. The average insecticide application also decreased from 10 units through 2 units to 0 units.

Farmers of South Africa too derived benefits by way of increase in yield in range of 23-26 per cent and reduction in use of pesticide spray by 4-6 per cent.

Biosafety Concerns and Risk Perception

Enhanced acreage, yield and return and growing demand notwithstanding, the reports of mixed response of their on-farm performance have kept the nation wide debate upon the use of transgenic crop alive. The transgenic technology faces criticism primarily for the bio-safety concerns, which include transgene movement to other varieties and wild relatives leading to possible development of super weeds, erosion of genetic diversity and ecological disturbances; widespread apprehension about toxicity or allergencity induced by transgenic products to humans and animals; its adverse impact on non-target organisms; emergence of more virulent forms of pests and pathogens; etc. Besides biosafety concerns, IPR, ethical and moral issues form the basis of debate towards adoption of transgenic crops. There are cases of undesirable results of transgenic technology, which further strengthen the apprehension about such technology. For example, transgenic soybean genetically modified to increase its sulphur containing amino acids by incorporating Brazilian nut *2S* gene was found allergenic to serum of people who are allergenic to Brazilian nut *2S* protein. Report of death of cattle by grazing Bt. cotton plants in Andhra Pradesh also fueled the debate on use of transgenic crops. Though farmers at large are greatly fascinated with the benefits of *Bt.* cotton, risks perception among them is still a challenging hurdle to push through transgenic food crops like *Bt.* rice, *Bt.* maize, *Bt.* brinjal and tomato etc. Farmers are very forthright in conveying their conception that since *Bt.* cotton kills the insect, a living being, it will certainly kill animals and other living beings. Burning of *Bt.* rice and destroying of *Bt.* brinjal trial lots by farmers amply show the level of concerns of bio-safety issues related with transgenic crops and technology. Scientific knowledge and arguments to clear the clouds of misconceptions and promotion of public awareness about transgenic technology haven't really percolated among the public and farming community *per se*.

Knowledge and Skill Gaps: Missing Links and Disenchantment

Though farmers accepted the potential advantages of *Bt.* cotton hybrids, they were cautious in their fuller adoption. Mixed pattern of cotton cultivation *i.e.*, putting acreage under *Bt.* cotton hybrids, unapproved *Bt.* cotton hybrids (in case of Punjab) and non-Bt cotton hybrids was observed. Though the pace of adoption was phenomenal, the cases of disenchantment with *Bt.* hybrids, where the expectations were not met, were the concerns of significance to boost up the adoption and fuller utilization of the technology. As the public extension system is not in fray for dissemination of knowledge about transgenic technology, there was enormous void related to knowledge and skill about proper application of *Bt.* cotton technology. Since *Bt.* cotton was deployed by private seed agencies, the extension services related to it were largely carried out by them, whose major focus was on promotion of sale of seeds and as a result the issues of biosafety, agro-techniques and risk communication remained unattended. There was enormous lack of communication about Bt. technology and related issues. Farmers just knew that *Bt.* cotton was a technology, which saved crop against American bollworm without use of pesticide, but they hardly knew that all *Bt.* hybrids were not same and different hybrids were suitable for different situations. For example RCH 2 was suitable for light soil and MECH 162 was suitable for heavy soil. *Bt.* hybrids for rainfed and irrigation systems have to be different. However, in want of proper knowledge the farmers just in the name of *Bt.* cotton adopted *Bt.* hybrids irrespective of their situation and location specificity and as a result the performance was poor and disenchantment set in. Similarly some *Bt.* hybrids (*e.g.* MECH 184) due to nutritional deficiency suffered from reddening of leaves and yield was affected. Farmers lacked knowledge about agro-techniques of *Bt.* cotton cultivation. Even spacing and plant density varied which led to higher seed rate as farmers followed the same crop geometry as with earlier hybrids/varieties keeping in view their plant vigour. They were not communicated about importance and use of refuge lines. Many farmers just kept aside the packet of non-Bt hybrids, which were given to them for sowing as refuge lines with the opinion that these would invite the problem of American bollworm again and affect the productivity of *Bt.* hybrids. Many farmers poöled the packets of non-*Bt* hybrid seeds and sowed them in a separate plot to save *Bt.* plot from American bollworm. Realizing the gap in knowledge and skill among the farmers about the transgenic technology, information and training needs of the farmers need to be assessed and addressed through training, interface and information dissemination.

Farmers' Perception about *Bt.* Cotton and Transgenic Technology

Affirmation with the positive dimensions of *Bt.* cotton hybrids that they are high yielder and eco-friendly and they reduced pesticidal pollution and health hazards revealed the positive perception of farmers about *Bt.* cotton. The positive perception of the farmers using *Bt.* cotton was further demonstrated by their disagreement with respect to statements which present negative shades of *Bt.* cotton and related implications like inducing genetic pollution, dangerous to living beings, destroying soil microorganisms, ineffectiveness against American Cotton boll worm, inducing allergencity, increase in labour man days, picking being cumbersome, loss of farmers' autonomy for seed with enhanced dependency on seed companies and their monopoly, being pro-resource rich, failing in benefiting small farmers, etc. Farmers agreed to the statement that it was advantageous if public institutions would do seed production of *Bt.* cotton. They disagreed with the statement that *Bt.* cotton was pro-resource for rich farmers; instead they had an opinion that it benefited resource poor farmers also. Farmers were divided over the issue of transgenic food crops. Some showed willingness to cultivate transgenic food crops if they were profitable to them and if approved by scientists many of them told that they would consume transgenic food crops. They expressed that even if *Bt.* food crops could be toxic they would be far less toxic than what we consumed with heavy pesticidal residues due to the rampant sprays taken up for food crops.

However a majority of farmers showed indecisiveness in want of proper knowledge about the heath implications of transgenic food crops.

Suggested Policy Initiatives for Safe Deployment and Dissemination of Transgenic Technology

Notwithstanding the numerous field–level constraints like rampant sale of spurious seeds, unavailability of genuine seeds, black marketing of seeds at exorbitant rate, lack of advisory services, emerging threats from other pests like mealy bug, etc, the experience in cultivation of *Bt.* cotton has certainly established the importance of transgenic technology in terms of profitability and ecological benefits. Besides, it also strengthened the process of positive and favourable perception building among the end-users about the potential benefits of transgenic technology in general and *Bt.* cotton in particular.

It reflects that farmers will accept cultivation of transgenic food crops if convinced of being profitable and devoid of health hazards. However, still not only educational drive is necessarily required to upgrade the knowledge and skill of farmers and other end users as well as the anti-transgenic social activists but also what is desired is availability in public domain of rigorous scientific experimentation based evidences and explanation about the facts of biosafety and health related concerns. Concerted efforts of educational campaign, scientists and end-users interface, extension literature and aids, large-scale demonstration and on-spot scientific explanation of biosafety concerns and misconceptions will go a long way in developing favourable perception and enhanced acceptance of transgenic technology.

Further, efforts should be initiated to address the field level and infrastructural constraints to facilitate development of supportive and caring climate for the end-users. The major suggestions to overcome the deployment and dissemination concerns include: recommendations of *Bt.* hybrids to be made according to their suitability to agro-ecological conditions; standardization and dissemination of agro-techniques; development of communication content and dissemination of extension literature related to *Bt.* cotton cultivation, *Bt.* technology and related biosafety concerns as well as organization of scientists-farmers interface to bring clarity about the transgenic technology and to remove unfounded apprehension about the risks of transgenic technology among the farmers and end-users; and ensuring easy availability of quality and authentic seeds of *Bt.* hybrids.

Emphasis should be laid upon augmenting the role of public extension system in deployment and dissemination of *Bt.* technology. Since the farmers find public extension system as the most credible and reliable source of information, it is essential to develop programmes for capacity building of extension workers in transgenic technology so that they could put in the desired educational efforts, which often is overlooked by the seed agencies at present.

The nation-wide network of KVKs must be utilized in deployment and dissemination of transgenic technology through involvement in field trails and extension programmes.

Necessary initiatives are required in areas of Biosafety regulation mechanism, capacity building of stakeholders for prudent compliance of biosafety measures and involvement of farmers and extension workers in deployment of transgenic to harness the potential benefits of transgenic technology with adequate biosafety.

Since the country is soon going to embark upon deployment of transgenic food crops like Bt. Rice and Bt. Brinjal, etc, it is high time to ensure flow of information with scientific reasoning and evidences to foster development of positive perception and wider acceptance among public. Many a social activist groups are campaigning against transgenic technology as public lack valid and genuine reasoning about the biosafety concerns. Labelling of transgenic food crops and regulation of markets with strict compliance of biosafety measures will be highly essential. Hence, foolproof policy needs to be developed and implemented.Public-private partnership should be emphasized upon research and development of transgenic technology to safeguard the farmers against the market monopoly of trade of seeds of transgenic crops.

Chapter 7

Conceptualizing Agricultural Extension Management for Rural Development

☆ ***A.K. Singh[1], Shantanu Kumar[2], Uma Sah[2] and S.K. Singh[2]***

Introduction

The present systems of Agricultural Extension in India are the outcome of several efforts and reforms made over the years for improving the rural life of the country. Prior to independence, the alien rule under East India Company was largely concerned with maintenance of law and order rather than rural development in sincere sense. It is a historical fact that recurrent famine from 1875 to 1901 forced the colonial government to think about agricultural and rural development. The famine commission of 1901, amongst other things, recommended the appointment of experts capable of applying scientific method to the improvement of agriculture. The government of India act 1919 empowered the transfer of all the departments closely connected with the rural development to the major provinces and agriculture became a state subject. The Royal Commission on Agriculture in India (1928), observed that agricultural research can be of no help to the cultivators until its results are given to them in a form in which it becomes a part of their agricultural practices. The commission made some

1 Zonal Coordinator, ZCU, ICAR, Kanpur (UP).

2 Indian Institute of Pulses Research, Kanpur.

important recommendations for action by the department of agriculture, which paved the way for development of agricultural extension in India.

The reforms and policies of independent India introduced many programmes and projects essentially focused to help farmers in providing technological inputs, information, knowledge and skills in order to bring overall development of agriculture in the country. The experiments indicate that efforts initiated needed far more thrust on effective planning, implementation, monitoring and evaluation. Albeit, it is important to know and identify what are the important management components, which are most responsible for making any extension, programme a success. The article at hand, therefore, attempts to address such issues with succinct focus on planning, monitoring and evaluation of agricultural extension programmes.

Antecedents for Effective Management of Agricultural Extension Programmes (AEP)

The success or partial success or complete failure of any development effort including AEP depends by and large on the extent to which it is being managed effectively. Management, literally, means the process of decision making regarding action to be undertaken in due course of time, resources to be allocated for any activity, having vigil on the performance of that activity so as to keep it on the right direction and finally, to see whether the activity under operation has been performed to the desired extent or not. Technically, management refers to the continuous process of planning, deciding, organizing, budgeting, coordinating, monitoring and evaluating the organizational activities through individual or group approach.

Agricultural extension programme involves multi-facet activities. Broadly, it includes educational efforts, input supply and related technical services, and rural institutions building both in agriculture and allied fields. For example, in dairy development Intensive Cattle Development Project (ICDP) attempts to ensure the timely delivery of adequate and good quality technical inputs like frozen semen, vaccines, medicines, fodder seeds, etc. in the form of technical services like artificial insemination, vaccination, treatment of diseased animals, demonstration of technical know-how related to improved fodder cultivation, etc. Similarly, operation flood programme intends to empower dairy farmers through sensitizing and organizing them for cooperative institutions building at various levels. Under the banner of these institutions, several activities like meeting, camp, campaign, demonstration, milk procurement, society and union election, etc. are undertaken. Similarly, dairy development efforts channelized by non-government organization like, *Goshalas*, voluntary agencies, private sectors, etc., also include multidimensional activities. The successful execution of all the activities related to above quoted example of dairy development and also related to agricultural development being carried out by the concerned organizations could be guaranteed only if that are well planned, properly monitored and objectively evaluated. This is equally applicable to the agricultural extension programmes.

More clearly, planning of the activities of AEPs would help to decide whether the programme is need based or not, what type and what amount of resources including human resource are required and what work plan or activity-schedules are to be followed. The precision and objectivity of such decisions will be the determinant of effective planning. Similarly, monitoring and evaluation is the action-oriented management process and instrument. It helps to gather information that assists in determining the relevance of planned pathway, effect and consequences of activities as systematically and objectively as possible. Findings of monitoring and evaluation help to improve concurrent and future activities in terms of proper planning, decision-making and programme execution. Monitoring assists the managers in guiding the programme when it is in execution and implementation stage. Evaluation, on the other hand, intends to see whether the programme created desired impact or not. Therefore, whereas monitoring is ongoing or concurrent, evaluation is generally ex-post or terminal.

All the above three important dimensions of management in relation to AEP have been deliberated in details under the subsequent sub-heads. In addition, attempt has also been made to elucidate the concept of participatory management for more effectiveness of agricultural extension programmes.

Planning of Agricultural Extension Programme (AEP)

This section of the article highlights the conceptual and theoretical base of development programme planning which is based on the Bos's decision-making model (1974). This model offers the extension managers two-pronged pathways towards knowledge and choice in order to arrive at the decisions to be taken either by themselves or in collaboration with other actors in development. On the pathway of knowledge and choice for planning, agricultural development functionaries can either collect the information needed for planning the AEP and activities therein in reciprocal cooperation with cattle owners or they can try to gain the knowledge by themselves or with the help of experts such as bank officials, market research bureau, etc. Both the pathways move through goal and target group analysis. The goal analysis would help in making choice whether the department is interested in educational and awareness activities or farmers' empowerment through rural institution and capacity building or to ensure adequate, good quality and timely supply of technical inputs and related services. Similarly, target analysis provides information on the aspects *viz.*, current/existing behaviour of the target group in the area where agricultural extension organization operates, what do the members of such group consider to be their problems, what knowledge, attitude and skills do the target group members have towards their problems and also towards the problems for which development personnel consider extension help is desirable. In addition, following points are must while analyzing the target group:

1. Which members of the target group make decision, which decision and who influences these decisions?
2. Which communication channels do target group use now and which one could they use in future?
3. What resources do different categories of the target group have access to?
4. What is the target group's situation and how does this influence the members' behaviour?

Similarly, management ought to make choice from the varieties of programme goals, segmented target group, multi-dimensional extension contents and methods to be used, etc. Based on the information obtained after the above analyses, extension managers can easily decide about the quantity and quality of resources required and can make their allocation as per the programme activities. Besides, media or channels to be used and *modus operandi* to be employed in the programme can be successfully selected for effective management of AEP.

Monitoring of Agricultural Extension Programme (AEP)

Monitoring is derived from the Latin word *'to warn'* and it is seen as the management technique wherein development managers gather data on the way in which the development programme is implemented and the constraints, if any, it faces in trying to stay on the right track. In this way, the concerned management is soon informed if the programme deviates considerably from the planned pathway and at the same time, it gives sufficient reasons for such deviation. This, in turn, helps management to take rapid necessary and corrective measures to implement the original plan or to adjust them if they were not realistic. For example, if the village development agent in a particular programme is expected to contact, say, ten farmers in a day who are representative of general agricultural farmers in that area. After monitoring, it may be found that poor transport situation made it impossible to visit not more than five farmers in a day. Moreover, the existing local social dynamics further made it obligatory first to visit the village head. Based on this information of monitoring, associated agricultural development authority will have to take the measures either to implement the original plan by facilitating better transport or to adjust the original plan to the field situation. This type of monitoring is also called on-going or concurrent evaluation. However, such type of monitoring is possible, only when the agricultural development programme is relatively less complex and the development agents have to deliver nearly same kind of message to all categories of farmers. In other words, when the agents are expected to cover large number of farmers with their specific problems, such monitoring is a bit difficult to execute. However, for monitoring any agricultural development programme/project, it would be better if all possible indicators on the several dimensions *viz.*, physical target, financial status, manpower and necessary infrastructure status and finally the benefits actually accrued to the beneficiary dairy farmers as perceived by them are taken into account (Van den Ban and Hawkins, 1998; Kumar *et al.*, 2001).

The format and frequency of monitoring will depend on the nature, size, complexity and structure of the implementing organization as well as the programme under operation. But in general, as suggested by Singh (1986), monitoring system should be linked with the responsibility center in the organization concerned. A responsibility centre, as operationalized by him, is simply a unit or a sub-system (*e.g.*, stockman centre, dairy cooperative society, dairy development center, veterinary dispensary, co-operative union, etc. in case of dairy development) headed by a responsible manager who is accountable for the work done by the unit. These managers of the responsibility centers should be provided with all the information relevant to their needs in planning, decision-making, control and remedial action. Further, it could also be argued for institutionalization of some kind of formal mechanism to monitor the agricultural development programme activities right from village level, block level, district level to the state as well as national level in order to ensure effective management of the same.

Evaluation of Agricultural Extension Programme (AEP)

It is usual to distinguish between evaluation and monitoring (Casley and Kumar, 1988). In the preceding sub-head also, both these management tools have been briefly differentiated. The current section of this article, therefore, attempts to highlight further theoretical clarification on evaluation, its importance and the level and criteria for evaluating any development programme with particular reference to AEP.

Evaluation in its broader sense means judging the value of something formally or informally. All evaluation process essentially involves three basic elements namely, (*i*) observing or collecting some information, (*ii*) applying some standard or criteria to the observed facts and finally, (*iii*) forming some judgment and drawing some conclusions (Reddy, 1993). Evaluation is used to determine whether any development programme has achieved its goal or whether these goals could have been achieved more effectively in different way. This is equally true for evaluating the agricultural extension programme. Darcie *et al.* way back in 1959 had identified two broad degrees of evaluation that included informal and formal evaluation. Informal evaluation contains the segments like, casual-everyday evaluation, self-checking evaluation and do-it-yourself evaluation. Formal evaluation, on the other hand, is more complex and uses more scientific approaches. Most of the extension/development studies (ex-post facto researches) and other scientific researches leading to determine cause-and-effect relationship come under the domain of formal evaluation. Evaluation of AEP may assume either of the degrees depending on the size of programme, nature of the implementing organization and purpose of the evaluation.

Evaluation may have several functions for development agents including agricultural extension managers as identified by Van den Ban and Hawkins (1998). It can provide information on which these agents/managers may base their decisions. In addition, evaluation information helps agents to set more realistic goals and assists them to discover more effective methods to reach these goals. One can begin on a small scale with a pilot project and evaluate this thoroughly so that work can be made more effective when starting a new programme on large scale. Moreover, evaluation of a complete programme provides no useful information for improving the programme, but the experiences gained in this way can be valuable for planning future programme or improving the other programme in pipeline. For example, in dairy development programmes, there could be several aspects *viz.*, input delivery system (Kumar *et al.*, 2001), health and breeding service delivery system (Ahuja *et al.*, 2000), beneficiaries' participation in dairy development programmes (Sharma and Rao, 1998), impact of dairy cooperatives (Shah *et al.*, 1995), etc., that could be evaluated ex-post. However, for proper evaluation of any AEP, it is essential to identify the levels of evaluation and also to ascertain the criteria corresponding to each level that ought to be used as the operational measurement of each level. Bennett (1976) attempted to delineate several levels while analyzing the impact of extension programme in general. The details have been furnished in Table 7.1. It is evident from the table that any extension programme must be evaluated right from programming and planning level to its consequences on target and society as a whole. Table also shows that there are many intermediate levels on which evaluators have to focus their attention. Higher the level becomes, the more accurately they will be able to state how far the goals have been reached. At the same time, higher is the level, it is more difficult to identify the suitable measurement criteria. In addition, choice of the level is very important for selection of evaluation criteria to be used by the evaluator. Though the enclosed model of evaluation has been advocated for general extension programmes, it can be adequately and specifically used by the development managers of agriculture, dairying, fisheries, poultry etc., as well by making appropriate modification, if any required, according to their circumstances and programmes.

Table 7.1: Levels and Criteria to be Used for Evaluating Agricultural Extension/Development Programme

Sl.No.	*Levels*	*Criteria*
1.	Programming of the activities	(*a*) Initial situation
		(*b*) Desired situation
		(*c*) Target group selection
		(*d*) Contents and methodology to be used
		(*e*) Organization and staffing of the programme
2.	Implementation of the programme (Usually called monitoring)	(*a*) Manpower status
		(*b*) Resource status
		(*c*) Direction of the programme
		(*d*) Performance status (if inputs and services are delivered)
3.	Farmers' participation in the programme	(*a*) Extent of participation in meetings, demonstration, camps, campaign, etc.
		(*b*) Extent of request made by the participating farmers
		(*c*) Types/categories of farmers participating in the programme
4.	Farmers' opinion about the programme	(*a*) Contents of the programme
		(*b*) Methodology used
		(*c*) Compatibility to their situation and needs
		(*d*) Understandability of programme contents
5.	Covert changes taken places in the participants	(*a*) Knowledge gained
		(*b*) Attitude changed/formed
		(*c*) Motivation developed
		(*d*) Group norms created
6.	Changes in participants' overt behaviour	(*a*) Extent of adoption
		(*b*) Level of non-adoption
		(*c*) Technological skills acquired
		(*d*) Reasons for technology rejection
		(*e*) Constraints in adoption
7.	Consequences for the target group	(*a*) Desirable/direct/functional consequences
		(*b*) Undesirable/indirect/dysfunctional consequences
8.	Consequence for the society	(*a*) Diffusion effect
		(*b*) Socio-economic up-liftment of farmers

Source: Bennett (1976).

Participatory Management of Agricultural Extension Programme (AEP)

Participation of clients in any development programme is often considered instrumental in making the programme successful, especially for addressing the problems of resource poor farmers. This is equally true in case of AEP. This section of the article attempts to highlight conceptually what, why and how of the peoples' participation in AEP as distilled from the works reported by Weitz (1971) and Van den Ban and Hawkins (1998).

The word 'participation' connotes differently to different development professionals. Some of them find farmers' participation simply as the cooperation extended by them in implementing the programme by participating in meetings, demonstrations, by asking the agents' questions, etc. Others perceived it as the organization of implementation of development activities by the farmers' group. Currently, however, the most acceptable meaning of people participation in development is given as the ways by which rural people could provide information necessary for deciding and planning the programme, identifying the target group, devising the messages and methods suitably, monitoring and evaluating the such programme activities and also sharing the cost, to certain extent, involved in the programme.

Similarly, there are several valid reasons for the desirability of farmers' participation in the decision-making regarding development programmes. Firstly, farmers have accurate and adequate information about their goals, preferences and priorities, situations and circumstances, resource set-up, perception about the technologies and about the social structure and social processes. This information is certainly crucial for effective planning and execution of the programme. Secondly, farmers' motivation could be sufficiently aroused if they are allowed for sharing the responsibility related to their welfare. Most importantly, there are many development agenda that could no longer be addressed through individual decision-making. Therefore, in such collective decision process, participation of the target group becomes mandatory.

It is, however, more important to conceptualize how the participation of beneficiaries could be guaranteed in development programme related activities. It has been experienced that there could be formal or institutionalized as well as informal approaches for the same. Formally, farmers' representatives could be chosen through farmers' organization, elected local members, village headmen, learned elders, etc. These representatives could be made accountable for their decisions to farmers they represent. Participation of farmers can and often be achieved informally. The development agents can carefully listen to the needs and aspiration, goals and opportunity of the farmers belonging to different types in a given area. This information could be importantly placed while planning and designing the development programmes. Similarly, success as well as failure experiences of the farmers with any technology could also be positively utilized in identifying new interventions, messages, etc. and may work more effectively than a formally organized participation system.

The above conceptual aspects of participatory plan and decision making process though appear very general in nature, but offer sufficient scope for their consideration while managing the activities of AEP. The status of beneficiaries' participation for the success and growth of dairy development organization was also empirically reported by Sharma and Rao (1998). They found that the participation of dairy farmers was poor in planning and decision stage than at execution stage. However, their involvement in managing the dairy development affairs was relatively better in well-managed dairy cooperatives than the intensive cattle development projects. Similarly, Kumar (1999) reported that the mutual strength of functional linkages between farmers and dairy development personnel was highest in supply and services (50 per cent) and it was relatively poor in joint planning and decision making (17 per cent) and implementation and evaluation (22 per cent). These empirical evidences definitely reflect that in most of the cases, dairy farmers were the passive recipients of inputs and services rather than the active actors in the management of dairy development programmes. These two examples from the field of dairying also help to infer the poor status of participatory management of agricultural extension programmes as well.

Conclusions and Implications

Based on the theoretical approaches for management of development programmes including AEP as deliberated in above sub-heads, it could be unequivocally concluded that planning, monitoring and evaluation of AEP are the cardinal pillar for successful and sustainable agricultural development in India. The contents of the present article are expected to adequately sensitize the agricultural extension institutions, agents, researchers and extension managers for effective management of agricultural extension endeavours being carried out in their respective areas. Further, it is also revealed from above discussion that beneficiaries' participation in the management of AEP ought to be institutionalized formally. However, care must be taken to strengthen the informal participation of farmers in the programmes as and when such situation is encountered. It could be also suggested for the extension/development researchers to undertake intensive investigative works for empirical comprehension of management processes being followed by various agricultural extension organizations in India and accordingly, some suggestive measures for effectiveness of AEP could be intervened.

References

Ahuja, V., George, P.S., Ray, S., McConnel, K.E., Kurup, M.P.G., Gandhi, V., Deininger, D.U. and Haan, C.De., 2000. *Agricultural Services and the Poor: Case of Livestock Health and Breeding Services in India*. Publication by IIM, Ahmedabad (India), The World Bank and Swiss Agency foe Development and Cooperation.

Bennet, C.F., 1976. *Analyzing the Impact of Extension Programmes*. USDA Extension service, Washington DC.

Bos, A.H., 1974. *Oordeelsvorming in Groepen*. Veenman, Wageningen, The Netherland.

Casley, D.J. and Kumar, K., 1988. *The Collection, Analysis and Use of Monitoring and Evaluation Data*. John Hopkins University Press, Baltimore.

Darcie, *et al.*, 1959. *Evaluation in Extension*. USDA Publication, Washington DC.

Kumar, S., 1999. A study on linkages among research, extension and dairy farmers in Haryana. *Unpublished Thesis*, NDRI (Deemed University), Karnal (Haryana).

Kumar, S., Sah, U. and Kumar, Ram, 2001. Input delivery system: Key to dairy development. *Indian Dairyman*, 53(2): 11–16.

Reddy, A.A., 1993. *Extension Education*. Sree Lakshmi Press, Baptla, Guntur, A.P., India.

Sharma, N.K. and Rao, S.V.N., 1998. Beneficiaries' participation: An approach for success and growth of dairy development organization. *Ind. J. Dairy Sci.*, 50(1): 12–17.

Shah, T., Ballabh, V., Pratima, B. and Talati, J., 1995. Institutional structure for dairy development: India's post-independence experiences. *Indian Dairyman*, 47(1): 1–36.

Singh, K., 1986. *Rural Development: Principles, Policies and Management*. Sage Publication India Pvt. Ltd., p. 354.

Van den Ban, A.W. and Hawkins, H.S., 1998.. Evaluation and monitoring. In: *Agricultural Extension*, CBS Publishers and Distributors, New Delhi, pp. 202–217.

Van den Ban, A.W. and Hawkins, H.S., 1998. Participation of farmers in extension programmes. In: *Agricultural Extension*, CBS Publishers and Distributors, New Delhi, p. 218–232.

Weitz, R., 1971. *From Peasant to Farmer: A Revolutionary Strategy for Development*. Columbia University Press, New York, Ch. 9.

Chapter 8

Indian Agriculture: At the Crossroads

☆ *D.K. Sujan*[1], *K. Ghadei*[2] *and S. Panigrahi*[3]

"Agriculture is the basis of all our development. If we fail in Agriculture, It doesn't matter what else we have got"

– Pt. Jawaharlal Nehru

Introduction

Agriculture, the lifeline of Indian economy is known for its multifarious approach of providing livelihood nutritional and ecological security, besides employing 70 per cent of workforce. India has witnessed new strides in agriculture in post independent era. Still share of industry and agriculture in GDP shows utter disparity with opposite trends. Is not it a strange paradox.

Gone are the days of chronic food shortage and famine. India has transformed from a 'begging bowl' to a self-sustaining to an exporting country in agrarian front. Over the ages our food grain productivity has increased 3.3 times, fruit 1.6 times, vegetables 1.8 times, fish 5.8 times, milk 1.8 times and eggs 4.8 times (Yojana, Jan-2002). But keeping in view, the booming population, this rate is not satisfactory in both quality and quantity aspects. Share of agriculture is gradually decreasing without

---

1 Professor, Department of Extension Education, Institute of Agricultural Sciences, BHU, Varanasi.

2 Lecturer in Department of Extension Education, Institute of Agricultural Sciences, BHU, Varanasi.

3 Scholar in MABM, Department of Agricultural Economics, Institute of Agricultural Sciences, BHU, Varanasi.

proportionate displacement of active workforce. In this context an eye to past efforts will enrich us for future progress.

Pre-Green Revolution

Only with a dismal 51 m T production from 81 m Ha area, agriculture had 52 per cent share in GDP at independence (Pratiyogita Darpan-Indian Economy-p 65-67). In every front we were lagging behind the global standards. From food grains (Rice 22 mT, Wheat 11 mT) to tuber crops (potato 1.45 mT), from rubber to coconuts to fertilizers, agricultural production was in a sorry state with per capita availability of food grains at 395 g and milk at 124 g (Dutta-2002).

Some community development programmes namely Sevagram attempt by Mahatma Gandhi, Works at Shantiniketan by R N Tagore, Marthandam project by Spencer Hatch, Gurgaon Experiment by F L Brayne etc. were notable. After setting up of Department of Agriculture in 1882,the Imperial Agricultural Research Institute (IARI) came into existence in 1905. 'Grow more Food' campaign was started in 1942.But the poor response towards this resulted in formation of community development block comprising 100-200 villages. Etawah Pilot Project in UP in1948 by Albert Mayer and CDP by Govt. of India in 1952 covered 25260 villages and 64 lakh people (India Today-Saran-2002). It was better equipped with more finance and technical manpower for wholesome development of rural community and renamed as National Extension Service in 1953.

'Panchayati Raj' was introduced in 1958 basing on the recommendation of Balwantrai Mehta Committee to promote local leadership in planning and implementation process. Agricultural Universities were started in various states, starting with Gobind Ballav Pant, University of Agriculture and Technology (GBPUA&T) at Pantnagar in 1960; to impart a more coordinated research, teaching and extension.

Post-Green Revolution

Massive food grain production and rapid diffusion of semi dwarf wheat varieties brought about not only self sufficiency in production but also turned to be a driver of growth for agriculture. As a result of which now India is able to produce 221 m T of food grains from 113 m Ha of area. Rice alone has jumped to 90 mT production from 45 m Ha area. Similar is the case with wheat, sugarcane, cotton, rubber and coconut. Significant increase in fruits (4.96 m Ha area and 49.29 mT production) and vegetables (6.75 m Ha area and 101.43 mT production) and milk production has resulted in better nutritional security (The Hindu-Survey of Indian Agriculture-2007).

Various ambitious schemes of socioeconomic development programme were started after green revolution. High Yielding Varieties (HYV) program in 1964-65 has given Indian agriculture a new dimension. Drought Prone Area Programme (DPAP) to mitigate the impact of chronic drought was started in 1970-71. To overcome the demands of input and credit, Small Farmers Development Agency was started in 1974-75, presently representing 19 per cent of total area and 12 per cent population. Krishi Vigyan Kendra and 'Training and Visit' Programmes were launched during 70s'.For upliftment of poor people Jawahar Rojgar Yojna (JRY) took off in 1989.For proper dissemination of innovative technology; Lab to Land Programme (1989) and Institution Village Linkage Programme (1994) were commenced. To conserve natural resources National Agricultural Technology Project (NATP) started in 1998. Several other ambitious programme like IRDP, TRYSEM, DWCRA, SITRA were initiated by Government. Ganga Kalyan Yozna was reconstructed to Swarna Jayanti Gram Swarojgar Yojna in 1999.Food for Work in 2001,ATMA, Pradhan Mantri Gramoday Yojna were some of the steps taken to curb the menacing Unemployment. Sampoorn Grameen Yojna and Pradhan Mantri Gram Sadak Yojna aimed both at rural employment and connectivity. Khetihar Majdoor Vima Yojna and Jay Prakash Kalyan Rojgar Guarantee Yojna during 2001-2003 were also notable regarding rural upliftment. Newly framed employment generating schemes include the National Rural Employment Guarantee Scheme, better known as NREGA.

Challenges Ahead

The growth rate of yield for rice and wheat, pulses have decreased over the years. The over all food grain production had shown a fluctuating and erratic trend in recent past. Still we need to produce more food grains to feed our enormous and ever increasing 1.2 billion strong populations by 2010 and 1.4 billion by 2025(India-2007).

Domestic Demand Contrast and Need for Employment

Green Revolution has created self-sufficiency along side coexistence of food surplus and mass poverty, malnutrition, hunger death and unemployment. Some implications of implementation of WTO norms in India are as follows

- ✰ Ever increasing Unemployment
- ✰ Increasing inter industry share in rural employment
- ✰ Decreasing self employment at expense of increasing actualizations

Supply Side Constraints

Agricultural GDP growth rate is not at par with total GDP. Increase in milk, eggs, wool and fish production and decrease in forestry, inadequate infrastructure and diversification of agriculture along with negligence to allied activities etc all these point to a system failure of teaching, research and extension.

Factor Market Distortion

Absence of broad based lease market, labour policy for unskilled labourers and failure of research and extension are resulting in decline in gross capital formation in agriculture.

External Factor Constraint

Fast falling international prices, high subsidies through direct and indirect foreign investment and non-tariff barriers imposed by developed countries coupled with lack of awareness, organizational research and development support are worries of Indian agricultural export front.

Vision for Future

We have failed miserably to translate our latent demand into effective demand. Decreasing poverty with a nearly steady growth rate of 8 per cent and nutritional security backed by friendly Government policy and corporatisation of agriculture may do the trick.

Some new initiatives like Accelerated Irrigation Benefit Programme, National Water Resource Development Project, and diversification of crops through National Horticultural Mission have been taken to double up our production. Small farmer agribusiness consortium is meant for good governance and proper credit disbursement to agribusiness. Crop insurance through Rastriya Krishi Vima Yojna and Farmer Income Insurance Scheme for income protection is among the other initiatives for mitigation of market risk. Seed bank for production and distribution of seeds, Kisan Credit Card with help of NABARD and 24 hour agricultural help line at 1551 as Kissan Call Centre are some of the farmer centric initiatives by the government. Under Krishi Upaj Yojna the dutyfree entitlement for export of fruit along with free trade and warehousing zones have been planned. Special Agriexport Zones, in the lines of SEZ,are soon to come up in our country.

This alarming situation calls for opening new vistas in agrarian sector. To tap the vast opportunity in agriretailing and food chain marketing some corporate houses have come forward.This corporate involvement will only result in better alignment of consumption and farming pattern.Some of the big guns of corporate sectors in agribusiness are.

ITC: e-Choupal

It has computer with VSAT internet and solar battery in about 30000 villages in 6 states reaching 3.1 million farmers. Its 5 core mandates are dissemination of information, imparting farming know-how, helping purchase of farm inputs, helping in selling the produce and development of rural community with affiliated NGOs.

TATA Kisan Sansar

Farmers are provided with inputs and services through loans from ICICI bank in Uttar Pradesh, Haryana and Punjab. Same thing is practiced by RALLIES in Madhya Pradesh.

Reliance

Rilcare-Starting from Navsari(Gujarat) mainly thriving on medicinal and herbal plants like Aloevera, Geranium and Aswagandha used in alternative medicinal therapies.

Bharti Group: Field Fresh Foods

Outsourcing agricultural produce from north Indian states to Europe by collaboration with Rothschild inc.

Godrej: Agrovet and Goldmohar

Equipping farmers with farm inputs backed by Adhaar, its retail chain and Manthaan, its retail chain for dairy and poultry industry.

Mahindra: Subhlabh

Retailing agri inputs besides growing oilseeds, cereals, flowers and providing loans to farmers in minimal interest.

DSCL: Haryali

These are the kissan bazaars coming within city proximity for the benefit of farmers dealing in agri inputs.(DCM Shreeram Consolidated Limited)

Conclusion

India now needs to increase productivity to meet the ever-increasing demand of food grain. India needs to come out from its cast and change its conventional approach towards agriculture. It should keep itself in pace with global trends and advancements like hybrid technology, efficiency and quality improvement with greater resilience and diversification. The key components may be biotechnology, information and communication system, water and energy management and policy analysis backed by cutting edge technology. Now the focus should be shifted from food security to prosperity. There is a need of holistic natural resource management, value addition and product diversification. Lastly the technology, services, supply and policies should be properly aligned to lead a sustainable growth towards a secure future, free from hunger and poverty.

References

Dutta, Sameer K., 2002. Indian Agriculture: Retrospect and Prospect, *Yojana,* January.

Foreign Trade Policy, 2004–09.

Pratiyogita Darpan, Indian Economy, 2004-05: Agriculture, p. 63–75.

Ray, G.L. *Extension Communication and Management: System of Extension in India.*

Saran, Rohit, 2004. Rural Market. *India Today*, p. 48.

Saran, Rohit, 2004. New Deals for Rural India. *India Today*, p. 56–59.

Singh, S.S. *Handbook of Agriculture.* Kalyani Publications.Agricultural and Rural Development Schemes, p. 721–731.

Survey of Indian Agriculture, 2007. The Hindu Group of Publications, Chennai.

Union Budget, 2004–05.

Chapter 9

Strengthening Research and Extension Linkages in Agriculture

☆ *P. Adhiguru*[1]

Introduction

Agricultural research and Extension are two important facets of agricultural development and these two spheres cannot function in isolation. It is worth pointing out the role of research system does not end with the release of improved technologies. For getting information on the agricultural production and to understand various technical, socio-economic and infrastructural constraints and for offering solutions, the research system will have to work in unison with extension. The formulation of plans and setting of priorities for agricultural research, generation, adaptation and modification of technologies require coordinated efforts. Effectiveness of research and extension linkages cannot be judged by the number of committees, meetings and monthly workshops but by the extent to which research and extension could influence and reinforce each other.

Functions and Types of Linkages

Input and Feedback

Expression of farmers' needs to research, extension, and have farmers' reactions to new technologies.

---

1 Senior Scientist, National Centre for Agricultural Economics and Policy Research (NCAP), DPS Marg, Pusa, New Delhi – 110 012.

Testing and Adaptive Research

Technologies must be adapted to local conditions and needs of different types of farmers through on-farm testing and demonstrations.

Regulations

Involving formal decisions on whether and when to release a new technology.

Communication of Research Results

The process of releasing research results to extension workers and farmers.

Training

Role of training programmes in transmitting research results *i.e.* new technologies and skills, also as a tool to achieve more effective interaction among researchers, extension personnel and farmers.

Types of Linkages

Production Linkages

To maintain regular flow of information from its source to the client system through govt. and non-govt. agencies including input supply and services.

Technology Generation Linkages

To develop new technologies to meet the emerging problems faced by the client system as well as contingency measures to manage pest outbreaks and natural calamities

Post Production Linkages

With marketing agencies both for domestic and export markets to dovetail the products and services to suit the consumer preferences. Linkages will be made with storage, transportation, packaging and agro-industries for value addition.

Causes for Weak Research-Extension Linkage

Weak research-extension linkages may be due to political, technical or organizational reasons.

Political reasons include pressure groups, such as government, farmers, donors and industry. Limited resources can lead to competition for funds between research and extension, and can seriously impair established linkages. Traditionally, extension has in general received less funding than has research, and has not been able to recruit people of high caliber.

Technical reasons include an inability to undertake location-specific diagnosis of problems and adaptation of technology; lack of infrastructure support, resulting in choice of inappropriate technologies; and lack of flexibility in developing different links appropriate to develop and developing countries.

Organizational reasons include the orientation and work style of scientists, who are more oriented towards the general scientific community and have a longer time horizon compared to extension staff, who are oriented towards farmers and pressed with immediate concerns; poor assignment of specific responsibilities; loss of credibility as a result of the diffusion of inappropriate technology; centralization of authority; and institutional incompatibilities. Other structural and organizational reasons are; No one assigned to perform functions such as adaptive research or provide feedback to researchers; Linkage activities assigned to an inappropriate institute or department, or divided in such a way as to reduce effectiveness; Excessive centralization or de-centralization; Insufficient authority to ensure that institutions coordinate their activities and perform their responsibilities; Institutional incompatibilities, such as research by commodity and extension by region; different clientele; or different time schedules for planning and budgeting. Motivation and incentive reasons include, Individuals may have little incentive from management to perform linkage functions; Maintenance of institutional autonomy may be reinforced; Rewards for journal publication may be higher than for technology transfer activities

Resource reasons may include Financial resources may be scarce for linkage functions such as publications, testing of research results and training of extension staff; Human resources may be overloaded and unavailable for linkage functions.

Communication reasons Value systems, educational backgrounds and communication patterns may differ widely between researchers and extensionists; Physical means of communication may be weak or non-existent in critical areas

Research-Extension Linkages Mechanisms

Under the reoriented extension system, systematic procedures have been established to promote and strengthen the necessary linkages by means of periodic meetings, workshops as well as communities. However, there is always scope for further refinement to improve the efficiency of linkages. So, it is worth to reflect upon the country's existing research-extension linkage mechanism. Efforts have been made to institutionalize research-extension linkages at national, regional, state, zonal and district levels. Such institutional linkages are being encouraged so that the ultimate objective of the delivering the technologies and services to the farmers arising out of harmonized efforts from the Research and Extension systems can be achieved. Following are some of salient efforts towards research-extension linkages.

DAC-ICAR Pre-Seasonal Interface

At the national level, under the ICAR-DAC pre-seasonal interface, joint meeting of the senior officers from Department of Agriculture and Cooperation (DAC), Department of Animal Husbandry, Dairying and Fisheries (DAH, D&F) and Indian Council of Agricultural Research (ICAR) have been organized twice a year to discuss critical research and development issues. This interface has facilitated evolving joint strategies in research and development sectors. In this pre-seasonal interface, 11 specific areas namely, Crops, Oilseeds and Pulses (TMOP), Seeds, Horticulture, Plant protection, Integrated nutrient management, Mechanization and technology, Rainfed farming system, Natural resource management, Animal husbandry, Dairying and fisheries and Extension. Subject matter-wise group is formed comprising officials/ Scientists both from DAC and ICAR. From ICAR, Assistant Director Generals (ADGs) from concerned subject matter areas function as nodal officers while the Divisional

Heads of all the subject matter Division which include Joint Secretaries and senior level officers' *viz.*, Deputy Commissioners/Additional Commissioners/of the respective divisions participate in the deliberations. The interface is steered by Joint Secretaries from DAC and Deputy Director Generals from ICAR for concerned subjects. The interface is conducted under the Chairmanship of Secretary (A&C) and Secretary, DARE &DG, ICAR provide overall leadership to the interface. Directors, Deputy Directors from DAC and senior level scientists from ICAR participate in the interface. Subject matter-wise recommendations are prepared and action taken reports of the previous meetings are reviewed. The recommendations of this interface are circulated in the National Conference on Agriculture for *Rabi/Kharif* Campaigns.

Proposed ICAR Framework for Technology Generation and Delivery

Technology development and delivery need to be in continuum and should be interactive. Keeping in a view the needed change, a framework for technology development and delivery is suggested. In the proposed Framework for Technology Development and Delivery Systems, it is envisaged that the ICAR institutes will substantially carry out basic and strategic research, while region-specific strategic research will be done by SAUs and AICRPs. The strategic research of AICRPs will also feed other areas of the country and a generic model for the purpose is presented in *Annexure 1.* Applied and adaptive research will substantially be carried out by SAUs and AICRPs. As regards private sector research, proprietary technologies can directly go to farmers and others as per agreements/understanding can also flow to SAUs/ATMAs for integration. Technology transfer will be through ATMAs located at the district level and bodies at district-subsidiary levels. In different models, linkages will vary depending upon the nature of crop/commodity/enterprise and technology service in involving relevant stakeholders to a specific technology-enabled zone/area/situation.

ICAR Regional Committees

At the regional level, the ICAR has established 8 regional committees in the country to review the status of research and first-line extension programmes in the ICAR institutes and SAUs located in the region as related to the development requirements of that region. These committees represented by the senior research and extension officers of the ICAR/SAUs and the state line departments; farmers' representatives and NGOs meet once in a year.

Krishi Vigyan Melas

The SAUs organize *Krishi Vigyan Melas* before each cropping season to enable an interaction between scientists, extension personnel and the farmers. The farmers fair include the exhibitions put up by the SAUs department, and agro-industrial organizations, competition of crops and produces by farmers, sale of certified seeds by various organizations and conducting of buzz session. Mandate includes (*i*) to attract and educate large number of farmers and field level functionaries about improved farm technologies, (*ii*) to arrange for farmers visit to experimental spots to demonstrate the most needed technologies for maximization of production and productivity, (*iii*) to provide opportunity to the visiting farmers to seek answers to their questions directly from the concerned scientists.

Agricultural Technology Information Centre (ATIC)

During the NATP, ATIC model has been conceived and established (about 40 centres). The Agricultural Technology Information Centre is a "single window" support system linking the various units of a research institution with intermediary users and farmers in decision making and problem solving exercise. The objectives of ATIC are as follows:

1. To provide a 'single window' delivery system for the products and species available from an institution to the farmers and other interested groups.
2. To facilitate direct farmers' access to the institutional resources available in terms of technology, advice, technology products, etc., for reducing technology dissemination losses.
3. To provide mechanism for feedback from the users to the institute.

The Policy Framework for Agricultural Extension (PFAE) and Schemes

The need for reforms in Agricultural Extension has been explicitly raised in the National Agricultural Policy. Accordingly to provide policy directives for extension reforms.

Support to State Extension Programmes for Extension Reforms

In line with Policy Framework for Agricultural Extension (PFAE) developed by Department of Agriculture and Cooperation, Ministry of Agriculture, Government of India, the Department is implementing a main scheme "Support to state extension program for extension reforms" so as to operationalize agricultural extension reforms across the country and the funding support shall be provided to the States/UTs.

The following key reforms are being promoted under this scheme.

1. *New Institutional Arrangements*: Providing innovative restructured autonomous bodies at the district/block level, which are flexible, promote bottom up and participatory approaches, are farmer driven and facilitate public-private partnership.
2. *Convergence of line departments'* programs and operating on GAP mode by formulating Strategic Research and Extension Plan (SREP) and Annual Work Plans.
3. Encouraging *Multi Agency Extension Strategies* involving inter-alia public/ private extension service providers.
4. Moving towards integrated *broad-based extension delivery* in line with farming systems approach.
5. Adopting *Group Approach to Extension* (Operating through Farmer Interest Groups (FIGs and SHGs).
6. Addressing *gender concerns* (mobilizing farm women into groups, capacity building etc.).

7. Moving towards *sustainability of extension services* (e.g through beneficiary contribution).

New Institutional Arrangement

Agricultural Technology Management Agency (ATMA)

Agricultural Technology Management Agency (ATMA) is a district level body created under the ITD component of NATP for pilot testing Agricultural Extension reforms. This is as an autonomous institution with participation of all the key stakeholders involved in agricultural activities for sustainable agricultural development (*Annexure* 2). It has the flexibility to receive funds directly (Government of India/States, Membership fees, beneficiaries' contribution etc.) ATMA has the main responsibility of all the technology dissemination activities at the district level. It has linkages with all the line departments, research organizations, NGOs and agencies associated with agricultural development in the district with a substantial representation of farmer organizations. Research and extension units within the district such as KVKs, ZRSs, Department of Agriculture, Horticulture, Animal Husbandry, Fisheries, Sericulture, and Marketing etc. are constituent members.

Local research and extension priorities are set through Strategic Research and Extension Plans (SREPs), which are developed by using PRA methodologies. The activities of ATMA are being carriedout/supported by a Governing Board (GB) (*Annexure* 3) and a Management Committee (MC) (*Annexure* 4). The programmes and procedures concerning district-wise R-E activities are determined by ATMA, GB and implemented by it MC.

In order to manage programme implementation at block level and below, ATMA has established a Farm Information and Advisory Centre (FIAC) at each block in the district. In effect the FIACs act as extension planning and operational arm of ATMA (*Annexure* 5). These are supported by two groups, one a group of technical officers at block derived from different functional areas termed as Block Technology Team (BTT), whereas, the others is a Farmers Advisory Committee (FAC) which is a body exclusively of farmers. While BTT develops the Block Action Plans (BAPs) in light of the SREP and is responsible for its implementation, the FAC plays a more proactive role by scrutinizing, improving and approving BAPs, before these are refereed to the ATMA GB for its final approval.

Commodity oriented Farmer Interest Groups (FIGs) are promoted at block/village level to make the technology generation/dissemination farmer driven and farmer accountable. These village level FIGs are ultimately federated at block/district level and represented in FACs and GB. In order to address the extension needs of these groups, ATMA has established closed linkages with various players *viz.* public, private, NGOs, para extension workers, input dealers etc. (*Annexure* 2).

In order to provide needed HRD support in innovative areas of extension delivery, a State Agricultural Management and Extension Training Institute (SAMETI) has been established either by strengthening one of the existing apex training institute in the state or by creating a new SAMETI in a State Agricultural University (*Annexure* 6).

Project activities, at state level are closely monitored by an Inter Departmental Working Group (IDWG) functioning under Chairmanship of APC or Secretary (Agriculture) of the state. A project Implementation Cell (PIC) created at the State Headquarter level provides support to the IDWG (*Annexure 7*).

Strategic Research and Extension Plan (SREP)

One of the first tasks of ATMA like institutions is to facilitate the preparation of Strategic Research and Extension Plan (SREP) of the district. The SREP is prepared through participatory methodologies (PRA) involving all the stakeholders and farmer. The SREP contains detailed analysis of all the information on existing farming systems in the district and research-extension gaps require to be filled-up. It also prioritizes the research-extension strategies within the district. It becomes the basis for development of work plans at block/district level.

State Extension Work Plan (SEWP)

Based on the research-extension strategies given in the SREPs, block/district level plans shall be developed by each ATMA like institutions. The State Extension Work Plan developed at state level shall contain a consolidated activity-wise plan incorporating all the District Agriculture Action Plans (DAAPs) in the state and state level activities to be carried out with activity-wise budgetary requirement as per the norms prescribed in the cafeteria.

Cafeteria of Activities

Under the cafeteria, activities to be undertaken at state and district level are categorized separately. The state level activities include support for upgrading state level training institutions such as SAMETI, HRD of extension functionaries, organization of state level Agri-exhibitions and Monitoring and Evaluation of the Scheme.

The district level activities are further categorized in four groups as given below.

1. *Farmer oriented activities*: development of SREP, mobilization of farmer groups, training/exposure visit of farmers, field demonstrations.
2. *Farm information dissemination*: local level agricultural exhibitions, information dissemination through printed materials and development of technology packages in electronic form are covered.
3. *R-E-F linkages*: organization of Farmer-Scientist Interaction at local level(at district level 25 farmers for 2 days), organization of Field days and *Kisan Goshties* (1 per block in each season) and support for local level researchable issues emanated from the SREP.
4. *The administrative expenses* under district level activities provide support for running ATMA like institutions and a few block level Farm Information and Advisory Centres.

Coverage of Districts

During 2007-08 ATMAs have been constituted in 303 districts as against the target of 300 new ATMA districts.

Lessons learnt

1. The project interventions have improved the R-E-F linkage and feedback process, which began when they cooperated in carrying out the field assessment for and preparing the SREP.
2. The ATMA Governing Board and Management Committee have provided common platforms for regular and personal interaction among scientists, extension administrators and farmers. On one hand, it has improved the awareness level of farmers, while on the other it has enabled the scientists and extension administrators to more clearly understand the farmers' needs and problems.
3. A large number of extension professionals are being trained with a more useful knowledge and skill set, plus an attitude to carryout these extension reforms.
4. ATMA is expected to become the single, most vibrant institution providing useful connection between all concerned stakeholders.
5. The shift from top-down, central planning to a bottom- up, farmer/ stakeholder-involved planning strategy is being accomplished.
6. Shift to a more market-driven extension system takes place.
7. There is usefulness and energy in the ATMA model, as channels are forged and connections made among stakeholders who serve better than before (Reddy and Swanson, 2006).
8. Integrated package of exposure visit, training and demonstration has resulted in better technology adoption.
9. Public research system has become more responsive to development of location specific demand-driven technologies. The adaptive research sponsored by ATMA has been successful in providing location specific solutions to many problems (Sulaiman, 2003).
10. ATMA could successfully solve some of the research-extension linkage problems as it is free from many bureaucratic and time-consuming procedures and this provided ATMAs the much-needed flexibility to quickly respond to demands from the field.
11. Mechanisms such as SREP, FAC and block action plans supported with adequate funds contributed to making ATMA demand driven. But performance of ATMA varies widely across states and districts and the reasons behind this differential impact need to be understood.
12. While ATMA has brought in some additional resources for extension activities to be decided at the district level in consultation with farmer representatives, ATMA has not addressed any of the fundamental institutional problems constraining extension performance.
13. Extension continues to be centrally planned, implemented and evaluated without much operational freedom at the local level. While farmers require a wider range of support to address the emerging challenges, extension mainly functions as an agency for technology dissemination.

14. Marketing extension has been a recent addition but is understood and implemented mostly as provision of output price information in various markets. This is highly inadequate when it comes to addressing challenges in marketing (10-14: Sulaiman, 2003).
15. Though SREP helped strengthening Research-Extension linkages some of the salient weaknesses include, wherever Project Directors are from University, the Research-Extension-Farmer linkages were better compared to PDs from State Department of Agriculture. There is no clear-cut role for the Deputy Project Director, which needs to be critically looked into while up scaling and institutionalization of SREP. The group dynamics and group approach is not adequately addressed during orientation training (Reddy *et al.*, 2006).
16. A field case study on Research-Extension-Farmers linkages of Agricultural Research Station (ARS), Mandore carried out by Rathore *et al.*, 2008 that scientists of ARS, Mandore were making varied efforts in order to find out the problems being faced by the extension personnel and farmers. Majority (64.28 per cent) of them were contacting the extension scientists and going through the official records of the research institute functioning in the zone or State Department of Agriculture or *Krishi Vigyan Kendras* (Table 9.1). Large proportion (92.85 per cent) of the scientists were having liaison with the State Department of Agriculture and research institutes functioning in the zone *viz.*, Central Arid Zone Research Institute and Defense Laboratory (Table 9.2).

Table 9.1: Efforts made by the Scientist in Finding Out the Problems being Faced by the Extension Personal and Farmers N=14

Efforts Made by the Scientists	*N*	*Per cent*
Through official sources	9	64.28
Directly from the extension scientists	9	64.28
From K.V.K. scientists	7	50.00
Directly from farmers	7	50.00
Through field level work	6	42.85
By joint diagnostic visits	6	42.85

Table 9.2: Liaison of ARS Scientists with Different Agencies

Agencies	*N*	*Per cent*
Department of Agriculture	13	92.85
Research Institutes	13	92.85
Krishi Vigyan Kendras	9	64.25
Non-government organizations	7	50.00
Soil and Water Conservation Department	7	28.57
Banks, Fertilizers, Seeds and Pesticide companies	4	28.57

Strategies for Strengthening Research-Extension Linkages

Strengthening research-Extension linkages is a continuous process; the following current issues emerged from the DAC-ICAR interface gives some ideas on periodical suggestions received for improving linkages.

1. State level interface between the concerned state departments and research system are to be strengthened and made more effective by due emphasis on the following:
   (*a*) Regular interface between SAU and State Department of Agriculture including other line departments.
   (*b*) Meetings of Zonal Research Extension Advisory and Formulation Committee once in a season.
   (*c*) District level diagnostic team consisting of specialists from SAUs and officers of the State Department
2. It was noted that, at most places KVKs are actively involved in preparation of SREPs and other activities of ATMAs. A suitable system should be evolved by the ICAR for channelizing the information on research issues emerging out of SREPs.
3. The information/knowledge available with the ICAR system should be digitized at the earliest so that it could be accessed at the info-kiosks at grass root level. IT Divisions of Department of Agriculture and Cooperation and ICAR may hold a separate meeting on their e-Governance activities.
4. KVK should be an active partner to operationalize the ATMA model. There should be no duplication between the activities of KVKs and extension activities to be taken up under the scheme. ICAR should ensure that research priorities of KVK are fully aligned with and sensitive to SREPs at the District level.
5. ATMA should be part of an integrated extension system consisting of farm schools in the field of farmers, KVKs, agri-clinics and agri-business centres.
6. SAUs should provide services of Subject matter specialists to State Agricultural Extension Management and Training Institutes (SAMETIs) in conducting training programs.
7. SAUs shall be fully responsible for providing technical backstopping to KVKs as well as training.
8. Kisan Call Centres are to be located in ATICs/SAUs.
9. KVKs are to involve ATMAs in planning and conducting FLDs.
10. In the Scientific Advisory Committee of KVKs, Project Director, ATMA is to be included as special invitee.
11. The deployment of Scientists as DPD to ATMA may be considered by the SAUs/ICAR institutes subject to availability of manpower and not from the KVK.
12. A committee of inter-disciplinary group of experts of SAU may be constituted under the Chairmanship of Vice-chancellor of SAU to provide technical backstopping for Mass Media Scheme.

13. Farmer–Scientists' interaction should be organized in the village itself for better linkages. At the same time scientists should get an opportunity to meet the extension personnel in farmers' field.
14. Farmers' problems should reach the scientists well in time either directly or through agriculture supervisors so that they may be solved at the earliest.
15. There should be more specialized staff at the field level for immediate solution to farmers' problems.
16. The presence of agriculture supervisor of the respective area should be made mandatory in any training or meeting organized for farmers.
17. Farmers' fair should be a regular feature in *rabi* as well as *kharif i.e.*, at an interval of six months to provide the farmers with latest know how.
18. There should be emphasis on quality of demonstrations, mini-kits and training along with quantity (13-18: Rathore *et al.*, 2008).

Conclusion

Strengthening Research-Extension linkages is a continuous process. In brief, the following aspects would be critical for improving the effectiveness.

1. Farmer driven funding: Institutional mechanism is to be developed such that in funding agricultural research, farmers have a voice to direct the funding to need based researchable issues.
2. Farmer driven output: In addition to greater emphasis on publications as research outputs, solving farmers 'problems should be necessarily reflected in the output of scientists.
3. Paradigm shift in who listens whom: In the past, research and extension system disseminate information and we expected farmers to respond. In the new paradigm, we need to receive information from farmers and have to respond favorably to solve their farm problems.
4. Capacity strengthening and change in Mind set: As part of human factors, these two pillars are essential for integral functioning of research and extension systems.
5. Continuous functional adjustments: Stakeholders in the institutional arrangements should be flexible enough to introspect accommodate changes for bringing harmony in research and extension linkages.

References

GOI, 2006. *Proceedings and Recommendations of Pre-Kharif Interface Between DAC and ICAR*, 6th and 16th February, Directorate of Extension, Ministry of Agriculture, Government of India.

GOI, 2006. *Proceedings and Recommendations of Pre-Rabi Interface Between DAC and ICAR*, 24th and 30th August, Directorate of Extension, Ministry of Agriculture, Government of India.

GOI, 2005. *Xth Plan Centrally Sponsored Scheme Support to State Extension Programmes for Extension Reforms, Operational Guidelines 2005–2007*. Ministry of Agriculture, Department of Agriculture and Cooperation, Government of India.

NATP, 1998. *National Agricultural Technology Project (NATP) Main Document.* ICAR, New Delhi.

Pal, Suresh, 1997. *Agricultural Research and Extension in India: Institutional Structure and Investments*. Policy Paper, NCAP, New Delhi.

Reddy, M.N and Swanson, B.E., 2006, Strategy for up-scaling the "ATMA" model in India. In: *22nd Annual Conference Proceedings, AIAE.*

Sulaiman, R., 2003. *Agriculture Extension: Involvement of Private Sector, Occasional Paper-*29, Department of Economic Analysis and Research, National Bank of Agriculture and Rural Development, Mumbai

Reddy, G.P., Sontakki, B.S., Rao, V.K.J. and Shenoy, Sandhya, 2006. Assessment Strategic Research Extension Plan (SREP) methodology for upscaling and institutionalization of r-e-f linkages, International Association of Agricultural Economists Conference, Gold Coast, Australia. National Academy of Agricultural Research Management, Rajendra Nagar, Hyderabad, India. http: icar.naarm.ernet.in.

Sulaiman, R.V., 2003. *Innovations in agricultural extension in India, National Centre for Agricultural Economics and Policy Research*. Indian Council of Agricultural Research, www.fao.org/sd/2003/KN0603a2_en.htm–44k.

Sulaiman, R.V., 2007. *Farmer First or Still Last? Uneven Institutional Development in the Indian Agricultural Innovation System*. Centre for Research on Innovation and Science Policy (CRISP), Hyderabad, India.

Surya Rathore, Intodia, S.L. and Singh, R.P., 2008. Analysis of Research-Extension-Farmer Linkage in the Arid Zone of India. *Indian Res. J. Ext. Edu.* 8(2&3), May and September.

Annexure 1
Proposed ICAR Framework for Technology Generation and Delivery

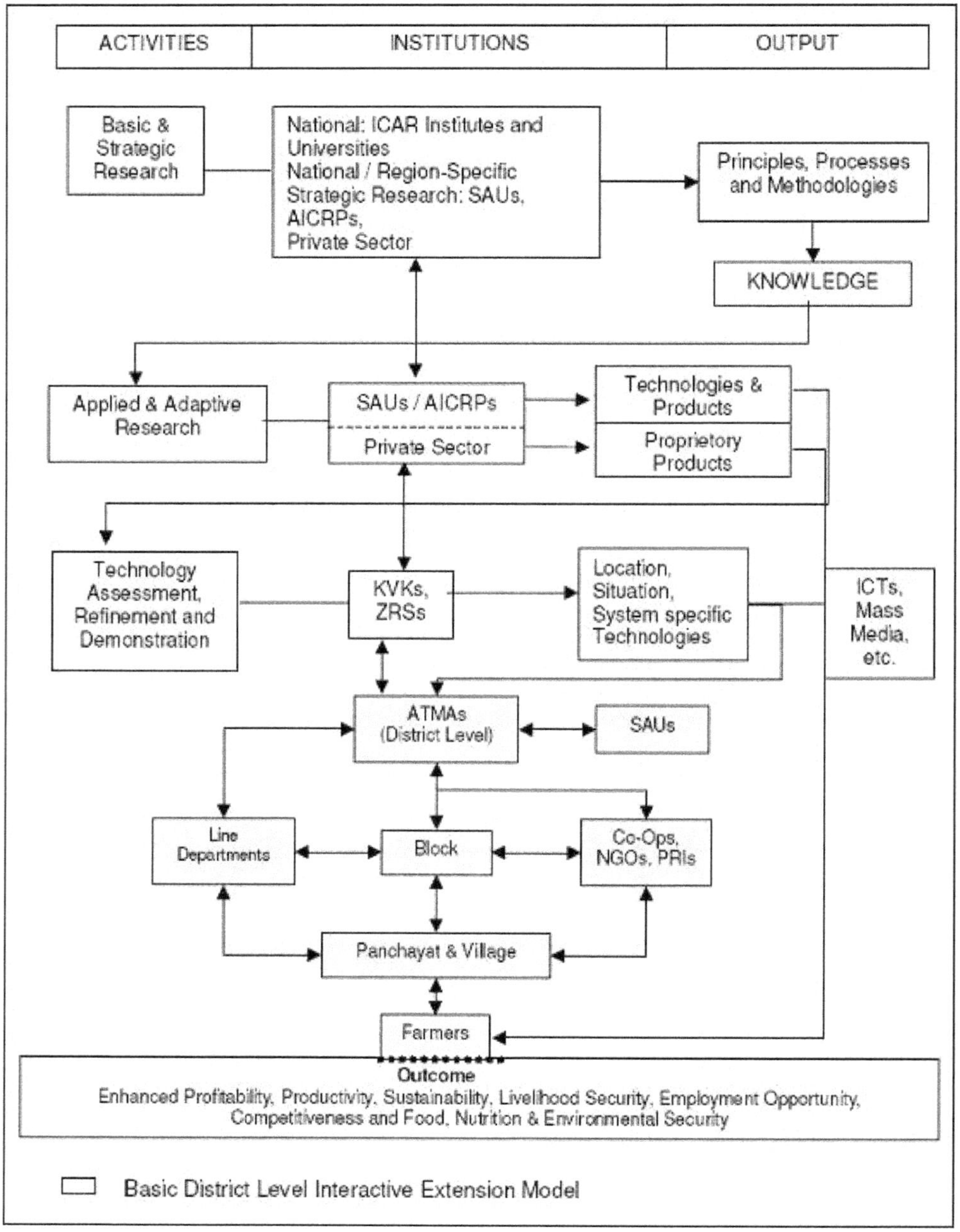

Annexure 2
Organizational Structure of ATMA

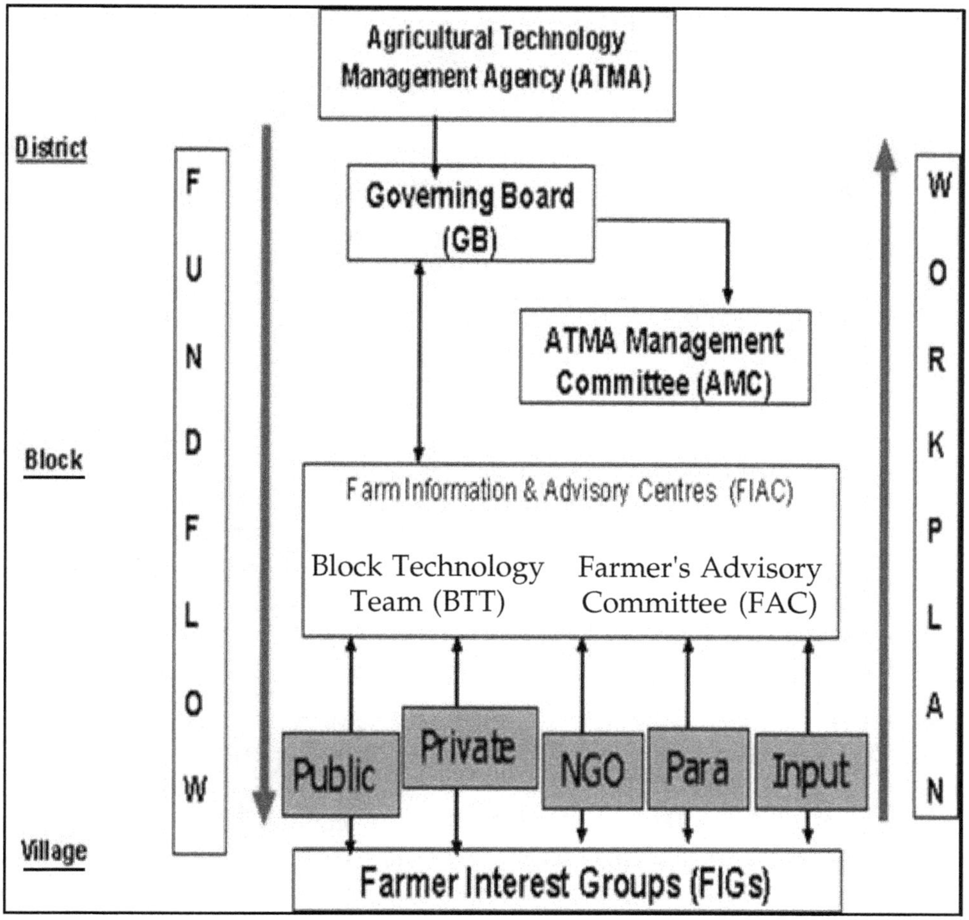

Annexure 3
ATMA Governing Board

Representatives of technology generation/refinement and dissemination systems, line departments, the farming community and other stakeholders who are members of its Governing Board (GB) govern ATMA. The District Collector with the Project Director serving as its Member Secretary heads the GB. The ATMA Governing Board is a policy making body and provide guidance as well as review the progress and functioning of the ATMA.

Key Functions of the ATMA Governing Board

1. Review appropriate strategic and annual work plans.
2. Receive and review annual reports carried out by the participating units and providing feedback to the participating units.
3. Receive and allocate funds to carry out priority research, extension activities.
4. Foster the organization and development of Farmers Interests Groups (FIGs) and Farmers Organizations (FOs).
5. Facilitate the greater involvement of private sector in providing inputs, technical support, agro-processing and marketing services to farmers.
6. Encourage agriculture lending institutions to increase the availability of capital to resource poor and women farmers.
7. Encourage each line departments, plus the KVK and ZRS, to establish farmer advisory committees to provide feedback and input into their respective R-E programmes
8. Enter into contracts and agreements as appropriate to promote agricultural development activities.
9. Establish revolving funds, arrange for the periodic audit of ATMA.
10. Adopt and amend the rules and by-laws for the ATMA.

Annexure 4
ATMA Management Committee (MC)

The ATMA Management Committee (MC) would be responsible for planning and executing the day-to-day activities of ATMA. Project Director of the ATMA is the Chairman. District Heads of line departments, Head, KVK, Head ZRS, representatives of NGOs and farmers constitute members.

Key Functions of Management Committee

1. Carryout periodic Participatory Rural Appraisal to identify the problems and constraints faced by different socio-economic groups and farmers within the district.
2. Prepare an integrated, strategic technology plan (STP) for the district that would specify short and medium term adaptive research as well technology validation and refinement and extension priorities for the district.
3. Prepare annual work plans that would be submitted to the ATMA Governing Board.
4. Maintain appropriate project accounts for submission to technology dissemination unit (TDU) for audit purposes.
5. Coordinate the execution of these annual work plans through participant line departments, ZRSs, KVKs, NGOs. FIGs/FOs and allied institutions.
6. Establish coordinating mechanisms at the block level, such as farmer advisory centres (FACs), that would integrate extension and technology transfer activities at the block and village levels,
7. Provide annual performance reports to the Governing Board.
8. Provide secretariat to Governing Board and initiate action on policy direction, investment decisions and other guidance received from the GB.

Annexure 5
Farm Information and Advisory Centre (FIAC) at Block Level

Under each ATMA, FIACs are created at the block level. It consists of two bodies namely Farmer Advisory Committee (FAC) and Block Technology Team (BTT). The FAC is a body of the farmer representative (11-15 members representing various enterprises and socio-economic strata). The BTT on the other hand is a group of technical advisors operating at block level representing agriculture and allied sectors. FAC and BTT, taken together, involved in planning and operational management of ATMA.

Block Technology Team (BTT)

It is an Inter Departmental Team of agriculture and line departments operating at block level. An indicative composition of BTT is given below.

Composition

Block level officers of Agriculture, Horticulture, Animal Husbandry, Fisheries, Plant Protection, Veterinary Science, Soil Conservation, Extension, Sericulture, and Marketing etc. are the members. The senior most officials irrespective of the participating departments head the Block Technology Team as BTT convener.

Key Functions of Block Technology Team (BTT)

1. Operationlize the SREP in each block and move towards single window extension system.
2. Help district core team in upgradation of SREP
3. Prepare Block Action Plan detailing extension activities to be undertaken
4. Coordinate the implementation of extension programs detailed in the Block Action Plan
5. Facilitate formation of FIGs/FOs at the block level and below.

Farmer Advisory Committee (FAC)

The Farmer Advisory Committee consists of 11-15 members covering different categories of farmers under the given block, with due representation to women farmers and weaker sections of the society. Chairman shall be elected out of the members on rotation basis. BTT convener also acts as Member Secretary to FAC.

Key Functions of FACs

1. Act as an agency for providing farmer feedback mechanism
2. Help set block extension priorities and recommend resource allocation across programme areas
3. Recommend Block Action Plan for approval of ATMA GB.

4. Review and provide advice to each implementation unit at block level
5. FAC shall meet once in a month during the season and quarterly in lean season
6. Help information of Farmer Interest Groups at block level and below.

Annexure 6
State Agricultural Management and Extension Training Institute (SAMETI)

The State Agricultural Management and Extension Training Institute (SAMETI) would be strengthened by way of providing training managerial and equipments, communication support, program cost. SAMETIs are proposed to be autonomous institutes with greater flexibility in structural and operational aspects.

Key functions of SAMETI

1. To provide capacity building support in Extension Management related areas to the extension functionaries both from public and private sector.
2. To provide consultancy in the areas like project planning, appraisal, implementation etc.
3. Develop and promote the application management tool for improving the effectiveness of agricultural extension services through better management of human and material resources.
4. Organize need based training programmes for middle level and grass-root level agricultural extension functionaries;
5. Develop modules on Management, Communication, Participatory Methodologies etc., as sequel to the feedback from training programmes.

Annexure 7
State Level Inter Departmental Working Group (IDWG)

In pursuance of the number of mechanisms of built into the project design and to ensure effective coordination among the departments like agriculture, animal husbandry, fisheries, horticulture, soil conservation etc., it is proposed to constitute a state level inter-departmental working group under the Chairmanship of the Agriculture Production Commission/Secretary agriculture. The secretaries of other line departments, Vice-Chancellor(s) of SAU(s), Secretary (Agri) constitute members.

Key functions of IDWG

1. To provide a mechanism for interactions with the Technology Dissemination Management Committee (TDMC) of the DAC, GOI guide the human resource development activity and to monitor the district level technology dissemination programme.
2. To oversee and support Agricultural Extension Research activities begin undertaken by ATMA and to make policy interventions on inter departmental matters including issues related to Women in Agriculture and coordination thereoff with the following composition.

Section II:
MARKET LED EXTENSION AND COMMERCIAL FARMING

Chapter 10

Issues and Emerging Challenges of Agricultural Marketing: Extension and Export Policy

☆ ***Ram Pratap Singh***[1]

Introduction

The supply pool for agricultural commodities is no more confined to domestic production alone. The countries having competitive edge in cost of production are looking for opportunities to dump their output anywhere in the world. Despite our complex and diversified agriculture, it is yet to pass the test for comparative advantage in terms of cost effectiveness for many important commodities. Unless, we succeed in inducing cost effectiveness having locational advantage. It may not be possible for us to harness the benefits of WTO through export of agricultural commodities. Besides, we have to face a number of emerging threats including unrealized opportunities in agricultural marketing unwarranted imports by the countries, under exploited export opportunity for Indian products and distortion in domestic markets.

Since, India has entered in global market and signed WTO, it is necessary to change our outlook on agricultural marketing system, especially in view of the Exim-policy and the existing dynamics in domestic markets and the new agricultural policy. Market-oriented agricultural extension is the need of the time, which is the real challenge for our public sector, dominated agricultural extension system.

---

1 Former Vice-Chancellor, MPUAT, Udaipur, Rajasthan.

Components for a Sound Agricultural Marketing Extension

Farmer-Agro-Industry-Consumer Linkage

There is a need to have strong linkage between agricultural marketing and agricultural products system as 'what to produce' how much to produce', how to produce and for whom to produce'.

Shift in Physical Output Value Realization

Ensuring shift in the approach form increased physical output to increased value realization by the produces in the need of the time.

Agro-processing and Value Addition of Agricultural Produce

India has great potential for expanding its export of processed products in view of the increasing production of fruits and vegetables, dairy products, meat and marine products etc. the practice of contract farming is coming up around such processing centers. More and more processing facilities for such commodities need to be augmented.

Grading and Packaging

In view of tremendous potential for export of fruits and vegetables, it is necessary to give greater attention to grading and packaging of these commodities by creating required infrastructure support.

Diversification of Agriculture

India has enormous inherent potential for diversification and scope to introduce new areas in agricultural production such as hi-the horticulture, precision farming, organic farming, etc.

Augmented Infrastructure Facilities

A strong infrastructure support has to be developed in terms of cold storage chains, transport, credit support, market information and insurance to fully exploit opportunities for export.

Cost Effectiveness

The country's agricultural has to be come more cost effective to meet the growing challenges and opportunities arising out of WTO agreements and the consequent globalization impacts. For this, future growth of agricultural has not only be yield based but should be titled towards ensured demand in the national and international market.

Global Market Research and Information Centre

An apex centre at the national level for market research, International price analysis, global demand, availability and also to pass on this information to the concerning authorities is very much required.

Involvement of Private Sector

Efforts should be made to involve private sector in creating necessary marketing infrastructures.

Marketing Information Network

Easy access to timely and relevant information to farmers, traders, policy makers and other marketing agencies so as to enable them to have proper marketing decisions is very much.

Marketing of Agro-inputs

For the marketing agro-inputs like seeds, it is necessary to streamline and simplify seed certification procedures, modify the seed Act and enforce if strictly. Greater participation of private sector, Cooperatives and NGOs in the production and distribution of seeds is required.

Newly Emerging Areas in Agriculture for Export

There are several global issues and global dimensions which the Indian agricultural sector has to confront as a consequence of globalization and liberalized made regime. It is a reality that our grains through export of agricultural commodities depend on the promotion of those commodities for which we have a comparative advantage in terms of cost of production and quality aspects. Some of the new areas for the extension system to mould itself to perform in an effective manner are listed below:

Green Food Production

It refers to organically grown crops, which are not exposed to any chemical right from the stage of seed treatment to the final post harvest handling and processing.

Hi-Tech Horticulture

It is the deployment of any technology, which is modern, less environment dependent, capital intensive and has the capacity to improve the productivity and quality of horticultural crops.

Precision Farming

It implies to the application of technologies and principles to manage special and temporal variability associated with all the aspects of production and improving crops performance and environmental quality.

Sustainable Agriculture

It is the aspect of agriculture which encompasses the use of natural resources land, water and genetic endowment in such a manner that is technically feasible, economically viable, environmentally non-degrading and socially acceptable.

Farming System Approach

The farming system refers to the activity mix of food crops, oil seeds, fodder crops, horticulture crops, livestock and other high value commodities that would increase farm income while ensuring the goals of stability, sustainability, diversification and commercialization of agriculture.

Protected Cultivation

It is the method of harvesting the natural resources through the creation of an environment to incash crop productivity by checking the vagaries of nature a weather variation often tend to have adverse effect on the yield and production of the crop.

Cold Storage/Cool Chain

It refers to the application of refrigeration technique in the preservation and storage of perishable agricultural products like fruits, vegetables and flowers to facilitate marketing system and to ensure higher return.

Micro-Propagation

It is a proven means of producing millions of identical plants under controlled and uniform condition, independent of seasonal constraints. It ensues economy in time and space and also gives greater output.

There are numerous other newly emerging concepts and techniques of crop production. The extension system will have to resolve these problems through its network in the year to come.

Agriculture Export Policy Issues and Emerging Challenges

There is a great need to orient our agricultural production systems for enhanced export promotion by considering followings:

1. Promoting of eco-friendly inputs, precision agriculture practices and participatory technology developed for sustainable increase in productivity and production, in order to increase the exportable surplus.
2. Exploit the agro-ecological diversity of the country, for product ional export of agricultural produce from different regions during the off-seasons in importing countries. The APEDA and SAD's can work on this and the state Department of Agriculture, Horticulture, Agricultural Marketing and State Agricultural Marketing Boards can implement this.
3. Enhance the competitiveness of the Indian agricultural produces; cost of production must be brought down by increasing yield per unit of inputs through facilitating adoption of latest practices and reducing the yield gaps.
4. Training must be given to farmers on hi-tech modules, pre-harvest management and post-harvest handling of produce to match importing country standards.
5. Agriculture must be treated as agri-business and all incentives must be extended for its promotion and value added products from agricultural commodities in general and dry land crops in particular.
6. Credit availability must be enhanced and all procedural bottleneck must be removed the institutional lending, besides, creating awareness among framers to use crop insurance programme available.
7. Harness benefits of new scientific advances such as biotechnology, cloning, remote sensing, modeling, information and IT for farmers and rural communities through Internet aided extension networks have to be materialized that too in local language.

8. Farm Science Centers (KVKs) should be strengthened as nerve centre for dissemination of technical know-how and promoting the market/export oriented quality produce of the commodity.
9. Formation of commodity specific extension kiosk on indigenous food, horticultural crops, livestock, poultry, beekeeping, mushroom, medicinal and aromatic plants, sericulture, tea and other value addition enterprises.
10. Distance education for farmers and small Agri-entrepreneurs need to be explored and strengthened through television, interactive audio and video system, besides, print and learning materials which would be the distinguishing feature of extension teaching and learning process.
11. The knowledge on freely exportable and importable commodities, restricted as well as non-exportable/importable and various levels as it has implications on the marketing of agricultural products, which needs application of Exim-Policy.

HRD in Extension

There is a need to train the farmers in the areas like products planning, marketing information, preparation of produce for marketing, improved marketing practices, rules and regulations, input marketing, etc. We have strong network of extension in the form of ICAR extension system compromising of SAUs Extension system of ministry of Rural Development and extension work by the Non-Government organizations (NGOs), business houses and voluntary agencies. The rapid advancements in the technological front necessitate that these should reach to the extension scientists in time so that they in turn could disseminate it among the farming community. The transfer of technology being a complex and continuous process, there is no alternative but to develop human resources to achieve desired goal. The first line as well as development extension agencies both at centre and in the states, involved in Extension Education activities are required to pay added attention to the following aspects of HRD to meet the challenges in the future.

1. Professionalism in extension has to be brought about by augmenting subject matter training with training in extension education.
2. The Extension education programme has to focus still greater attention in future on small and poor farmers.
3. Indigenous knowledge system and farmers 'traditional practices need to be integrated with formal R and D by approaches, which have system' perspective and participatory orientation.
4. The public sector should embrace, besides technology transfer, other roles like human resources development, broad basing and farming system perspective and gender differentiated strategies.
5. The participation in agricultural technology development and dissemination process should also empower the local groups to exert demand-pull on research and extension system.

6. A substantial training effect is needed to upgrade the technical, managerial and organizational skills, knowledge and qualifications of field and junior extension personnel.
7. The extension efforts must concentrate more on group approach in the form of 'self help group" which have proved more effective than the individual approach due to several limitations.
8. The extension scientists should be trained in areas of modem information and communication technology. The access to extension-by-extension functionaries must be ensured.

Suggested Model of Agricultural Extension

It is beyond the capability of any extension system to come out with a unified model to resolve all the problems of such a complex and diversified agricultural system prevalent in the country. It is recognized that approaches like farmer's participation, institutional linkages; system management approach, policy reforms, capacity building, empowerment of farmers and farm women, use of media and information technology and a host of other suggestive approaches may go a long way in making our extension system more vibrant. We can perceive the market oriented extension system as a sub-set of overall economic development strategy as shown below:

Economic Development Strategy

↓

Agricultural Development Strategy

↓

Extension System

↓

Market Orientated Approach

The agricultural extension strategy initiated to achieve self-reliance in agriculture has to be re-oriented and made market centered. Such a shift in extension strategy has become inevitable owing mainly to strong backward linkage that agricultural marketing has with agricultural production as 'what to produce' and how much to produce, ensuing shift in the approach from increased physical output to increased value realization by the producers; changes in agro-processing and value addition for agricultural products; potentials for diversification and new areas in agricultural production such as hi-tech horticulture, precision farming organic farming etc: emergence of glut in market in the wake of surplus production certain agricultural commodities; complex clauses on WTO agreement; declared EXIM policy (2002-07) by the Government of India listing freely exportable and importable commodities (restricted as well as non-exportable/importable commodities) and lack of accessibility to information technologies due to vast area, illiteracy and the like.

Research and Extension linkage has been considered as an intrinsic aspect for the development of agriculture in the country. It is well recognized that research and extension personnel have complementary roles in agricultural development the success of each group is determined largely by the effectiveness of linkage activities. Efforts should be directed to improve the existing linkage between research, extension and farmers. Technology transfer will be more effective if all of them show more commitment on their respective fields and more co-ordination among them. Involvement of the farmers at planning level may help in better execution of the extension activities.

Thus, in a comptetive environment of today and in the wake of missing expectations from the consumers, the hall-mark of coming decades will quality, innovations and efficiency in every sphere of business activity. By setting standards and enforcing commercial laws, government has fostered economic growth and provided the framework within which a modern economy can function. It is high time that we take stock of the existing situation, improve upon it, adopt and adapt it according to the exigencies of the circumstances. We are running a race and we must run faster in order to improve our relative position in the right path for meeting the global competitiveness in the agricultural sector.

Chapter 11

Demand-Driven Extension Services for Farmers

☆ ***C.S. Arneja[1] and Shabana Yasmin[2]***

Introduction

The emergence of Demand Driven Agricultural Advisory (Extension) Services is a vital element of the array of market and non-market entities and agents that provides critical flow of information that can improve farmers' and other rural peoples' welfare (Anderson, 2007).

The term "Demand-Driven" refers to the economic concepts of supply and demand. In economic theory, "Demand" refers to the amount of good or service that a consumer is willing and able to buy at a given price or it may be defined as what people ask for, need and value so much that they are willing to invest their own resources such as time and money in order to receive the services.

In extension context, the concept of Demand Driven Advisory or Extension Services emphasizes the need to provide services that meet the needs and priorities of farmers.

The main principles for Demand Driven Service Delivery Systems include:

1. Services shall be driven by user demand.
2. Service providers shall be accountable to the users.

---

1 Department of Extension Education, PAU, Ludhiana.

2 Research Scholar.

3. Users shall have a free choice of service providers (Neuchatel group, 2006) and the objectives are:
   - ☆ To increase agriculture income.
   - ☆ To improve the household food security of small and medium farmers by providing access to services that have the content and quality they ask for.

Agricultural Extension Services play an important role in promoting agricultural development by supporting and facilitating people engaged in agricultural production to solve problems and to obtain information, skills and technologies to improve their livelihoods.

It plays an important role in meeting the new challenges agriculture is confronted with: changes in the global food and agricultural system including the rise of supermarkets, growth in non-farm rural-employment and aggressiveness, constraints imposed by health problems that affect rural livelihoods and the deterioration of the natural resource base and the emerging need to cope with climate change.

It also plays an important role in promoting the spread of farmer-based innovations. In doing so, the extension agents must see whether a particular technology suits the socio-economic and agro-ecological conditions of the people.

Criteria for Success

The success of Demand Driven Extension Services depends on:

Extension Services Should Conform to the Demands of Different Stakeholders

The demand of farmers may consist of food safety standards and other quality standards, as well as official regulations related to environment and natural resource management.

For *e.g.*, farmers who want to sell their products with an organic premium must comply with the standards for organic production and are therefore likely to demand advisory services to develop such systems.

Women farmers are often more responsible than men for food production for the household. A focus on female farmers' demand may contribute more to household food security than a focus on male farmers.

Services Should be Related to Immediate Benefits Rather than Pursuing Long-term Goals

Farmers demand services that serve their private interests. They are most likely to demand services that relate to immediate benefits rather than pursuing long-term goals such as the protection of the environment and natural resources. For *e.g.* if a new technology or practice gives more yield or income or saves time, labour and cost; or has less risk than the existing one; has more relative advantage are likely to be adopted more quickly.

Demand for Services Require Motivation and the Capacity to Formulate Demand

Farmers demand services when they are motivated to develop their agricultural production. Their motivation is closely linked to the existence of market opportunities and prospects for improved working conditions and when they have the motivation and potential to develop their production, they require adequate capacity to articulate their demand for advice and training as well as organization necessary to formulate their demands and negotiate with potential service providers.

Demand Driven Supply of Services Require the Availability of Service Providers

Advisory/Extension services can only be demand driven if there is a functioning market for service provision. A choice of advisers must be available who are able to offer quality advisory services at an appropriate price. This in turn requires that the services are financially viable as a business for the providers. It also requires that the farmers are well informed about the different services and service providers.

Sustainability of Advisory Services Require the Existence of Backstopping Institutions

Long-term sustainability of Demand Driven Advisory Services requires continuous capacity building of the farmers, their organizations and their advisers. Institutions are required which can offer training to farmers and advisers.

Guiding Principles Pertaining to Farmer Demand Driven Extension Service

Decentralization

The extension services delivery system and its activities should be decentralized to the district and below. This implies that decision-making and management of resource will be decentralized.

Farmer and Organizational Empowerment

Farmer and organizational empowerment will feature prominently in the reformed extension service system. Farmers and their organizations should be empowered in terms of training, financing and organizational facilitation.

Targeting

Targeting of extension services will concentrate on categorization of farmers into various groups based on enterprise, resource availability, gender, environment, cultural affiliations and prevailing norms. Issues of who needs subsidies and who will be able to pay will feature during targeting.

Cost Effectiveness

The extension services should consider issues of costs against expected outputs. Agricultural production will consider economic aspects in order to make agriculture commercial and profitable.

Plurality

There will be many actors besides the public services. This means the involvement of the secondary and territory sectors *i.e.,* private sector, NGOs in delivering the services.

Subsidiary (Networking)

The new system should promote the participation of grass root organizations. It will facilitate group formation and networking.

Demand Driven Technology and Utilization

The services will be planned and implemented in a participatory way. The services/messages prepared for utilization will depend on demand by the end users (ASDP, 2003).

Steps to Follow during Implementation of Extension Services

Identifying Leaders

1. Identify the potential stakeholders
2. Empower farmers to demand service and negotiate with Advisory Service Providers.
3. Mobilize villagers into action and negotiate partnerships with a variety of Advisory Service Providers.

Consultative Workshops

1. Explore interest among stakeholders and build a common vision.
2. Promote self-organization for interested stakeholders at different levels.
3. Promote vision based planning by competent local planning staff.

Utilize Communication and Information System

The communication media will include mass and print media (radio, television, newspapers, pamphlets, leaflets etc.). Besides, leaders at all levels will organize public meetings.

Piloting and Programme Consolidation

At the beginning of the programme implementation, baseline surveys will be conducted to determine current levels of agricultural development in the selected districts/villages. The programmes will be conducted in the selected districts and villages. Experiences gained from pilots will be consolidated and scaled up to move into new districts and villages.

Review Institutional Set Up

There is a need to review the current institutional set up with a view to prepare a dynamic management and organization structure for the reforms. The proposed institutional set up should be the one that enhances decentralization of extension services at village and district levels but without compromising quality (ARD, 2002).

Agencies Providing Advisory Services to Farmers

1. Farmer Organizations
2. NGOs
3. Marketing and input supply companies
4. Community based services
5. Public Services
6. Public-Private Partnership (PPP) *e.g.* Agricultural Technology Management Agency (ATMA)

Farmer Organizations

Farmer organizations with good financial management frequently provide services to their members. These organizations either employ staff or private professionals to deliver services. The service providers are the elected representatives of the farmers. An example of Farmer organization is Foutah Djalon Farmers Federation in Guinea (FDFF). This organization provides services such as:

1. Credit for input supply.
2. Facilitation of negotiation between producers and local brokers for delivery contracts.
3. Farm management advice
4. Machinery and equipments for bigger farmers

NGOs

Implementation of NGOs usually involves the delivery of goods and services. An NGO can be engaged in providing services to its clients through programmes such as health care, credit, agricultural extension, legal advice or emergency relief, non-formal education programmes etc.

It also strengthens the already existing public delivery system through research and training services to upgrade the knowledge and skills of the government staff.

Marketing and Input Supply Companies

They often deliver free advice along with their supply of input or marketing. They get their income from farmers but the costs of advice are included in the price of the inputs.

Community Based Services

Farmer-to-farmer services can be viewed as an initial step towards the development of private advisory services in remote rural areas. They are relatively inexpensive since the advisers provide the services for their neighbours on a part time basis. The price and transaction cost are appropriate for subsistence and small-scale farmers (Neuchatel Group, 2006).

Public Services

Public services are the services provided by the government to its citizen, either directly (through the public sector) or by financing private provision of services. Public services often include:

1. Broadcasting
2. Education
3. Electricity
4. Health Care
5. Public Transportation
6. Telecommunication
7. Waste Management
8. Water Service
9. Food Pantry

Public-Private Partnership (P.P.P.)

To exploit the opportunities of decentralization and to overcome the problems of implementation, some countries have tried to combine decentralization with other reform models.

For example; Agricultural Technology Management Agency (ATMA) model is proposed as the key concept of decentralizing decision making at the local level. ATMA promotes coordination among government agencies and coordination among the public, private and third sectors (NGOs). (Swanson and Samy, 2003) This model was originally introduced under the National Agricultural Technology Project (NATP) funded by the World Bank. It is an innovative extension mechanism for improved participation and accountability of all the stakeholders involved in agricultural activities for sustainable agricultural development.

Strategies for Enhancing the Development of Demand Driven Services

Demand Services

Strengthen Farmer Organization

Farmers are very heterogeneous, but groups and organizations of farmers with similar interests can secure better and more responsive service provision and more efficient use of public resources than individuals. They can have stronger negotiation power with private providers than individual farmers as their capacity to purchase advice as groups makes it more interesting.

Poor Farmers Need Empowerment in Order to Effectively Demand Services

The limited power and voice of the poor in influencing services, markets and policies are the most important constraints for the poor farmers to be able to realize new agricultural opportunities. The service providers must focus on their empowerment by strengthening their voice and negotiating power with the service providers and policy makers.

Promote Access to Markets for Farmers

Access to markets is a pre-condition for succeeding in development of Demand Driven Agricultural Advisory Services:

Improvement of access to and benefits from market requires:

1. Access to market information
2. Availability of infrastructure
3. Communication (telephone, radios etc.)
4. Appropriate credit facilities
5. Organization for improved bargaining power
6. Appropriate market regulation.
7. Risk management measures (*e.g.*, insurance schemes)

Consider Development of Demand as a Process

Development of Demand Driven Advisory Services consists in principle of five steps:

1. Formulation of demand among farmers.
2. Dialogue and negotiations with service providers
3. Preparation of services.
4. Delivery of services
5. Evaluation of services received

The first step (formulation of demands) and the fifth step (evaluation of services received) entirely owned by the owners.

The second step is an interface between demand and supply; both users and service providers are involved.

Third and the fourth steps are related to the capacity and availability of the advisers.

Public Interests

Demand Driven Advisory Services are not suitable to enforce public sector interventions if they are not of immediate benefit to the farmers. However, different kind of incentives can increase the farmer's interest. If the incentives make the intervention profitable to the farmers, they will demand services that are necessary for implementation. *e.g.*, Tree planting promoted with public incentives such as subsides for trees or directly for planting. This increases the demand for tree seedlings and eventually also for training and advice on tree management.

Supply Services

Enhance the Availability of Services

Small-scale farmers in most countries have very little choice of advisory service providers. Both public and private advisers exist but their capacities and geographical presence are often limited. The service promoters are hesitant to invest in providing services to farmers in remote areas because of long distance and poor infrastructure which makes the cost of delivery high. Further, the farmers in these areas often lack the market opportunities that should be the basis of demand for advisory services. Moreover, small scale farmers often demand services where the content is different from what the service providers can offer. Many advisers are well qualified and focused on technical services, whereas the farmers are more interested in improved farm management, marketing advice and information, business development and legal support. This means that due to the increasing choice of the services requires accessing of different type of advisers rather than those who are usually associated with extension.

Several types of agents provide advisory services to farmers:

1. Specialized fellow farmers
2. Farmer organizations
3. Input supply companies
4. Independent private service providers such as private advisers, veterinary, animal health workers etc.
5. Public services such as extension, research institutes etc.

Promote Appropriate Approaches to Quality Assurance

The most important factor for assuring quality services is the market. If farmers have a choice, they will choose the advisers that they trust and who deliver the services according to the farmers' own quality criteria-such as timeliness, efficiency, innovative advice, respectful attitudes, strong networks etc. A local forum consisting of all stakeholders with a strong representation of the users should handle assessment of quality of service providers.

Services Related to Private Interest, Quality Assessment should be Based on Users' Criteria

The assessment should be based on the users' criteria for quality services, which should be defined in the demand formulation process.

Services Related to Public Interest, Public Authorities should Define Realistic and Appropriate Standards

Public authorities should define realistic and appropriate standards for handling certain goods such as handling of veterinary drugs, pesticides and communicate them to the users and to local forums (Neuchatel Group, 2006).

Conclusion

The emergence of Agricultural Advisory (Extension) Services is a vital element of the array of market and non-market entities and agents that provide critical flow of information that can improve farmers and other rural peoples welfare. Apart from their conventional function of providing knowledge for improved agricultural productivity, these advisory services have been successful in fulfilling a variety of new functions such as listing new technologies on farm, linking small holder farmers to high value and export markets, promoting environmentally sustainable production techniques etc.

In the near future, advisory services can also play an important role to meet the new challenges agriculture is confronted with: changes in the global food and agricultural system, including the rise of supermarkets and the growing importance of standards and labels; growth in non-farm rural employment and agri-business.

References

Anderson, J.R., 2007. Agricultural Advisory Services. Background Paper for World Development Report 2008, Agriculture for Development Washington D.C. World Bank.

ARD, 2002. *Agricultural Extension and Rural Development: Converging Views for Institutional Approaches*. Workshop Summary–Washington D.C.

ASDP, 2003. *Framework and Process Document MAFS*, Das ES, Salaam.

National Agricultural Extension Project II (NAEP), 2000. A mid term review report on the failure of Past Extension Approaches : NAEP.

Neuchatel Group, 2006. Demand Driven Agricultural Advisory Services. http:///www.neuchatelinitiative.net/english/documents/Demand Driven Agricultural Advisory services.pdf.

Swanson, B.E. and Samy, M.M., 2003. Decentralization of agricultural extension systems: Key elements for success. In: *Proceedings of Regional Conference on Operationalizing Reforms in Agricultural Extension in South Asia*. May 6-8, New Delhi, India.

Chapter 12
Market-led Agricultural Extension

☆ ***Virendra Kumar Yadav[1] and C.Y. Manikanhaiya[2]***

Introduction

India is basically an agriculture country and agriculture continues to contribute about 25 per cent of national income and remains a major sector that employs about 60 per cent of the labour force in rural India. Likewise it accounts for more than 13 percent of Indian export (Hindu Survey of Agriculture, 2005). India was depending on import of food grains in earlier days to meet the needs of expanding population. But with better management of agriculture, India has increased its food production manifold. However, the Indian farmers are subjected to the vagaries of nature as well as to the market fluctuation to a great extent. When the market is upbeat with very high price offers for certain products, the farmers wherever possible switch on to that crop and by the time the crop is ready, the prices come down so sharply creating utter glut in the market. In spite of the support price guaranteed by the Government, the farmers are left with no other alternative, but to sell at the available prices. This is the case with both food crops and cash crops.

With the opening of the market under the W.T.O. agreement, the farm products have become more volatile and the farming community is at crossroads wondering how to face the future shocks in the market. The market is already witnessing arrival of horticultural products and other food items from other countries (Subrahmanyam, 1998).

---

1 Scientist, Directorate of Maize Research, Pusa Campus, New Delhi – 110 012.

2 Ph.D. Scholar, Division of Agriculture Extension, IARI, Pusa Campus, New Delhi – 110 012.

In India, although at village and farm level agricultural production extension services exist but market-led extension work designed to benefit farmers and other market functionaries do not exist. Overwhelming number of States do not have even separate market-led extension cell to undertake regular extension activities. In the absence of emphasis on market-led extension, technology transfer in the field of agricultural marketing has been sluggish. As customer based or market driven strategies is now replacing the days of the mass production and mass marketing, an effective market-led extension service is need of the hour. This has added significance in the light of post WTO scenario. If the Indian farmers have to withstand the possible onslaught of international competitors both in domestic as well as overseas markets the market-led extension would be an effective instrument to safeguard farmers interest through proper education and guidance on regular basis. The market-led extension service to assist small and marginal farmers in solving the problems faced in marketing their produce is, therefore, a sine-qua-non in the free trade environment.

Market-led agricultural extension envisages educating the farmers, traders and consumers regarding marketing to bring desired changes in their knowledge, skills, attitude and behaviour towards adoption of improved techniques with regard to product planning, securing market for farmers, group action, direct marketing, marketing credit, marketing information, adoption for improved market practices, processing and value addition, problem solving methods and marketing capacity building for realizing better returns of their agricultural produce. There is a need of regular extension service in agricultural marketing.

Role of Agricultural Marketing Extension

For efficient disposal/marketing of agricultural produce, the farmers need guidance regarding:

- ✰ What to produce
- ✰ How much to produce
- ✰ When to produce
- ✰ When to sell
- ✰ Where to sell
- ✰ In what form to sell
- ✰ At what price to sell

There is no tailor made solution to these problems. Every question has to be answered taking account of market demand-supply, scenario, cost of production, storage facilities and several other personal and situational factors by using participatory methods. The decisions regarding marketing of agricultural produce may vary from place to place, demand pressure, supply pressure, financial pressure, family pressure and variety of other reasons.

Extension Issues in Informal Marketing

Informal marketing may be described as marketing transactions without any formal terms and conditions between farmers–sellers and traders–buyers. Small sized holdings coupled with smaller marketed surpluses in our country have made "informal marketing" as most common method of marketing of agricultural produce in producing areas. A whole set of constraints which prevents farmers participation in organized marketing system constitute extension issues in informal marketing. It includes (*i*) Advise on product planning, (*ii*) Securing market for farmers, (*iii*) Group action, (*iv*) Direct marketing, (*v*) Marketing credit, (*vi*) Marketing information, (*vii*) Advice on improved market practices, (*viii*) Processing and value addition, (*ix*) Problem solving methods, (*x*) Marketing capacity building essential.

Extension Issues in Cooperative Marketing

Cooperative marketing is composed of two words–Cooperative or Cooperation and Marketing. If the word Cooperation is split up, one finds that it is composed of two Latin words "Co" and "Opus". The word "Co" means to work and "Opus" means together. Cooperation thus means working together or jointly. In simple words, the cooperation can be defined as a form of association of people to work together in order to achieve a particular end.

The word marketing is derived from a Latin word *Marketus* that means to trade. Marketing, in this sense means the act or technique of buying and selling. Marketing in the modern sense connotes all business activities involved in the flow of goods and services from the point of production until they are in the hands of the consumers.

Cooperative Marketing is an extension of the principles of cooperation in the field of marketing. It is a process of marketing through a cooperative association formed voluntarily by its members to perform one or more marketing functions in respect of their produce. In other words Cooperative Marketing is the marketing for the producers and by the producers that aims at eliminating the chain of middlemen operating between the producers and the ultimate consumer and thus securing the maximum price for their produce.

Extension Issues of Cooperative marketing are (*i*) Marketing Information System, (*ii*) Grading and Standardization, (*iii*) Credit Facility, (*iv*) Processing and Cool Chains, (*v*) Coops Role in Inputs Supply.

International Agricultural Trading-Extension Issues and Strategies in Agricultural Marketing

In order to come up with suitable extension strategies to make Indian agriculture internationally competitive we should know our strength and weaknesses. To achieve this, in this paper SWOT analysis was used as the main tool.

Strengths

The main strength of our agriculture sector to compete in the international markets are (*i*) Varied and diversified agro-climate base, (*ii*) India is a leading producer in some commodities, (*iii*) Cheap and abundant labour and (*iv*) Huge scientific and technical manpower.

Weaknesses

- ☆ Lack of good quality produce suitable for exports
- ☆ Lack of infrastructure facilities like pre-cooling chambers/cold storage units at production centres
- ☆ Too many marketing intermediaries resulting in less returns to the producers
- ☆ High taxes/transport costs
- ☆ Inefficient domestic marketing resulting in huge post-harvest losses
- ☆ Lack of good and cheap packing material
- ☆ Lack of awareness among the farmers regarding the benefits of grading, cold storage, proper packing, etc.
- ☆ Lack of grading and sorting facilities in market yards
- ☆ Lack of information on buyer specification with respect to importing countries.

Opportunities

With more awareness and demand in developed countries on nutritional foods and organic foods, a vast scope exists for our exports. The signing of GATT and WTO also provides good access to international market.

Threat

Other developing countries and developed countries are also posing a stiff competition to us. For example, Sri Lanka in the case of tea, Brazil in the case of cashew, Pakistan, Mexico, etc. in the case of mangoes, Chile in case of grapes. These countries are also selling at lower prices compared to our country.

Extension Strategies

The extension strategies devised should overcome our weakness and build the capacity of our cultivators to take a plunge in the international market and compete with other countries.

- ☆ Identifying the farmers/group of farmers interested in export of agricultural commodities
- ☆ Creating quality consciousness among the farmers
- ☆ Training in cost reduction strategies
- ☆ Dissemination of the information about the financial and other assistance available under various government schemes
- ☆ Giving practical training to the farmers in grading and packing, storage and preservation aspect
- ☆ Creating, sorting and grading facilities at village level and at assembling markets

- ☆ Development and screening of video-films showing different steps involved in preparing the produce for exports
- ☆ Creating awareness about the scope of export of different commodities
- ☆ Giving information through electronic mass media T.V. and internet
- ☆ Investment in infrastructure and value addition initiatives
- ☆ Formulation of enterprise based Self Help Groups (SHGs).

Conclusion

Globalization is compelling Indian agriculture to shift its focus from being 'supply- driven' to 'market-driven'. In ability to adapt to this changing need may limit farmer's market and/or income opportunities. To facilitate such shifts market–led extension is not only necessary but also essential. Before the extension personnel's are asked to help the farmers to reorient production and marketing activities to adjust to trade, the extension personnel's need to be well equipped in terms of trade information. For this purpose, the training of change agents in the area of agricultural trade and continuous flow of information to them about changes in trade scenario are essential. Globalization should be viewed as an opportunity and not as a threat. The farmers should go for high quality products, not only for exports but also for the domestic market. Simultaneously, the facilities for quality control, storage, packing material, transport services etc. will have to be created and extended to convert the WTO provisions in our favour.

References

The Hindu Survey of Agriculture, 2005.

Subrahmanyam, K.V., 1998. Marketing of horticultural crops in Karnataka: Present status and future requirements. *Indian Journal of Agricultural Marketing*, 2(1): 15–22.

Chapter 13
e-Choupal and Agricultural Marketing

☆ *Samir Rajan Dash*[1]

Introduction

Agriculture forms the backbone of Indian economy and contributes a major share (23 per cent) in GDP of the country, feeds a billion people, and employs 66 per cent of the workforce. The agricultural exports accounts for 21per cent. Because of the Green Revolution, India's agricultural productivity has improved to the point that it is both self-sufficient and a net exporter of a variety of food grains. Yet most Indian farmers have remained quite poor. The causes include remnants of scarcity-era regulation and an agricultural system based on small, inefficient landholdings.

In India agricultural marketing operates basically according to the forces of demand and supply in the private sector. Farmers in private sectors sell a major part of agricultural produce in India to the moneylenders (to whom the farmer might be indebted) or to the village traders. The other ways of selling agricultural produce would be the weekly village markets, or in a irregularly held market in a neighboring village or town or in a mandi. Traditionally, a farmer sold his agricultural produce to a small trader referred to as a 'Kaccha adat.' The small trader then sold the products to a larger trader called the 'Pakka adat', who sold the produce in a 'mandi' (market place).

The purpose of regulation of agricultural markets was to protect farmers from the exploitation of intermediaries and traders and also to ensure better prices and

---

1 KVK , Kandhamal, G-udayagiri Orissa University of Agriculture and Technology, Bhubaneswar.

timely payment for his produce. Over a period of time these markets have, however, acquired the status of restrictive and monopolistic markets, providing no help in direct and free marketing, organized retailing, competitive trading, information exchange and adoption of innovative marketing systems and technologies. Farmer cannot sell his produce directly in bulk except on retail basis to the consumers. Farmers have to bring their produce to the market yard. Exporters, processors and retail chain operators can not get desired quality and quantity of produce for their business due to restrictions on direct marketing.

Origins of e-Choupal

ITC initiated its e-choupal project to streamline its dealings with Indian farmers in 2000. This is a project on a massive scale that ultimately aims to cover every sixth Indian village. Each choupal covers around eight villages and 15,000 villages have been covered to date in Madhya Pradesh, Uttar Pradesh, Maharashtra, Karnataka, and Andhra Pradesh. e-Choupal is an initiative of ITC Limited to link directly with rural farmers for the procurement of agricultural/aquaculture produce like soybeans, wheat, coffee, and prawns. e-Choupal was conceived to tackle the challenges posed by the unique features of Indian agriculture, characterized by fragmented farms, weak infrastructure and the involvement of numerous intermediaries. Traditionally, these commodities were procured in 'mandis' (major agricultural marketing centres in rural areas of India), where the middleman used to make most of the profit. These Middlemen used unscientific means to judge the quality of the product to set the price. Difference in price for good quality and inferior quality was less, and hence there was no incentive for the farmers to invest and produce good quality output. With e-Choupal, the farmers have a choice and the exploitative power of the middleman is neutralized.

The Benefits

1. Enhanced relationship with farming community across 15,000 villages
2. Reduced transaction cost for its Agri-commodity purchases.
3. Information on inventory retained by the farmers that can improve the quality of trading decisions.
4. Provided means to expand the reach to the rural markets through the cross-selling of the company's products and services.

Due to the e-Choupal services, farmers have seen a rise in their income levels because of rise in yields, improvement in quality of output and a fall in transaction costs. Customized knowledge is offered to the farmers despite heterogenity. Farmers can get real-time information despite their physical distance from the mandis. The system saves procurement costs and the model is quite different from the other models, as the farmers do not pay for the information and knowledge they get from e-Choupals. The company has maintained a role for the traditional commission agents by making them Sanyojaks (coordinators) in this project who manage physical flows in the supply chain and collect price information from local rnandis and maintain records. Thus, the company has chosen to re-intermediate, and not disintermediate (Sawhney, 2002).

There are 6,500 e-Choupals today. ITC Limited plans to scale up to 20,000 e-Choupals by 2012 covering 100,000 villages in 15 states, servicing 15 million farmers.The e-Choupal model demonstrates that a large corporation can play a major role in recognizing markets and increasing the efficiency of an agricultural system, while doing so in ways that benefit farmers and rural communities. e-Choupal' also unshackles the potential of Indian farmer who has been trapped in a vicious cycle of low risk taking ability–low investment–low productivity–weak market orientation–low value addition–low margin–low risk taking ability. This made him and Indian agribusiness sector globally uncompetitive, despite rich and abundant natural resources. 'e-Choupal' makes use of the physical transmission capabilities of current intermediaries–aggregation, logistics, counter-party risk and bridge financing–while disintermediating them from the chain of information flow and market signals.

Such a market-led business model can enhance the competitiveness of Indian agriculture and trigger a virtuous cycle of higher productivity, higher incomes, enlarged capacity for farmer risk management, larger investments and higher quality and productivity.

Chapter 14

Market Led Extension System in Tribal Zone: An Experiment

☆ *C. Satapathy*[1] *and Sabita Mishra*[2]

Introduction

Transfer of technology in the field of agriculture, horticulture, fishery, forestry and animal husbandry essentially means to benefit farmers in terms of income, social prestige, exposure to outside world and sustainable interest for development. Extension approach made up to the end of the last century was confined to research relating to technology and human behaviour. Marketing was kept aside assuming it as the function of non-extension agencies or existence of good marketing network. With rise in production system, the hypothesis has failed to prove that the expansion of technology is conditioned by marketing net work. In other words, market decides extension strategy whether to give emphasis on kind of production. Extension background and achievement so far have not taken marketing as an integral part in planning extension project. With change of situation, taste of people, global change, general consciousness, the need for market led extension has come to discussion among the scientists concerned. With this background, the paper entitled "Market led Extension System in Tribal zone–An Experiment" is an attempt to explain how to internalize idea of marketing in extension system.

The study conducted in tribal area is based on the realities that tribal women are producer, promoter, retailer, trader and purchaser of agri-products. It is a matter of observation that women dominate tribal markets. The male members participate to a

---

1 Director, AMITY Business School, Bhubaneswar, Orissa.

2. Senior Scientist (AE), NRCWA, Bhubaneswar, Orissa.

limited extent in marketing of agri-products in tribal areas. Keeping these considerations in view the objectives of the study were delineated as follows.

Objectives

1. To study the role of tribal women in production, promotion and pricing of agri-products and NTFP.
2. To find out awareness of tribal women about market, price of different produce sold in the locality.
3. To ascertain techniques followed by tribal women retailer in fixing price of the produce.
4. To find out marketing behaviour of tribal women in terms of livelihood management, empowerment, decision-making and future aspiration.
5. To investigate into facilities available at market place to sale the products which directly decide the rate of the produce?

Methodology

The study was undertaken in Koraput district covering two Panchayats and eight villages of Sadar block. The samples of the study consisted of 130 tribal women who were very much regular in attending three weekly markets to sell and purchase of agri-products. The randomly selected samples were interviewed with help of a structured schedule and observations were recorded at market place.

Results

Participation of tribal women in farming was studied with respect to crop grown in the area as per demand of the market. According to agro-climatic zone, taking rainfall, temperature and available technologies into consideration, the preference for growing of crops was examined as follows.

Market Led Production System

With increase in market demand, the tribal women who are involved in both production as well marketing have a change in their farming practices. Farming of yesterday and today witnesses changes in production preference.

Table 14.1: Market Oriented Production System

Sl.No.	*Crops*	*Preference Over Past (Score)*	*Rank*
1.	**Cereals**		
(*a*)	Paddy	2.35	I
(*b*)	Ragi	1.20	III
(*c*)	Maize	2.00	II
	Average	**2.00**	

Contd...

Table 14.1–Contd...

Sl.No.	Crops	Preference Over Past (Score)	Rank
2.	**Pulses**		
(*a*)	Green gram	1.10	I
(*b*)	Black gram	0.95	III
(*c*)	Red Gram	1.05	II
(*d*)	Horse gram	0.62	IV
	Average	**0.93**	
3.	**Oilseeds**		
(*a*)	Groundnut	2.65	I
(*b*)	Niger	1.15	III
(*c*)	Castor	0.70	II
(*d*)	Sesamum	1.10	IV
	Average	**1.40**	
4.	Vegetables		
(*a*)	Bean	2.55	III
(*b*)	Cabbage	2.00	V
(*c*)	Tomato	2.65	II
(*d*)	Brinjal	2.90	I
(*e*)	Radish	2.45	IV
	Average	**2.51**	

Now in the tribal areas of Koraput district, there is change in cropping pattern. As per demand of the market and profitability, vegetables receiving priority over oil seeds, cereals and pulses. Market led extension system can be very well planned to cover these crops keeping watch on market.

Marketing of NTFP

Besides field crop, the tribal women are also involved in marketing of NTFP since generation to generation. The sample women involved in procurement and marketing of NTFP is observed as contained in Table 14.2.

Table 14.2: Marketing of NTFP

Sl.No.	Name of NTFP	Women Involved (n=130)	Per cent
1.	Tamarind	73	56.15
2.	Lac	63	48.46
3.	Araroot	55	42.30
4.	Mahul	67	51.53
5.	Kusumi seeds	62	47.69
6.	Silk cotton	52	40.00
7.	Honey	43	33.07

The facts contained in table reveal the marketing of NTFP by tribal women. Since it is their way of living, certain training programmes should include marketing in extension activities.

Place of Marketing

In tribal districts, weekly markets are most common. Keeping a considerable distance in locations, the weekly markets are held in all locations. These markets sit in morning hours and over by afternoon. Some women of distance places used to come day before marketing stay at market place for next day marketing. They face problems in market place and marketing. The information given in Table 14.3 reveal the problems encountered in marketing by tribal women.

Table 14.3: Problems in Market Place

Sl.No.	*Problems in market place*	*Frequency*	*Per cent*	*Rank*
1.	Sitting arrangement	84	64.61	II
2.	Safety	72	55.38	IV
3.	Rainy days	63	48.46	V
4.	Night halt	76	58.46	III
5.	Transportation	100	76.92	I
6.	Tax at market	48	36.92	VII
7.	Seasonal effect	51	39.23	VI

The problems in marketing are many folds. The women of tribal areas are proned to such problems. Transports, sitting arrangement, night halt, safety and shelter in rainy days are some of the problems which need attention.

Techniques of Price Fixing

It is interesting to observe that the tribal women have their own techniques in quantifying products per unit. The pricing of product is fixed on oral conversation, looking into quantity of arrival, price of last market day and next market day of their locality. The factors that they consider are given in Table 14.4.

Table 14.4: Techniques of Price Fixing

Sl.No.	*Techniques of Price Fixing*	*Frequency*	*Per cent*
1.	Quantity of arrival	120	92.30
2.	Grading of produce	90	69.23
3.	Unsold material	45	34.61
4.	Bargaining	98	75.38
5.	New produce of the season	92	70.76

Quantity of arrival, bargaining, grading and freshness and unsold produce of last marketing normally are followed to fix the price of produce. Above all, the price goes up when new produce of the season are coming to market for sale.

Marketing Behaviour

The marketing behaviour of tribal women is regulated by livelihood system, extent of empowerment, freedom to take decisions, facilities and next date of marketing. The marketing behaviour is regulated by market information and quantity of production. Market behaviour is a complex concept. The parameters determining such behaviour of women involved in marketing are, (*a*) regularity in marketing (*b*) procurement from own and neighbouring field (*c*) making calculation of profit and loss and (*d*) known in the locality as good sellers. Correlating these dependent variables with that of behavioural parameters, the following results were obtained.

Table 14.5: Factors of Regulating Market Behaviour

Sl.No.	*Variables*	*'r' Value*
1.	Livelihood	0.753 **
2.	Empowerment	0.521 **
3.	Independent decision	0.492 **
4.	Facilities in market place	0.328 **
5.	Future aspiration	0.612 **

**: Significant at 1 per cent level of probability.

Results reveal that livelihood system, empowerment, independent decision, facilities and aspiration do influence the marketing behaviour of women.

Market Information

The above result warrants a market led extension strategy to empower tribal women for efficient marketing. The women to be armed with marketing efficiency desire constant information on the following aspects.

Table 14.6: Need for Marketing Information

Sl.No.	*Information*	*Score*	*Rank*
1.	Price of produce	2.62	I
2.	Appropriate demand time	2.20	II
3.	Expected demand in local market	1.79	IV
4.	Place of sale	1.93	III

Sample women desire information on price of produce in nearby locations preferably in district, appropriate demand of the produce, place of sale and expected demand on weekly market.

Conclusion

The study entitled "Market Led Extension System in Tribal Zone–an Experiment" conducted with 130 randomized sample reveal:

1. Now there is shift in farming pattern. The vegetables are replacing traditional crops in view of profit and market demand.

2. Tribal women are involved in procurement and marketing of NTFP and they need marketing skill in these areas.
3. The problems faced by women at market places like transport, sitting arrangement, safety, night halt and rainy day shelter need attention.
4. In fixing price of produce they follow the traditional methods. Training on techniques of price fixing will help tribal women to do efficient marketing.
5. The marketing behaviour of tribal women is conditioned by their livelihood system; extent of empowerment, available facilities and future aspirations.
6. The tribal women involved in marketing need information on price of the produce, market demand, place of sale and weekly market demand.

Chapter 15

Farmers' Driven Participatory Pulses Seed Production (FDPSP) for Ensuring Seed-Sufficiency at the Village Level

☆ *S.K. Singh*[1], *Shantanu Kumar*[1], *S.N. Nigam*[2] *and A.K. Singh*[3]

Introduction

Seed security is a pivot to food and nutritional security. Quality seed of improved varieties alone can cause significant improvement in crop productivity leading to increased availability of pulses, particularly to the poorer sections of the society. Further, improved varieties provide an efficient vehicle for transfer of other improved crop management technologies. In spite of significant achievements made in pulses through crop improvement research and development, traditional varieties and landraces still dominate the farming in the country. The vertical seed replacement rate in pulses remains low as compared to other major crops for a variety of reasons. Limited access for farmers to improved legumes varieties is one of the major factors contributing to low farm level productivity. Without improving access to improved seed through effective, farmer-friendly seed system, this scenario is not likely to change.

---

1 Indian Institute of Pulses Research, Kanpur.

2 ICRISAT, Hyderabad.

3 Zonal Coordination Unit (ICAR), Kanpur.

Compared to cereal crops (700-800 in pearl millet and 200 in sorghum), the seed multiplication ratio in pulses is low (chickpea: 10-20, pigeonpea: 20-30). No single agency can, therefore, fulfill the demand of quality seed of improved varieties of legumes with the existing practice of seed production and supply. The strategies devised by the Government agencies in the past to address these issues have either failed to be initiated or were unsustainable. The quality and integrity of the seed bought in local markets is most often uncertain and inferior resulting in low productivity of legumes. It is imperative to revisit the seed issues and consultation with relevant stakeholders, to develop strategies and appropriate approaches for addressing the constraints relating to sustainable multiplication and distribution of improved legumes seeds.

The sustainability of rainfed pulses cultivation is heavily dependent of the availability of seed of locally adopted, farmer- preferred varieties at the right time in required quantities and at the right price to meet the demand of resource poor farmers. Both informal as well as formal seed sector are essential and complimentary to ensure an effective seed security strategy. But it is important to place farmer-based seed system at the center of the seed security. Against this context, ISOPOM funded ISOPOM Project Development and Popularization of "Modal Seed system (s)" for Quality Seed Production of major legumes to ensure Seed sufficiency at Village Level is being implemented in partnership with ICRISAT, Hyderabad among fifteen villages of Fatehpur and Kanpur Dehat district of Uttar Pradesh with following basic objectives:

- ☆ To evolve a replicable 'Model' seed system (s) to achieve seed-sufficiency of farmers farmers' preferred improved varieties of pulses
- ☆ Knowledge empowerment of farmers in pulse seed production, processing and storage technologies and integrated crop management practices to promote seed production enterprises at the village level.
- ☆ To strengthen and sustain seed production chain in public sector seed system to improve availability of seed of improved varieties of pulse crops to farmers.
- ☆ To establish linkage between informal and formal seed sectors and with local village institutions to enhance the availability of seeds of farmer-preferred improved varieties of pulses to farmers at the village level.
- ☆ Research backstopping for further improvement of varieties for resistance to pest and diseases and traits preferred by the farmers and traders in the areas.
- ☆ To suggest policy and institutional changes, if needed, to ensure supply of quality seed to farmers.

Methodology

Baseline survey as well as informal discussion with individual farmer and in groups was conducted to ascertain the prevailing cropping system, extent of space occupied by the pulse crops, sources and quality of seed utilized, level of pulses productivity, etc. It was ascertained about the traits of quality seeds as perceived by the partner farmers. The prevailing formal and informal seed system was also documented to identify the deficiency therein. Based on production situation present, Farmer Participatory Varietal Evaluation Trails (FPVETs) were laid out. The yield related attributes like plant height, number of branches, etc were collected for the varieties under FPVETs. Besides, the actual productivity and profitability of various pulse crops under different production situation were also analyzed. Farmers' perception of the qualitative attributes and appropriateness of the varieties were ascertained. Based on the outcome of FPVETs, large-scale seed production was initiated at the farmers' fields. The total production, consumption and disposal pattern for various pulse seeds were documented. A functional model for implementation of this project was also evolved.

Results and Discussion

Analysis of Agro-Eco Situation of the Project Sites

Fatehpur

Five villages namely Mahuar, Alipur, Pahur, Godrauli, Sai, Aung and Manakpur in Fatehpur district were selected for the project. District Fatehpur is spread over an area of 4152 km^2 with 13 blocks, 132 Nayaypanchayats, 786 Gram Sabha and 1516 Villages. The district is supported with 170 nationalized gramin and cooperative banks and 8 cold storage houses. It receives 960 mm of annual rainfall that flows along the network of 1532 kms of canals. There are 549 and 53,161 tube wells under public and individual ownership respectively. The population density is 457 per km^2 and the scheduled cast and scheduled tribes contribute about 25.0 per cent of the total population. About 81 percent of the population is dependent on agriculture as farmers and agricultural labourers. Small and marginal farmers cultivate about 74.0 percent of total cultivable land of the district. District Fatehpur is distributed into four diverse agro-ecological situations. The major farming system of the district is crop based. Rice- Wheat, Rice-Chickpea, Til-Chickpea, Wheat-Sugarcane, Rice-Potato/ Chilies, Urd/Mungbean- Mustard, etc. are the main cropping sequences of district Fatehpur. Mixed cropping of pigeon Pea+Jowar+Urd/Moongbean+Til is also practiced. Chickpea is the important *Rabi* pulse crop grown in about 46662 ha of area with 45685.0 metric tons of total production. The productivity of chickpea in the district in 9.79 q/ha. The pigeon pea area in the Fatehpur district is 1634.9 ha and total production is 19150 metric tons. The average yield of pigeon pea is 11.71 q/ha.

Kanpur Dehat

In Kanpur Dehat district, chickpea is cultivated in the area of 25071 ha. Total production of chickpea is 34906 m tones. The productivity of gram is 12.01q/ha. Pigeon pea area is 15,750 ha with productivity of 13.51q/ha. Five villages namely Barahapur, Kuitkhera, Sariyapur, Bevan, Sangasiapur and Mohummedpur have been identified as project sites for seed production of pigeon pea and chickpea in the Kanpur Dehat. Sandy loam, loam soil with good drainage facilities are found in the project villages in Kanpur Dehat district. Area under pigeon pea cultivation ranges between 5 per cent to 30 per cent. Paddy, hybrid bajra, maize, urd, til and pigeon pea are the main *Kharif* crop. Farmers grow wheat, chickpea, potato, barley, mustard, during *Rabi* season on large scale.

Status of Formal and Informal Seed Systems Prevailing in the Area

Base line Survey was conducted to ascertain the existing informal and semi-informal seed system in the project villages of Fatehpur district. It was observed that farmer obtain seed from local sources and mostly use their own saved seeds. Besides, they also utilize relatives and other neighboring farmers, local traders, shopkeepers and open-air market grain traders as the sources for seed. The majority of farmers in the project villages depends on seed saved from their own harvest and continues recycling it as long as the harvest is "adequate" and most often they are able to keep some of them for subsequent sequences. Local traders play a critical role in usual communities by purchasing grain at harvest; storing it and selling it back to the same farmer, either as raw or processed grain or as seed at planting time. The prevailing informal seed system is not at all organized in the study area. However, some of the innovative/progressive farmers occasionally procure seed of Chickpea from Agriculture Research Station/Krishi Vigyan Kendra located nearby.

Conducting Farmer Participatory Varietal Selection Trials (FPVST)

FPVST of Pigeonpea

Farmer participatory Varietal selection trials (FPVSTs) were conducted to ascertain farmers' preferences along with and suitability of long duration pigeonpea varieties under the situation of loam soil and partially irrigated condition. Further long duration improved pigeonpea varieties *viz.*, N.A.1, Bahar, Amar, Azad, Mal 13 along with local varieties were included in each FPVST. On the basis of overall performance of FPVSTs on pigeonpea it was inferred that Narendra Arher1 was the highest yielder (2208 kg/ha) followed by Bahar (2020 kg/ha), Amar (1979kg/ha), Mal-13(1812kg/ha) and Azad (1804kg/ha). These improved varieties yielded almost 1.0 tha^{-1} more than the local variety.

FPVST for Chickpea

Farmer participatory on-farm varietal selection trials on chickpea involving six improved varieties *viz.*, DCP 92-3, JG 16, KWR 108, PG 186 KGD 1168 and JGK 1 and two local large and small seeded varieties were carried out in 10 (Ten) and 5 (Five) farmers field in Kanpur Dehat and Fatehpur district, respectively. The performance of six chickpea varieties in FPVS trials obtained from Fatehpur district was as fallows: JG 16 (grain yield 2.85 tha^{-1},100 seed weight 16g) DCP 92-3 (25.50 tha^{-1}, 13 g) KWR 108 (2.03 tha^{-1}, 16 g), PG 186 (1.70 tha^{-1}, 16 g), JGK 1(Kabuli 1.45 tha^{-1}, 32 g) and KGD 1168 (1.30 tha^{-1}, 22g). Farmers preferred JG 16 and DCP 92-3 for their high yield and PG 186 for its adaptability in late sowing (November to first week of December) (Table 15.1).

Participatory Seed Production, their Economics and Disposal Pattern

Pigeonpea

Seed production of short duration and long duration pigeonpea were also initiated in Fatehpur and Kanpur Dehat during rainy season 2007. Seed production of early pigeonpea involving variety UPAS 120 was undertaken in 16 ha area at 80 farmers field in Pahur village of Fatehpur district. Total 9250 kg seed of early pigeonpea was produced and out of which National Seed Corporation @ Rs. 2900/q. Farmers fetched net return of rupees 600/q and produced will be utilized as a seed in coming *kharif* season 2008 procured 3976 kg.

Table 15.1: Performance of Farmer Participatory Varietal Selection Trials in Fatehpur (Situation II)

Micro Farming Situation	*No. of Fields*	*Plot Size (ha)*	*Varieties*	*Productivity (q/ha)*		
				Max.	*Min.*	*Mean*
Loam soil, Partially irrigated, double cropping system (Sown in between 10th November to 1st week of December, 2007)	3	0.50	DCP 92-3	25.8	21.4	23.6
			JG 16	27.7	22.0	24.9
			KWR 108	18.6	14.1	16.4
			PG 186	25.1	19.8	22.5
			KGD 1168	23.2	17.4	20.3
			JKG 1	9.8	7.8	8.8
			Local (large seeded)	15.8	12.0	13.9
			Local (Small seeded)	11.7	7.9	9.8

Seed production of long duration pigeonpea varieties (N.A.1 and Bahar) was undertaken in two compact blocks in 12 ha by 46 farmers in Fatehur and 17 hectare area at 44 farmers in Kanpur Dehat district during *kharif* 2007. Certification of all the pigeonpea growers was done through U.P. Seed Certification Agency, Allahabad and Kanpur Dehat. Total 22800 kg including N.A.1 (16800kg) and Bahar (6000kg) was produced as seed of pigeonpea in Kanpur Dehat district. Some of the farmers preferred Bahar whereas others preferred Naredra Arher1. The economics of seed production for long duration pigeonpea has been computed and presented as follows in Table 15.2.

Table 15.2: Economics of Seed Production for Long Duration of Pigeonpea (Kanpur Dehat)

Sl.No	*Indicators*	*Varieties*		
		Local	*N.A. 1*	*Bahar*
1.	Average seed yield (q/ha)	11.34	21.72	19.13
2.	Increase in yield over control	-	51.46	33.40
3.	Cost of cultivation(Rs/ha)	9,240	12,530	12,530
5.	Gross income(Rs./ha)	35,850	65,160	57,390
6.	Net income (Rs./ha)	26,610	52,630	44,860
7.	B:C ratio	2.88	4.20	3.58
8.	Market rate for seed (Rs./q)	2500*	3000**	3000**

*: Prevailing market price for pigeonpea grain was taken at Rs. 2500/q.

**: Procurement price of pigeonpea seed @ Rs. 3000/q.

Chickpea

Seed production of chickpea varieties (DCP 92-3, JG 16 and PG 186) was done on 19 ha area at 96 farmers' fields in Fatehpur district during *rabi* 2007-08. Three improved varieties of chickpea *viz.*, DCP 92-3, JG 16 and KWR 108 were used in seed production at five villages of Kanpur Dehat in 20 ha area at 113 farmer's fields in the month of October and November, 2007. Farmers of the project villages in Fatehpur district sold 53 quintals chickpea seed produced during *rabi* 2006-07 and fetched 20 per cent higher price in comparison to prevailing market rate of chickpea grain. Variety-wise seed production and disposal pattern is given as follows in Table 15.3.

Table 15.3: Variety-Wise Seed Production and Disposal Pattern of Chickpea in Fatehpur District

Sl.No.	*Varieties*	*Total Prodn. (q)*	*Purchased by IIPR (q)**	*Sold to the Farmers **(q)*	*Sown in the Rabi (2007-08)*	*Used as Dal, Besan, etc.*	*Sold in Market as Grain****
1.	DCP 92-3	61.00	20.00	22.00	10.00	05.00	04.00
2	JG 16	51.00	20.00	17.00	08.00	03.50	02.50
3	KWR 108	50.00	10.00	13.00	06.00	02.80	18.20

*: Purchased @ Rs. 3100/- per quintal in May 2007 (Prevailing market rate was Rs. 2000/- per q).

**: Sold @ Rs. 2800/- per quintal.

***: Sold @ Rs. 2200/- per quintal.

Total 368.95 quintals seed of chickpea was produced as foundation category at five villages in Fatehpur district. Out of the total, 205.90 quintals was procured by National Seed Corporation @ Rs. 3500/q as foundation seed. Majority of farmers have kept seed of DCP 92-3, JG 16 and PG 186 for next year sowing, distribution and sale to other farmers in the same village or the neighbouring villages within the district and the farmers of the adjacent districts.

Table 15.4: Economics of Seed Production of Chickpea in Situation I (25 Oct–10 Nov, 2007)

Sl.No	*Indicators*	*Varieties*			
		Local	*DCP 92-3*	*JG 16*	*PG 186*
1	Average seed yield (q/ha)	13.56	22.63	18.34	15.72
2	Increase in yield over control (per cent)	-	66.89	35.25	15.93
3	Cost of cultivation(Rs/ha	12198	14200	14200	14200
5	Gross income(Rs./ha)	33900	79205	64190	55020
6	Net income (Rs./ha)	21702	65005	49990	40820
7	B:C ratio	1.78	4.58	3.52	2.87
8.	Seed rate (Rs./q)	2500*	3500**	3500**	3500**

*: Prevailing market price of chickpea grain was @ Rs. 2500/q.

**: Procurement price of chickpea seed @ Rs. 3500/q.

Coverage of Area Under Improved and Quality Seed of Pulses

As results of quality seed production, the area under high yielding variety increased. It could be seen from the Table 15.5, out of total area under chickpea in project villages (190.24 ha), about 65 ha area could be covered with the improved varieties, which is by any standard, is remarkable seed replacement scenario.

Table 15.5: Area Coverage Under Improved Varieties in Chickpea in Fatehpur

Villages	*Improved Varieties Area (ha)*	*Local Varieties Area (ha)*	*Total Area (ha)*
Mauhar	25.96	52.00	77.96
Alipur	8.00	10.64	18.64
Pahur	13.00	15.29	28.29
Harsinghpur	7.50	9.00	16.50
Sai	4.00	12.73	16.73
Godhrauli	6.12	26.00	32.12
Total	64.58	125.66	190.24

Conclusion

FPVST is an important tool for the capacity building of farmers for on-farm experimentations. Experiences of the study showed that all the recommended pulse varieties may not perform equally in the various micro-farming situations prevailing at the village level. Hence, identification of location specific pulses varieties followed by its participatory seed production is an important tool for pulse farmers' productivity and income augmentation besides ensuring the seed-sufficiency at the village level.

Chapter 16

Dynamics of Ajinkyatara Farmer's Co-operative: An Innovative Extension Strategy for Agricultural Development and Rural Prosperity– A Case Study

☆ ***P.G Khalache*[1], *H.P. Sonawane*[2] *and J.H.Gaikwad*[2]**

Introduction

Co-operative is an autonomous association of persons united voluntarily to meet their common economic, social and cultural needs and aspirations through a jointly owned and democratically controlled enterprise.

Floriculture is a rapidly growing industry in India. Floriculture has become an important agri-business. It is contributing to the nation's economy and development. It also is providing employment opportunities in sub-urban and rural areas.

---

1 Head, Department of Extension Education, MPKV, Rahuri. Dist. Ahmednagar – 413 702.

2 Assistant Professor, Department of Extension Education, MPKV, Rahuri. Dist. Ahmednagar – 413 702.

The Satara district of southwest Maharashtra is a leading district as regards flowers and vegetables cultivation under protected conditions. The total area of about 64 ha is solely under protected floriculture crops in the district. The total production of flowers from Satara district was 35 tons with an average yield of 0.547 tons/ha. Along with flower, crop like capsicum are also cultivated under protected conditions by the cultivators. These farmers also experience a number of constraints regarding the management of flowers and vegetables under protected conditions. Marketing of the produce of these crops is depending on the availability of the market and the market rates. Sometimes due to these dimensions the results is un-remunerative to the farmers. This cause a need of co-operative organization of such farmers those are cultivating flowers, vegetables and fruits crop under protected conditions. This can maintain a spirit of service as the profit-earning motive on the real basis of co-operative organization, with a view to protect the members from the intermediaries and to improve the economic status of its members. Co-operatives should, however, run efficiently and earn due profits in its transactions which are primarily meant to strengthen the financial position of the society with a view to render better and cheaper service to its members.

Ajinkyatara Farmer's Co-operative Society: A profile

The Satara is located in western ghat zone having dry and somewhat cool climate. The annual rainfall ranges from 700-1000 mm. Initially farmers in this area were interested in growing vegetables like tomato, chilli, cauliflowers, cabbage, brinjal, turmeric, ginger and fruits like mango, guava, sapota, grapes etc. However, the farmers were not getting satisfactory returns.

In 1997, the revolting of Hi-Tech farming took place and many of the farmers diverted towards cultivation of coloured capsicum in green houses. Initially they got much more returns from their produce but after one year they started suffering due to unmanaged planning and lack of market. This situation compiled them to divert towards taking flowers in green houses. However, there was no way with them to sale their produce at Delhi and Mumbai market by avoiding private traders. In this practice the farmers were suffering due to delay in payments and huge commission taken by the agents. Moreover, the transportation cost was more as the farmers were transporting their produce individually and in limited quantity of their produce. In such circumstances to overcome these problems and to strengthen the limitations that the cultivators were, facing there was a need of agency to help the problems faced by them in market and marketing transactions for their better earnings.

Ajinkyatara Farmer's Co-operative Society was registered on 28th September, 2001 having its area of operation of entire Satara district. Primarily the society was dealing with the marketing of fruits, flowers and vegetables produced by the member cultivators.

Area of Operation and Working Pattern of the Society

The Ajinkyatara Co-operative Society has a unique activity in green house and polyhouses at various Hi-Tech farms to train the farmers in ultra modern agricultural techniques and equipment not only in horticulture but also in overall activities. It is a marketing society established on a co-operative basis in the year 2001 vide its registration No. SAT/SAT/AGR/0/303/2001-2002/28-09-01 under the Co-operative Act, 1860. The members of the society are spread over in the Satara district of Maharashtra state in India. It also guides and helps to the producers by supplying quality cut flowers in time in the market. However, growers have the freedom of selecting their potential trader approved by the society. It assists the growers in minimizing risk and optimizing results with various options and through cultivation support.

In addition to this, the society provides guidance to its members for cultivation of flowers, vegetables and fruits in polyhouses and open field cultivation. Society also arrange the market their behalf for this purpose by undertaking the survey for their demands at Mumbai, Delhi, Hyderabad, Chennai and outside the country. The head office of this society is situated at Satara with well-furnished office and modern facilities like telephone, computer and e-mail with appropriate human resource. The society has provided a vehicle to collect flowers, fruits and vegetables from polyhouses and farmhouses at the producers and supplying up to Railway station. Moreover, fertilizers, pesticides, packing materials and boxes were supplied at door to door of the producers on reasonable rate so that farmers can spend maximum time in the field for proper management of their farm operations.

The meeting of Board of Directors is called every month to discuss on transactions of the society in which the decisions are taken unanimously about rates prevailed in respect of production cost for the benefit of members. Future plans and forecasting of rates are also assumed by the meeting. The society also arranges quarterly meetings of all members along with marketing agents, greenhouse erect contractors, irrigation system material suppliers and planting material suppliers in order to solve the problems of members if any. In this way, the society acts as a catalytic agent to run the business of the society for the welfare of its members.

Objectives of the Present Study

1. To study the mandate and activities carried out by the Ajinkyatara Farmer's Co-operative
2. To study the training facilities made available by the society to the respondent members.
3. To study the marketing behaviour of the respondent members.
4. To study the infrastructural facilities created by the society.
5. To find out the difficulties faced by the society, if any to fulfill the mandates.

Locale of the Study

The present study was conducted in Satara, Karad, Koregaon, Wai, Phaltan and Khatav tahsils of Satara district. 239 respondents from 42 villages who had members of the society were selected for the study. The data were collected with the help of well-constructed and pre-tested interview schedule.

Findings

Mandates of the Co-operative

1. To market fruits, flowers and vegetables produced by the members in local markets, metropolitan markets and international markets.
2. To construct office building, godown and cold storage for storage of agriculture commodities.
3. To process the agricultural produce in order to obtain maximum returns.
4. To maintain proper records of the rates of agriculture commodities and inform the members.
5. To provide technical guidance to the members in regards of cultivation of fruits, flowers and vegetables.
6. To provide necessary equipments, tools, fertilizers, insecticides, pesticides and packaging materials to the members at the site of production.
7. To arrange study tours of the members in order to study the hi-tech practice involved in production of fruits, flowers and vegetables.
8. To arrange quarterly meetings of the members with the market functionaries in order to design effective marketing strategy and,
9. To arrange farmers' rallies, exhibitions, seminars for transfer of technology for promotion of agriculture/horticulture.

The Activities Undertaken by the Society in the Sample Year

1. The society provided technical guidance to the members in regards to cultivation of fruits, flowers and vegetables.
2. The society had arranged seminars regarding green house, floriculture, water management, and package of practice for Hi-tech Horticulture plantation processing.
3. The society marketed various agriculture commodities *i.e.* carnation, gerbera, glad bird of paradise, capsicum produced by the members of Delhi, Hyderabad, Chennai, Ludhiana, Mumbai, Chandigarh, Kolkata markets etc. marketing of flowers to Singapore was also undertaken by the society.
4. The society had arranged exhibition of flowers, fruits and vegetables under the name of *Nisarg*.
5. The society provided necessary tools, equipments, fertilizers, pesticides and packaging materials to the members at the site of production.
6. The society had arranged quarterly meetings of the members with marketing agents, green house erectors, irrigation system suppliers and planting material suppliers in order to solve the problem of members.
7. The society arranged monthly meeting of the Board of Directors.
8. The society will arrange to increase the rose flowers as per demand in the market.

Turnover at a Glance of the Society

Now a days selling arrangement is done in the metropolitan cities of India *i.e.* Mumbai, Delhi, Chandigarh, Ludhiana, Hyderabad, Kolkata. Our farmers have option for to send their agricultural production anywhere in India and any traders as listed with our society. A payment was done in twice a month by cheque after deduction of actual expenses on transactions. The figures of sales shows the increasing trend as detailed under in Table 16.1.

Table 16.1: Trends of Sales of the Society

Sl.No.	Years	No. of Boxes	Total Sale Rs. (Lakhs)	Av. Sales per Month (Lakhs)
1.	2001-2002 (6 months)	2991	33.25	5.54
2.	2002-2003	9103	102.55	8.54
3.	2003-2004	20668	190.44	15.87
4.	2004-2005	37299	291.04	24.25
5.	2005-2006	49987	408.17	34.01
6.	2006-2007	71401	475.59	39.63

Training Facilities made Available by the Society to the Respondent Members

The Ajinkyatara Farmer's Co-operative Society being bulk supplier of flowers, fruits and vegetables the commission agents as better communicator to the members. The market intermediaries not only provided the interest free advance through society but also contacted members to pass on hints of market information useful to them. The society functioning whole-heartedly for its members inspired to improve both marketing as well as production. Better prices are realized for quality product that needs specific techniques of production. In view of this, society provided training facilities to the respondent members. The results pertaining to this are presented in Table 16.2.

Table 16.2: Distribution of the Respondent Members by their Training Received from the Society

Sl.No.	Training Imparted by the Society	Respondent Members (n=239)	
		Number	Percentage
1.	Undergone no training	43	17.99
2.	Undergone one training	119	49.79
3.	Undergone two trainings	77	32.22
	Total	239	100.00

From the above Table 16.2, it was observed that, trainings were mainly conducted by the society twice a year for the period of 3 days. Most of the trainings were related to greenhouse technology, production, packaging and marketing of the fruits, flowers and vegetables. The data further indicates that 49.79 per cent of the respondent members had undergone only one training, while 32.22 per cent of them had undergone two training programmes. Only 17.99 per cent of them had not attended any training programme.

It was also observed that the society is not arranging only training facilities to its members but also some extent providing services to the respondent members like technical guidance in regards to cultivation practices of fruits, flowers and vegetables. The society had arranged seminars regarding greenhouse, floriculture, water management, package practices to be followed for Hi-tech floriculture plantation. In all, the society arranged two seminars during the sample year. The society had arranged two study tours of the members to the research stations and the field of progressive farmers during last two years in order to study the hi-tech practice involved in production of fruits, flowers and vegetables. In addition, every year farmer's rallies, exhibitions were arranged for transfer of technology for promotion of protected cultivation technologies. The society also arranges quarterly meetings every year of all members' along with-marketing agents, greenhouse erectors, irrigation system suppliers, planting material suppliers in order to solve the problems of the respondent members.

Marketing Behaviour of Respondent Members

From Table 16.3, it is observed that 58.58 per cent of the respondent members sold their produce in out of state market *i.e.* Delhi, Hyderabad and Kolkata with an average quantity of 43,950 boxes of produce and received total returns of Rs. 290 lakhs, while 16.74 per cent of them had sold in within state market *i.e.* Mumbai (12,500 boxes of produce) and got the returns of Rs. 75 lakhs, while 14.22 per cent of them had sold their produce in the district market *i.e.* Satara (8500 boxes of produce) and received total returns of Rs. 25.59 lakhs and only 10.46 per cent of them had exported their produce at Singapore (quantity of 6551 boxes) and received Rs. 85 lakhs.

Table 16.3: Distribution of the Respondent Members by their Marketing Behaviour

Sl.No.	*Place of Market*	*Number of Respondent Members*	*Percentage*	*Total Produce Sold in Market (No. of Boxes)*	*Total Sale (Rs. in lakhs)*
1.	District market (Satara)	34	14.22	8500	25.59
2.	Within state market (Mumbai)	40	16.74	12500	75.00
3.	Out of state markets (Delhi, Hyderabad, Kolkata)	140	58.58	43950	290.00
4.	Export (Singapore)	25	10.46	6551	85.00
	Total	239	100.00	71401	475.59

Infrastructural Facilities Created by the Society

The Ajinkyatara Farmer's Co-operative Society was established to take up marketing of flowers, fruits and vegetables in local markets, metropolitan markets and international markets. The society provides necessary tools, equipments, fertilizers, insecticides, pesticides, hormones and packaging materials to its members. Also, arrange transport and communication facilities to them.

The Table 16.4 shows that the society provides marketing facilities to cent per cent to the respondent members. The fruits, flowers and vegetables produced by the members were marketed to Delhi, Kolkata, Hyderabad, Chennai, Ludhiana and Mumbai markets and at district market *i.e.* Satara. Export of flowers to Singapore was also undertaken by the society. The large quantity of produce of the members was sold in Delhi, Hyderabad, Mumbai and Kolkata markets. However, the less quantity produce was sold in district market *i.e.* Satara.

Table 16.4: Distribution of the Respondent Members Availed Infrastructural Facilities Created by the Society

Sl.No.	*Infrastructural Facilities*	*Respondent Nembers (n=239)*	
		Number	*Percentage*
1.	Marketing facilities	239	100.00
2.	Consumer's store		
(*i*)	Packaging material	239	100.00
(*ii*)	Tools and equipments	175	73.22
3.	Krishi Seva Kendra	194	81.17
4.	Transport facilities	220	92.05
5.	Communication facilities	239	100.00
6.	Head office	239	100.00

The society maintains proper records of the rates of agricultural commodities and inform to all the members from time to time regularly for their knowledge. The society provides necessary tools and equipments to 73.22 per cent and packaging material to all the (100.00 per cent) respondent members on reasonable rate from the society consumer store maintained for this purpose. Also fertilizers, insecticides, pesticides, hormones etc were supplied on reasonable rate to 81.17 per cent of the respondent members from its own Krishi Seva Kendra.

The society has afforded a vehicle to collect flowers, fruits and vegetables from greenhouse and farmhouse of 92.05 per cent of the respondent members and to transport the same up to railway station during the sampled year. The society have modern communication facilities like telephone, computer and e-mail with necessary human beings through which society inform the all (100.00 per cent) members about meetings, market rates of produce etc.

The head office of this society is situated at Satara with well-furnished building. From office, all the (100.00 per cent) respondent members received their bills of sold produce through the society from time to time and immediately.

Difficulties Faced by the Society to Fulfill the Mandates

A correct appraisal of the difficulties faced by the society with a view in mind the manager of the society and other staff members are approached to realize their opinion about difficulties faced by the society. At present, the difficulties faced by the society are as below.

1. Fluctuation of market rates
2. Lack of necessary infrastructural facilities such as cold storage and pre-cooling unit facilities
3. Un-availability of own processing plant
4. Inadequate capital (funds) with the society to meet the demands of the members
5. Unable to take diversified activities as expected by the members
6. Improper co-ordination between credit and marketing co-operative society to make available sufficient finance for marketing societies.

Conclusions

The infrastructural facilities created by the society to the respondent members were marketing facilities, consumer's store, Krishi Seva Kendra, transport facilities, communication facilities, main head office etc. Majority of the respondent members were received benefits from it. The maximum (58.58 per cent) of the respondent members have sold their produce in out of state market.

The society provided training facilities to the respondent members twice in a year for the period of 3 days. The trainings were related to production, packaging and marketing of flowers, fruits and vegetables, greenhouse technology. Nearly half (49.79 per cent) of the respondent members had attended one training. The 32.22 per cent of them had attended second training and only 17.99 per cent of the respondent members had not attended any training.

The society market fruits, flowers and vegetables produced by the all members (100.00 per cent) in local market, metropolitan markets and international market, also maintain proper records of the rates of agricultural commodities and inform to all (100.00 per cent) members. The society provided necessary tools and equipments to 72.33 per cent respondent members and packaging material and boxes to 100.00 per cent respondent members from own consumer store. Also fertilizers, insecticides, pesticides, hormones etc. were supplied to 81.17 per cent of the respondent members on reasonable rate. The society provided transport facilities to 92.05 per cent and communication facilities to 100.00 per cent of the respondent members.

The major difficulties faced by the society were fluctuation of market rates, lack of necessary infrastructural facilities such as cold storage and pre-cooling unit facilities and inadequate capital with the society.

Implications

The study would be of great utility in developing greenhouse through Farmer's Co-operatives with reference to flowers, fruits and vegetables sale-purchase. On the basis of the results of the study, the following implications are drawn.

1. The study indicated that respondent members had medium level of knowledge about greenhouse cultivation practices. It is, therefore, suggested that adequate number of demonstrations should be organized and more practical experience should be given to the trainees at their localities. Relevant literature be supplied to them at the time of training.
2. The study revealed that most of the respondent members availed seedlings of flowers, fruits and vegetables from private companies. It is, therefore, suggested that society should arrange to supply good quality of seedlings at reasonable rates.
3. It was observed that facilities of refrigeration transport vans and cold storage were not available with the respondent members of flowers, fruits and vegetables. It is, therefore, suggested that since flowers, fruits and vegetables are perishable in nature, the society should arrange to avail the facilities of refrigeration vans and cold storage to the respondent members for storing the transportation of flowers, fruits and vegetables during period of market glut.
4. The study also showed that the respondent members were not getting market information regarding rates of produce regularly. It is, therefore, suggested that society should provide regular information about market rates of produce to the members.

References

Barman, K.K. and Deka, H.C., 1995. Role of marketing institutions in market regulation in Assam : a critical evaluation. *Indian Journal of Agricultural Marketing*, 16(3): 147–158.

Desai, B.R., Girase, K.A. and Kokate, K.D., 1996. Appraisal of training programmes of Krishi Vigyan Kendra. *Maharashtra Journal of Extension Education*, 15: 74–79.

Deshmukh, S.J., 1990. A study of communication network between VEWs and farmers for agricultural information. *M.Sc. (Agri.) Thesis*, Dr. BSKKV, Dapoli (Unpublished).

Jadhav, V.B., 2004. A study of production and marketing constraints experienced by the cultivators growing crops under protected conditions from Satara district. *M.Sc. (Agri.) Thesis*, MPKV, Rahuri.

Kadam, S.S. and Nirban, A.J., 2004. Constraints in production and marketing of flowers. *Asian Journal of Extension Education*, 13(2): 178–181.

Prasad, Raj Kumar and Kumar, R., 1996. Role of state in agricultural marketing: An overview. *Bihar Journal of Agricultural Marketing*, 4(4): 335–339.

Shivalingaiab, Y.N. and Remesha, Y.S., 2002. Constraints in production and marketing of horticultural crops. *Current Research*, UAS, Bangalore, 31(3–4) : 39–41.

Section III:
ICT APPLICATION AND METHODOLOGICAL ADVANCES IN AGRICULTURE

Chapter 17

A Web Based Soil Health Card Application: Decision Support System for Agriculture Development

☆ *M.C.Varshneya[1], R.S. Parmar[2], Arun Patel[3], Kalpesh Patel[3] and Hiren*

Introduction

The Soil Health Card (SHC) system evaluates a soil's health or quality as a function of soil, water, plant and other properties identified by farmers. The Soil Health Card System is web based information system designed to run on a networked environment including intranet, internet and on Gujarat State Wide Area Network (GSWAN). This is a repository of agricultural information for the benefit of farmers, agricultural scientists and decision makers at Government level. The Soil Health Card System is a unique information initiative of its kind in India pioneered and initiated by the Government of Gujarat for the benefit of farmers at the grass-root level. In India, there is enormous amount of agricultural information available with the scientific community restricted to the labs and research stations across the State Agricultural Universities (SAUs). The farmers can get benefit, if this scientific

---

1 Vice Chancellor, AAU, Anand.

2 Principal, Extension Education Institute, AAU, Anand.

3 Project Coordination, IT, AAU, Anand.

agricultural information is made available to them to use practically in the crop fields. The Soil Health Card System brings together the scientific community in the field of agriculture, the information repository of latest tools, techniques and cropping practices, the farmers and the Government for the economic growth of the people at large. This system provides a Soil Health Card for each farmer with details about his land condition and cropping practice to be followed. The farmers get direct advice from agricultural scientists regarding alternative cropping pattern, fertilizer usages and other crop practices based on their soil analysis report and moisture index.

Features of SHC System

- ☆ Gujarati Web Based application designed to run on networked environment, intranet, internet and GSWAN.
- ☆ Secured access to the Planners, Scientist and other users based upon unique identification.
- ☆ User-friendly navigation.
- ☆ Auto generated built in alarms and messages.
- ☆ Uniformity of user interfaces for ease of understanding and operation.
- ☆ Various kinds of qualitative, quantitative and analytical reports.

Functionalities of SHC System

- ☆ Master data entry for Districts, Taluka, Villages, Farmers, Crops and Crops related details, parameters, labs, offices and users.
- ☆ Entry and modification of the Soil Health Card data.
- ☆ Processing of various transaction data for soil, weather, fertilizers, profitability etc. for generating various reports and recommendations.
- ☆ Generation of recommendation of the fertilizers needed for a particular crop based on a nutrition status of farmer's soil.
- ☆ Generation of recommendation of the alternative crops to a farmer for better crop production based on his cropping practice and weather condition in his area considering the facts like MAI, AWC and LGP.
- ☆ Generation of recommendation of the alternative crops with a generic ranking of crops in terms of profitability, looking into the various costs involved for producing that crop.
- ☆ Generation of the taluka wise plans and entry, correction and modification of various actions plan related data.
- ☆ Generation of various types of reports both textual and graphical.

Soil Health Card Perspectives

As mentioned earlier that, Soil Health Card System is crop productivity and income enhancing scientific information provider structure for the grass-root level farmers at their doorstep or any locations about their land, soil fertility status, suitability of their land to grow different crops, irrigation, scientific recommendation of crops, the best cropping pattern to be followed in their fields, guidance of agricultural scientists,

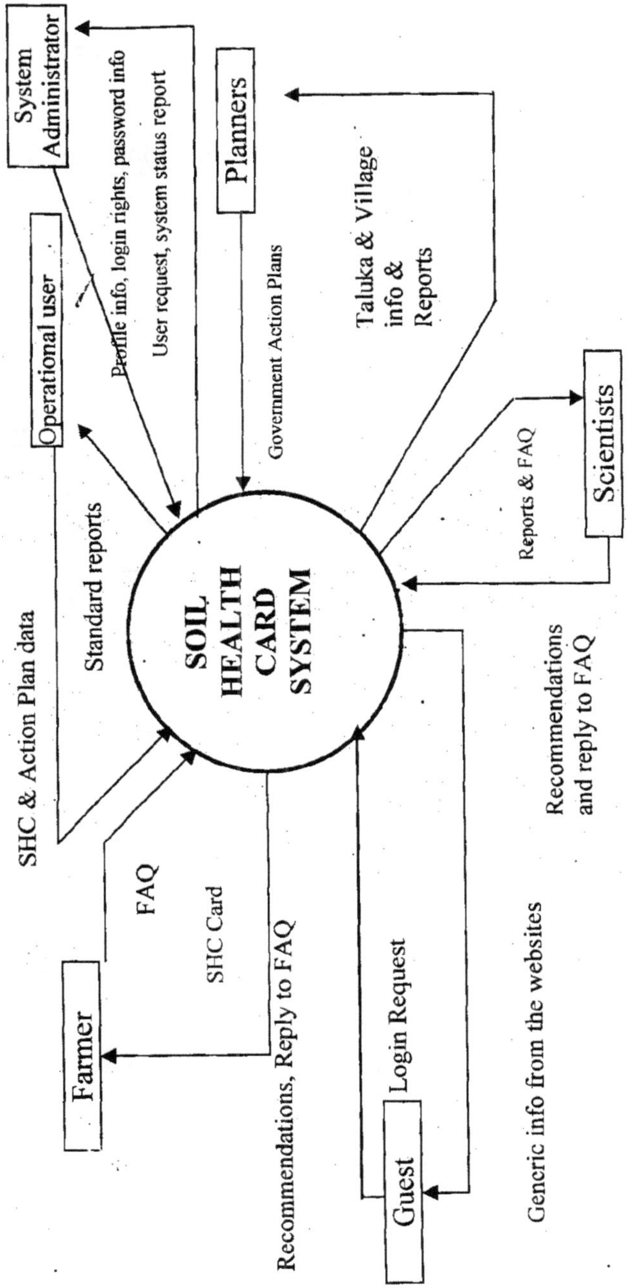

Figure 17.1: Data Flow Diagram of SHC

alternative cropping pattern and requirement if fertilizers considering soil analysis report and moisture Index of the area all the way through intranet, internet and Gujarat State Wide Area Network.

Schematic Flow for Accessing Soil Health Card and Other Information Through Button માહિતી

The Soil Health Card can be accessed on internet adopting the steps showing in the *Schematic flow of Accessing Soil Health Card.*

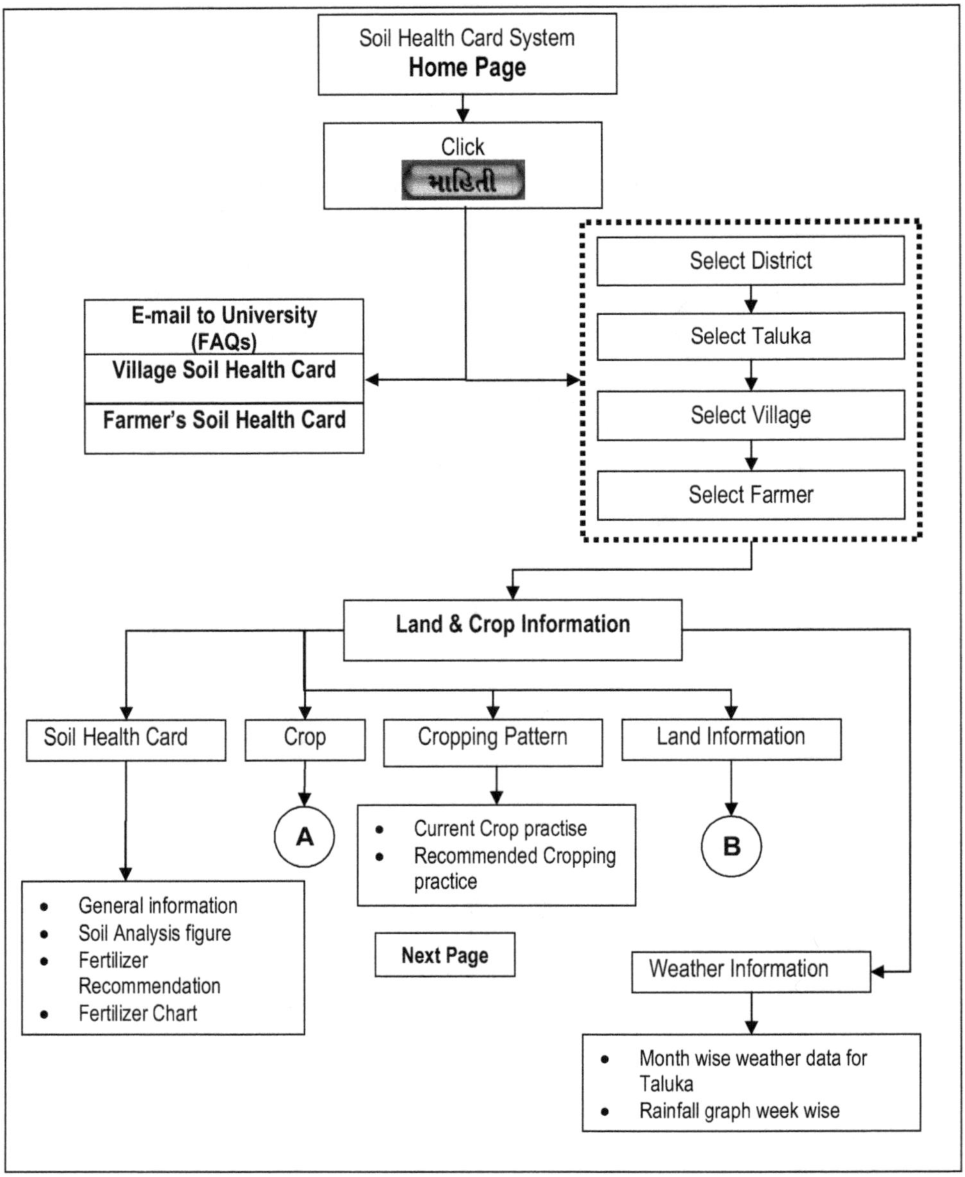

Figure 17.2: Schematic Flow of Accessing Soil Health Card and Related Information

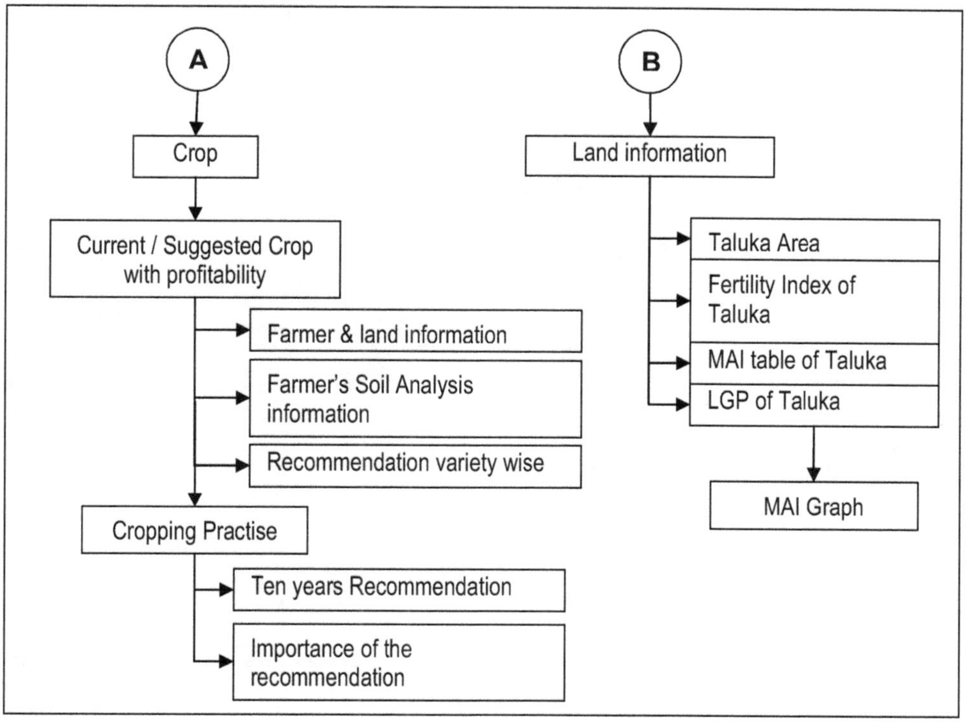

Figure 17.3: Schematic Flow of Accessing Soil Health Card and Related Information

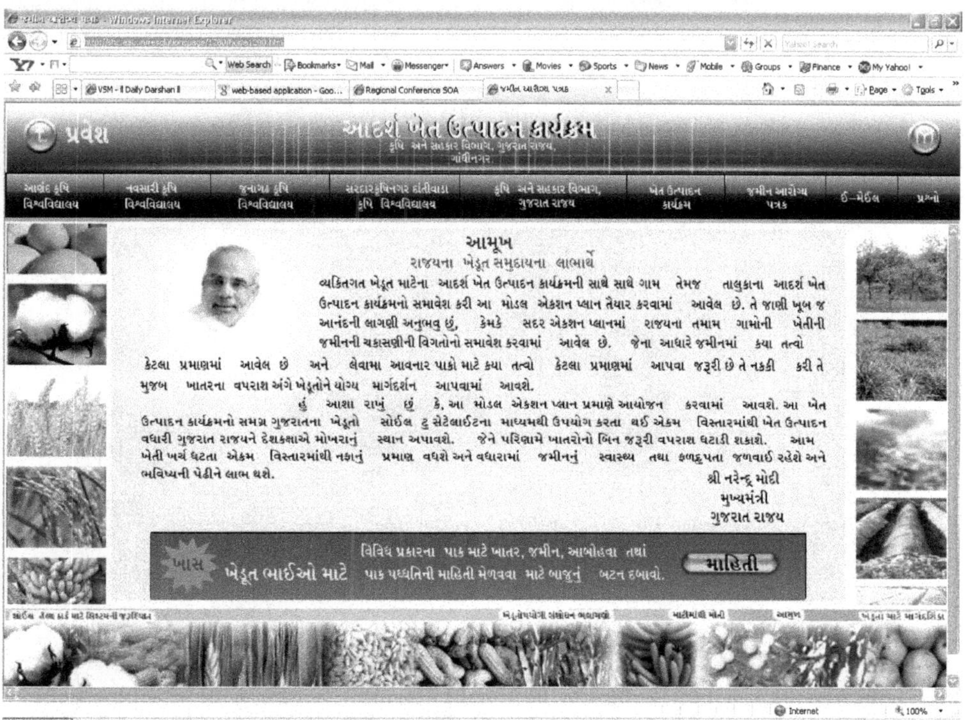

Figure 17.4: Home Page of Soil Health Card System

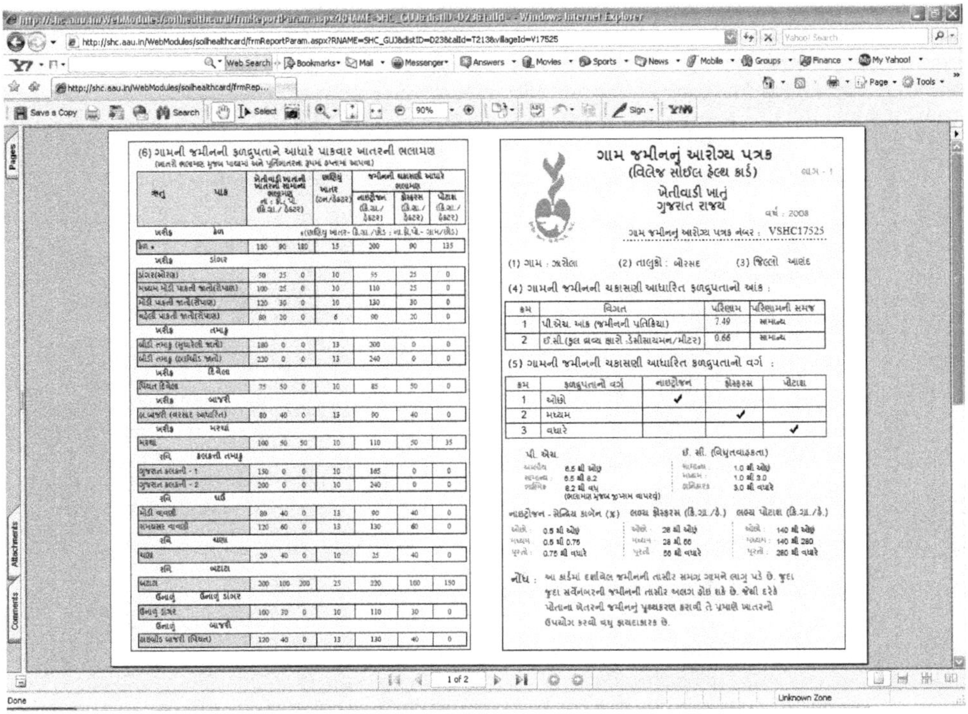

Figure 17.5: Farmer Soil Health Card

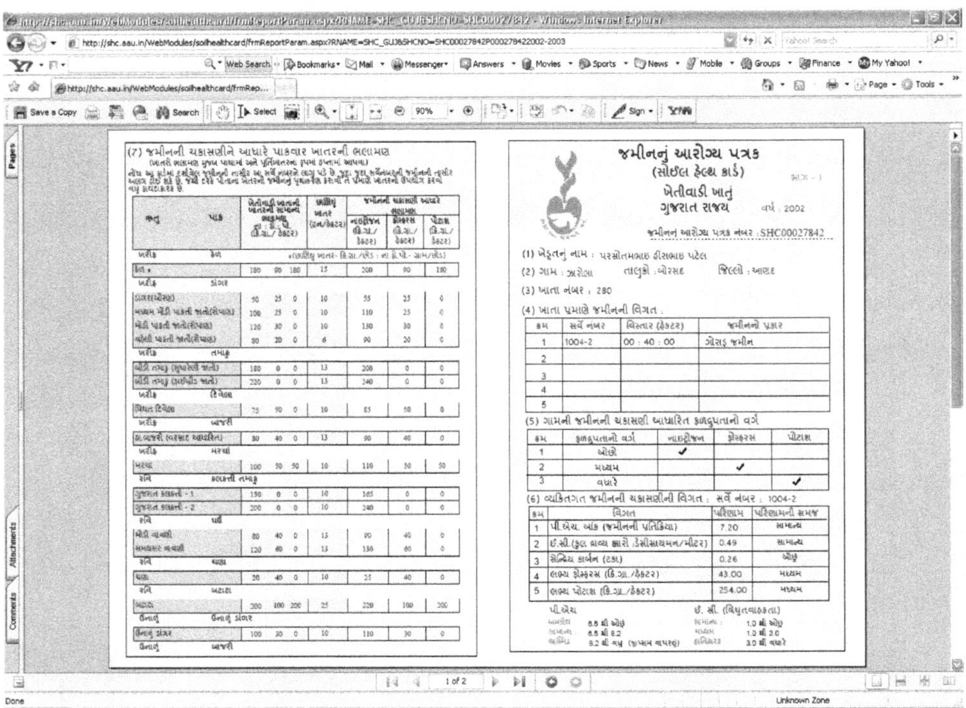

Figure 17.6: Village Soil Health Card

Taluka Action Plan–Information

The Taluka Action Plan Module produce all kinds of agricultural, demographic, geographic, climate and socio-economic Information/Analysis associated with a taluka so that these can further be used to decide the effectiveness of the recommendations given to the farmer or the scientists/planners so as to enable them for better decision making. The following are the list of Taluka Level Information/ Analysis can be retrieve:

- ☆ Total no. of farmers in current taluka
- ☆ Classification of farmers according to their amount of land
- ☆ No. of farmers belonging to SC/ST catgory
- ☆ Monthly rainfall, maximum and minimum temperature, Humidity and 10 yearly average rainfall for each month for the current taluka
- ☆ Total area of agricultural land, irrigated area, un-irrigated area, wasteland, forests, land used for cattle fodder etc.
- ☆ Types of soil and the area covered by them
- ☆ Fertility information of current taluka (soil pH, Organic Carbon, NPK status)
- ☆ Modes of irrigation and quality of irrigated water based on chemical analysis

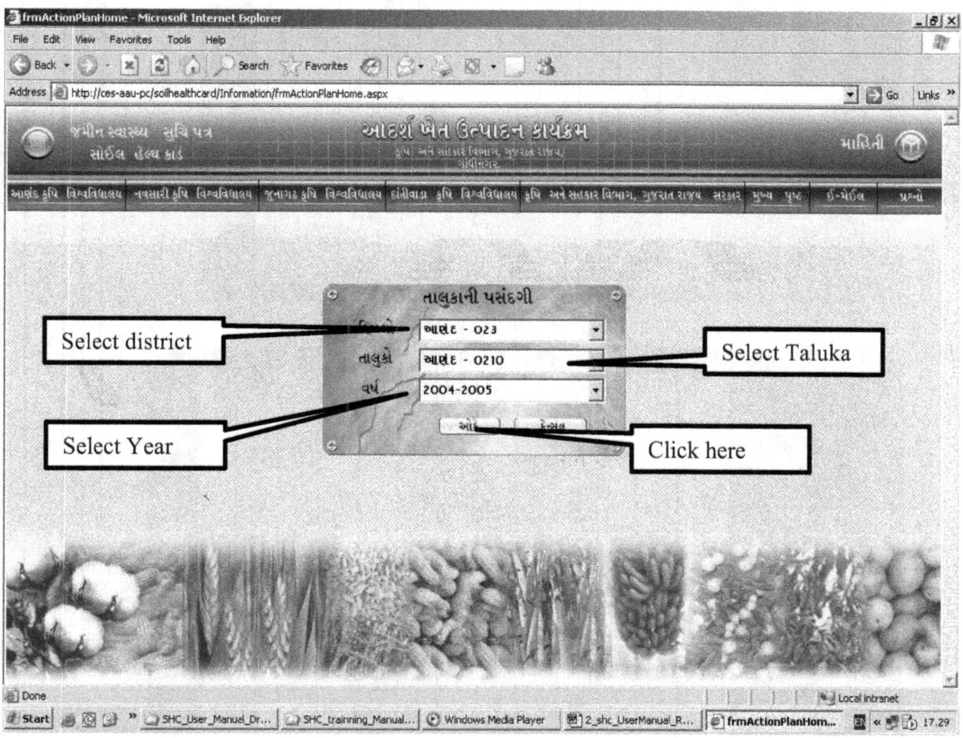

Figure 17.7: Taluka Selection page

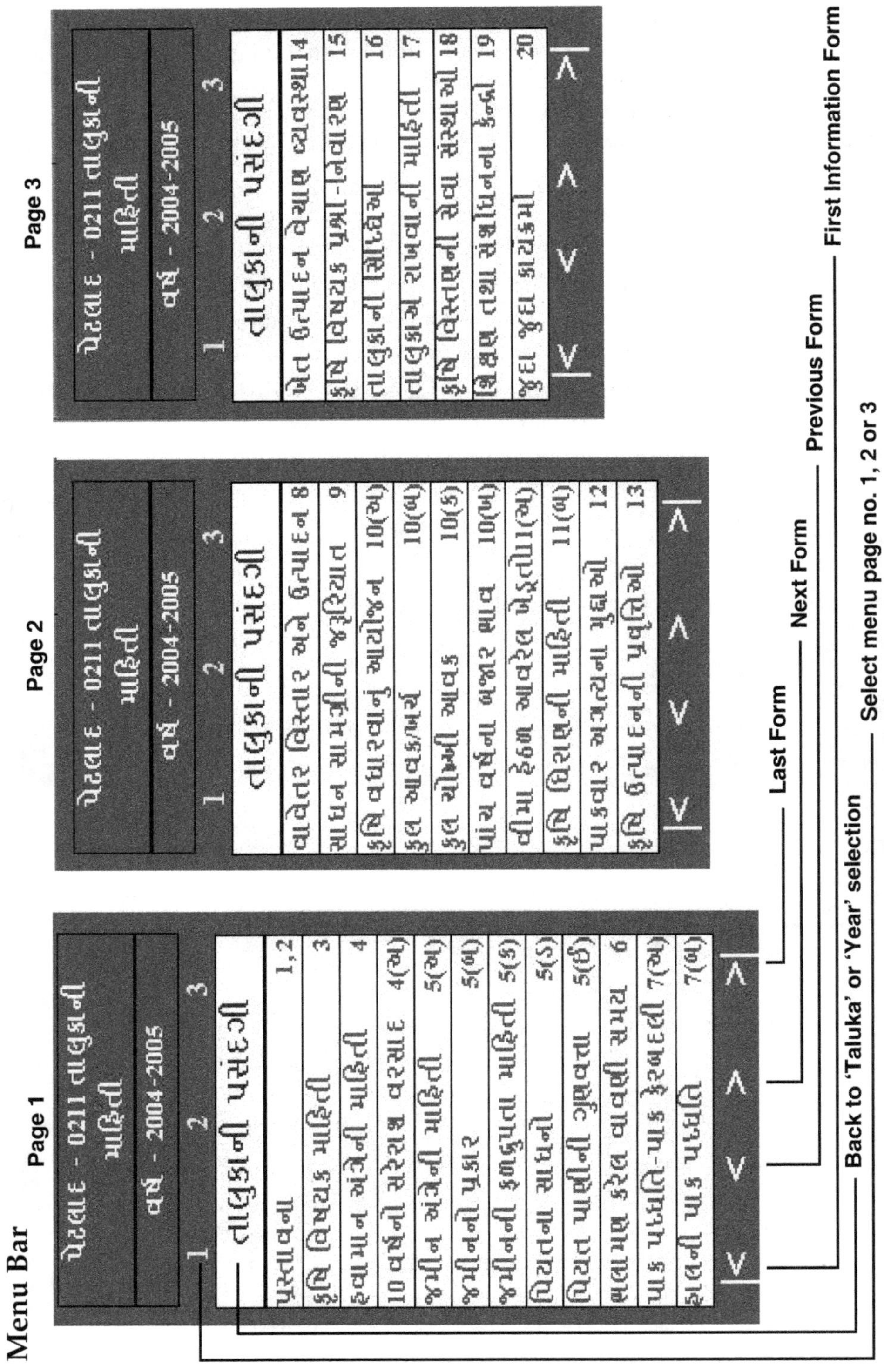

Figure 17.8: Taluka Information Menu Page

- Information on recommendation, crop name, current pattern, recommended pattern, percentage of farmers not accepting recommendation, reason for not accepting
- Farmer's demographic information
- Resources used in agricultural production

Soil Health Card System–Data Management

The Soil Health Card module handles the master data entry operations related to Soil Health Card System. It also generates the necessary reports on master data and Soil health Card that go to the planners and the scientists for further action. Following are the functions of this module:

- Capture and validate Soil Health Card information, District, Taluka, Village, Farmers, Farmer's Plot, Soil testing laboratory, Climatic Zone and Zone region data before saving to database
- Correct the existing data
- Prepare standard reports

Crop Management–Data Management

The crop management module handles the master data entry operations related to the crop. Master data for the crop group, crop type, crop variety, irrigated crops and un-irrigated crops are entered and maintained through this module. It also generates the necessary reports on master data that go to the stakeholders for further action and decision making. Following are the functions of this module:

- Capture and validate crop related data
- Correct the available data (if any)
- Prepare standard reports

Recommendation Module–Data Management

The Recommendation Module handles the master data entry and processing agricultural and climatic information which gives the Recommendation to farmer on Fertilizer to be used, Alternative and possible cropping system based on weather condition and Crop possibilities based on profitability. Following are the functions of this module:

- Capture and validate Daily Rainfall, Weekly PET, Available Water Capacity, Taluka wise main crop and Alternate Crop, Crop Profitability, Taluka wise existing Cropping and Possible Cropping information before saving into the database.
- Correct the existing data.
- Taluka wise Data processing for Moisture Availability Index, Length of Growing period and Fertility Index.
- Village wise data processing for Fertility Index.

- ✰ Crop wise data processing for NPK.
- ✰ Prepare standard reports.

Taluka Action Plan–Data Management

Taluka Action Plan Module handles the master data entry of all kinds of agricultural, demographic, geographic, climate and socio-economic data associated with a Taluka so that these inputs can further be used to decide about the effectiveness of the recommendations given to the farmer. Following are the functions of this module:

- ✰ Capture and validate under mentioned information before saving to database:
  1. General Information of Taluka
  2. Agricultural Information
  3. Information Regarding Weather Condition
  4. Rainfall Information
  5. Information Regarding Land (Ha.)
  6. Types of Soil
  7. Source of Irrigation
  8. Quality of Irrigation water
  9. Recommended Sowing Time and Soil Moisture Percentage at the time of sowing.
  10. Cropping System and Crop Rotation
  11. Current Cropping System
  12. Cultivated Area and Production (kg/ha) of Taluka
  13. Requirement of Inputs under Agriculture Production Programmes of Taluka
  14. More Planning for next year of Cultivated area and production for increasing crop production
  15. Total Income/Expenditure (In Rupees)
  16. Market Price
  17. Information about No. of Farmers covered under Insurance
  18. Information Regarding Agriculture Loan
  19. Information about distribution System of Agricultural Production of Taluka
  20. Program for Increasing crop wise Agricultural Production
  21. Agricultural Production Programme
  22. Market Facilities
  23. Agricultural Problems and Solution
  24. Achievements

25. Information should be kept at Taluka level
26. Agricultural Institute
27. Education, Research and Extension

☆ Correct the existing data

☆ Prepare standard reports

Administration and Security Management–Data Management

The Administration and Security management module handles all the administrative task such as Creation of new user, update user profile, delete user, view user profile, allow module permission provided to user, Create new Role (Database Role) and Database privilege which is assigned to role, allowing form permission to role, Backup and Restore of database, Export and Import data, Activity tracking and Tracing from log reader and Add/Edit Employee details as well as Look-up table updating facility. Following are the functions of this module:

☆ Create, Update and Delete User.

☆ Assign/Update Module Permissions to user.

☆ Assign/Update Operation Controls Access permissions to user.

☆ Assign Agro Climate Zone based Taluka level permissions.

☆ Add new Role to Database and Assign Farm level permissions to role.

☆ Backup and Restore Database.

☆ Export and Import data.

☆ Activity Tracking and Tracing System.

☆ Add/Edit Employee details.

☆ Look-up table updating facility.

☆ Prepare standard reports.

Summary

The Soil Health Card System is structure developed and available for the farmers to learn crop productivity and income enhancing scientific information at their doorstep or any locations about their land, soil fertility status, suitability of their land to grow different crops, irrigation, scientific recommendation of crops, the best cropping pattern to be followed in their fields, guidance of agricultural scientists, alternative cropping pattern and requirement if fertilizers considering soil analysis report and moisture Index of the area all the way through intranet, internet and Gujarat State Wide Area Network.

Implementation

In the present system of transfer of agricultural technology from SAU to extension functionaries of line departments of the states and in turn to the ultimate users consumes considerable time. It is therefore to reduce time lag to a considerable extent, "Soil Health Card System" type of facility should be made available for the farmers of other states of India under the technical support and guidance of State Agricultural Universities. To accelerate application of such facility, community based internet centers/kiosks should be established at village level with the collaboration of NGOs, schools, co-operatives and Government organizations.

Chapter 18

Community Radio: Looking Forward at Sustainable Development

☆ ***Suchi Gaur and Sarita Anand***[1]

Introduction

The UNDP report (2001) states, "the power of knowledge for development can be greatly enhanced by Information Communication Technology (ICT) if they are harnessed to improve access and break down barriers to knowledge because 'while education develops cognitive skills, information gives content to knowledge'."

There are nations in this world, which still use radio as a medium to communicate to the masses, at a very low cost. The relevance of 'New World Information Order' diversifies media access and encourages alternatives to the existing global media system dominated by industrial powers.

Looking at the concept and basis of Community Radio from all these angles, it can be termed as one which encourages participation in community, diversifies media access and brings out management of information system.

Many developments in the field of media for especially radio let to the range and scope being extended to huge levels be it local or community. In the area of radio, these were:

---

1 Lady Irwin College, New Delhi.

1. Development of FM (frequency modulated) radio transmission, which resulted in growth of local and community radio stations.
2. Increase of availability of low cost, portable radio receivers.

Various changes have taken place in the information and its role in developmental process due to this change in radio. Radio today has resulted in bridging the Digital divide between the so called have's and have not's. This concept has been harnessed in using radio in the field of agriculture apart from its usage today in other development issues too all over the world. Broadcasting of local level farm programmes has led to a lot of farmer's development and coming up of many such initiatives.

Defining the Community Radio

The definition of CR varies widely. Common elements of various definitions can be extracted by different definitions analysis:

1. Community radio is a social process where the local community members are involved in planning, designing and implementation of programmes related to education, agriculture, gender, health, entertainment, livelihood systems etc.
2. It is non-profitable and social welfare in nature maintained and managed by the local communities.
3. It looks forward to sustainable development by the use of participatory communication.
4. It fulfils the needs and interests of different segments of the population of the community.
5. It follows democratic principles in decision-making: content generated from the community.

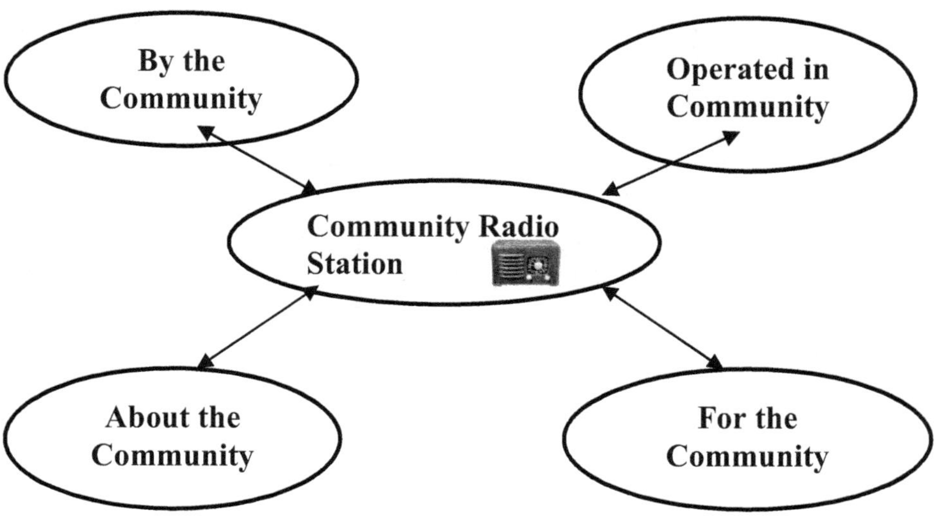

Figure 18.1: UNESCO Concept for Community Radio

Participatory Development Perspectives

When Wilbur Schramm (1964) and Daniel Lerner (1958) promoted models for Communication for development, the emerging Third World hoped to achieve in a few decades what the developed, wealthy, industrialized nations of west had achieved over centuries. India was among the first to use mass media sponsored by National Government through regulative planning to support development initiatives. Beginning with the use of rural radio forums to promote rural development, even satellite communication broadcast was promoted to thousands of villages. The agricultural extension programme and other development efforts were strongly influenced by Rogers (1962) Diffusion and Adoption of Innovation model. There was a strong belief in 'trickle down' theory. In agriculture, this model emphasis on top to bottom flows of information *i.e.* from agricultural universities to the farmers. Scientists and officials are seen as facilitators. In CR perspective, the whole community is seen to be involved in development and implementation of programmes and communication materials.

But largely the problems of third world remained unresolved. The model used was very dependent and knowledge, social and economic gaps between the 'haves' and 'have not's' became more. A number of writers have influenced the thinking on participation. Paulo Freire's orientation towards participation stems from his critique of existing pedagogy of extension education which was both paternalistic in its philosophy and basically non participatory in its operation. Freire proposed the abolition of 'transmission mentality' in education and communication, and its replacement with a more liberating type of communication education that would be more receiver-centered and more conscious of the social structure, making the receivers participants and critical reflectors of information and knowledge gained. Although, he never linked his analysis to the use of a particular media, it is implicit in his writings that communication, in order to be effective has to be participatory, dialogic and reciprocal. Various participatory communication projects in Community Radio production in India, Brazil, Chile, Jamaica, South Africa utilize Freire's perspective.

It is seen that learning is much more if the methods followed are participatory as it not just develops relations with the local communities, involves them, identifies stakeholders and problems, identifies communication needs, produces and implements plan but also helps in sharing of resources and monitoring and evaluation of the processes.

Advantages and Potential of Community Radio

Giving voice to the voiceless, community radio is a 'hot' medium that extends one single sense in 'high' definition. A CR has characteristics of identity, inclusion and innovation. They are extremely relevant for agricultural development programmes.

1. Identity refers to local clientele identification and establishment of rapport with them, developing a sense of belongingness and adherence to the community.

2. Inclusion brings up the efforts made to involve all the sectors and segments of community in developing and sharing common information relevant to needs and interests of the farmers.
3. Innovation refers to new opportunities, approaches and space that can be used by the CR in formulating the various programmes targeted at the farmer groups.

Just not attractive medium, a CR is available, accessible, affordable, acceptable and amazing medium of creating change in the grassroots level. There are farmers with no electricity and ones with even internet being made available; both can use the power of a Community Radio which along with community goodwill target encouraging effective listener participation.

Community Radio for the Farmers

A community radio aimed at producing programmes for farmers is not a new concept. The earlier concept of Rural Radio Farm Forums developed by the AIR aimed at promoting such programmes and internship. Though they promoted the government policy, today the CR developed by local organizations promotes farmers agenda and target their needs. Any CR aimed at producing programmes for the farmers should develop the following characteristics:

1. Collaborative and collective planning: defining and identifying the target groups, deciding the objectives of the programmes to be broadcasted to the farmers.
2. Designing and developing programmes content and support material top suit the needs and profile of farmers.
3. Developing and Imparting Skills for effective participation and presentation of messages targeted at the needs of the farmers.
4. Organizing internship groups of farmers for discussion about the contents of the programme and providing feedback.
5. Promotion and Publicity of the programmes and its impact on adoption and use of messages by the targeted listeners *i.e.*, the farmers.

Various issues that the Community Radio follows for Sustainable Development can be used for strengthening the paradigm of sustainable agriculture development. The 'green revolution' failed to sustain the level of development though it brought improvements in farm production. Today, Indian agriculture faces problem of fatigue in technology and extension.

CR as an effective tool can help in restructuring the Indian Agriculture scene and the information diffusion process.

- ✰ It can target at the local indigenous and grassroots technologies and networking in communities to enhance the production and improve the community local resources.
- ✰ CR can help in creating and sustaining linkages and networks between researches and field *i.e.*, the agricultural institutes and the local farmers.

- ✫ Help in conservation and protection of biodiversity, which is at stake.
- ✫ Providing appropriate markets to the farmers, helping them deal with issues of livelihood management and income expenditure.
- ✫ Promoting balanced consumption pattern and nutrition among the listeners.
- ✫ Catering to issues which cause problems for the farmers: pest management, markets, what to grow and how to grow, etc.

There have been in the past many such initiatives been taken up for the farmers. Namma Dhwani Community Radio in Kolar District of Karnataka is one such, which involves local people in programme production and message development. Also one case studies like the Alternative for Indian Development (AID) initiative in Daltongunj in Jharkhand reaching out to 45 villages? Others are Kutch Mahila Vikas Sangathan in Bhuj, Pastapur initiative by the Deccan Development Society in Andhra Pradesh, etc.

Community Radio: The Models

The various CR initiatives conform to two basic models:

Model 1

This refers to community participation in radio programming. Here, the institution buys time from AIR and broadcasts usually on weekly basis during a particular time slot. Community members take role of protagonists and take active role in feedback generation. Example can be AID in Jharkhand.

Model 2

Here, it has community management and ownership. Namma Dhwani is such where the radio is owned and managed by the local communities

Both models have deep roots in Participation and people's initiatives. Today the newly developed Community Radio Policy by the Government of India offers three approaches for establishing a CR station:

Institute–Community Approach

Institutes initiating the process of participation of community members.

Community–Institute Approach

Community member's request institues to take initiative like the Deccan Development Society.

The last being that which is focused on agriculture, which includes the KVK's, and ICAR initiatives.

A different form of radio being campus radio in which the mandate being that the 50 per cent content generated from the local community, with active participation of the surrounding communities is also well presented by various universities and colleges. Though, this is in question by people as in many cases the situation is a lot different. Krishi Community Radio Station 90.4 FM of Dharwad (University of Agricultural Sciences) is a form of campus radio station which develops majority of the programmes from the community and broadcasts programmes on horticulture, Irrigation, Food and Nutrition, Market intelligence, cultural aspects and weather forecasts. Even Banasthal CR is such.

Community Radio and ICT's

According to Rao (2002), there are 8 C's of Digital Economy, which are: connectivity, content, community, commerce, capacity, culture, capital and cooperation. For bridging the Digital Divide between have's and have not's, networking is very important. Community radio, which is key body of local information systems, has to use and transfer this knowledge available, *i.e.* the content. In agriculture, poverty of farmers can be escaped by use of this knowledge. Today, with communities not having much of knowledge being developed, people have come up with ICT's usage especially the internet for creating data base of contents on various issues. There are websites like *ek duniya anek awaz* etc where content sharing is taking place.

Access to new ICT's need not be understood to be the significant barrier to participating in an information society or even to using the internet for development. We should not underestimate what can be done with determination and imagination. ICT's are adaptable and if basic tools and knowledge are available, people will find a way to make the technology serve their local communication needs. Adaptability and decentralization are fundamental characteristics that have made the radio so enduring and effective because they have allowed for different approaches to its use in terms of range, interactivity and content, enabling it to integrate so effectively with existing social communication networks and practices (Bruce Girad, 2005).

One has to ensure, that technology fit's the user needs and not vice-versa. Use of ICT helps too transform generic information and render it locale specific. There is today, immense scope of covering community-based radio with different ICT tools including the World Wide Web *i.e.* the Internet. Experiences from the field indicate that extension of wireless technology, spreading use of cell phones, community radio services, use of audio, video and multimedia, establishment of tele-centers and information kiosks can help overcome the problem of using different scripts and languages and enabling bridge the digital divide. Internationally, many communities use the internet and other ICT techniques for and with Radio.

Conclusion

With increasing Globalization, where the local is going global and vice-versa, we can foresee a lot of scope in Community Radio for Development and preservation of our traditional farmers and farming strategies. Community Radio, just not aims at technological development but holistically at even social and economic one, making it target sustainable development. With the coming up of licensing policy by the Government of India, a lot of changes have been brought in. today, CR faces a lot of challenges from training of people, to getting license, to cost effective technology and content sharing. Newer models of CR are being explored and brought in. while the important emphasis is on community participation, inclusion of ICT's is also a facet to the issue. For convergent use of community radio in the field of agriculture, there is an urgent need for advocating and capacity building of grassroots thereby leading to a sustained empowered population.

References

Anthology on Community Radio: Practices and Possibilites, IGNOU, Delhi.

Capacity Building and Training for ICT, 2007. Workshop by Response Net, New Delhi.

Freire, Paulo, 1970. *Pedagogy of the Oppressed*. Herder and Herder, New York.

Girad, Bruce, 2005. *The Way Forward: Internet, Radio and Network Extension*.

Pavarala, V. and Malik, K.K., 2007. *Other Voices: The Sturuggle for Community Radio in India*. Sage Publications.

Rao, Madanmohan, 2002. *The Asia-Pacific Internet Handbook, Episode IV: Emerging Powerhouses*. Tata McGrawhill, New Delhi.

Sen, Ashish, 2003. *Waiting in the Wings*. Voices, March.

Singhal, A. and Rogers, E., 2001. *India's Communication Revolution from Bullock Carts to Cyber Marts*. Sage Publications India Pvt Ltd.

Servas, Jan, 2003. *Approaches to Development: Studies on Communication for Development*. UNESCO Communication and Information Sector.

Tabing, Louie, 2005. *How to do Community Radio*. UNESCO.

UNDP Annual Report, 2001. www.undp.org/annualreport/2001.index.html.

Chapter 19

Importance of Information Communication Technology in Rural Development

☆ *Prabhjot Kaur*[1] *and Sahil Garg*[2]

Introduction

The need for improved agricultural education throughout the world has never been greater. Agricultural technologies that can increase food security in the developing world while developing income-generating activities and conserving environmental resources do exist. Many of these technologies have existed for decades, yet famine, hunger, desertification, deforestation and droughts continue to plague the world's developing countries, which are least prepared to deal with these devastating problems. In developed countries, farmers are facing a changing atmosphere in which organic foods, new methods in raising animals, disease outbreaks, agri-business, crop insurance, and banking all continually present new challenges. All of these problems do have workable solutions, yet the global difficulty is getting the appropriate information to farmers. ICT's can benefit every agricultural aspect around the world, from research scientists in agricultural universities to the poor subsistence farmers of developing countries. It can benefit persons of all ages, all locations, and bridge the gaps between what is and what it should be. ICT in agriculture can assemble resources

---

1 Professor of Extension Education, Department of Extension Education, Punjab Agricultural University, Ludhiana – 141 004.

2 M.Sc. Scholar, Department of Extension Education, Punjab Agricultural University, Ludhiana – 141 004.

and knowledge from distant places that may otherwise be unobtainable. It can connect farmers with far away researchers and experts. Adoption of modern agriculture technologies depends upon the effective ICT to which the recipients, audience or the target group *i.e.*, the consumer of message, the farmers, are generally exposed directly or indirectly.

Importance of ICT in Agriculture

ICT has expanded the scope of information dissemination by covering market intelligence, weather forecast and post-harvest processing in a need-based and user-friendly mode. These can also support bottom-up articulation of development needs and perceptions of farmers. ICT offer enormous opportunities to improve communication potential, scope and efficiency. These enhance productivity, encourage competitiveness and accelerate growth in various business and social sectors.

Agriculture in India being dominated by the resource and information poor farmers, can get a boost through the use of ICT. Impact of multimedia extension programmes that promoted a variety of technologies varied across crops and region within the country. Allowing the farmer to identify the source of learning instructions for the technologies on which one relies can collect data. Multimedia technology is lower in cost per farmer in making contact and present viable alternative. The use of multimedia as a tool is advantageous to turn agriculture into industry and enhance turnover, reducing training costs, improving morale of an employee and increase the efficiency and effectiveness of training programmes and reducing the cost of delivery.

The adoption curve for adoption of an innovation usually follows a normal, bell-shaped curve when plotted over time on a frequency basis. If we draw the same curve for adoption of ICT in agriculture then, the curve will be as shown in Figure 19.1.

The farmers when exposed to different sources/media, they usually try to absorb the technology only on selective basis depending upon host of variables present. There have been number of innovative initiatives of ICT and the media engaged in the

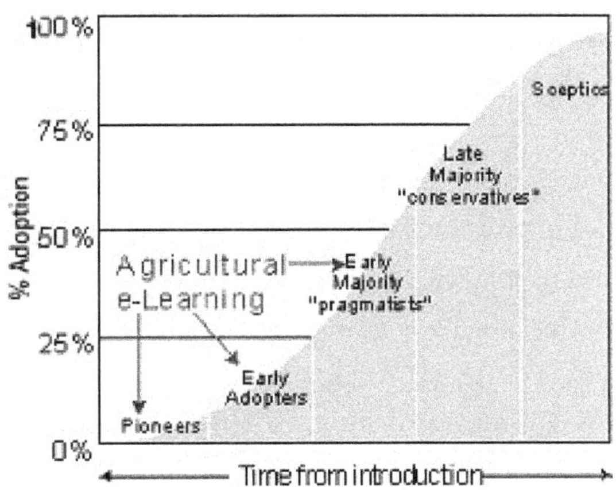

Figure 19.1: Adoption Curve for Adoption of ICT in Agricultural Sectors

communication of modern agricultural technology to the farming community which is discussed below.

Computer Aided Decision Support System

Computer aided decision support systems enhances the users' capability of making timely and knowledgeable planning, management and analysis decisions. But nowadays, research projects are to develop a handheld-based agricultural decision support system (HADSS) for advising the efficient management of irrigation, N fertilizer application and crop rotation in grain cropping systems.

With this expected HADSS, farmers will be capable of electronically recording soil and crop data, retrieving meteorological data through the internet, running and calibrating a series of widely-recognized empirical and biophysical models for agro-ecosystems, and making decisions in the field, for pursuing site-specific best management practices. The developed HADSS will greatly improve the farmers' capability to adopt the latest and most innovative agricultural research outcomes to help improve cropping management and therefore improve production and minimize losses. The HADSS will have an advantage over currently used CADSS (Computer Aided Decision Support Systems) because it can be used in the field rather than in the laboratory.

This research project has been embedded in a large ICAR project "Improving the management of water and nitrogen for agricultural profitability and environmental quality in the Yellow River basin, China" as an important means of delivering the proposed research outcomes to farmers and agricultural extension officers.

Expert Systems

"Expert Systems" is one of the important application oriented branches of Artificial intelligence. In the past decade, a great deal of expert systems had been developed and applied to many fields such as office automation, science, and medicine including agriculture. At the beginning of development of the agriculture expert systems, the areas selected are applications to diagnosis the diseases and pests of various crops. The complexity of problems confronting farmers like yield loses, soil erosion, diminishing market prices from international competition, increasing chemical pesticides costs and pest resistance and economic barriers hindering adoption of farming strategies necessitates that they become expert managers of all aspects of their farming operations. On the other hand, agricultural researchers need to address problems of farm management and discover new management strategies to promote farm success.

Expert systems are used to aid:

- ☆ Single point decisions. *e.g.* Planning
- ☆ Designing. *e.g.* Design of an irrigation system
- ☆ Selection. *e.g.* The most suitable crop variety or market outlet
- ☆ Diagnosis or identification. *e.g.* Of a livestock disorder
- ☆ Interpretation. *e.g.* Of a set of financial accounts

- Prediction. *e.g.* of extreme events such as thunderstorms and frost;
- A sequence of tactical decisions throughout a production cycle. *e.g.* plant protection and nutrition decisions, livestock feeding.

An Agricultural Expert System is a Decision Support System for Agricultural Extension Agents who have to decide what advice to be offered to farmers, who have to decide what action to take based on it. It is one of the most efficient extension tool to take the technology from scientists to the farmers directly without any dilution of content which normally creeps in because of the number of agencies involved in normal technology transfer systems.

Community Radio System (FM Radio)

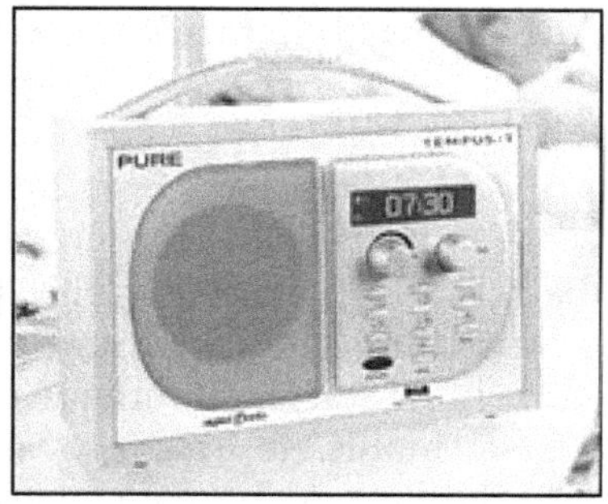

Community Radio is a powerful medium to reach the unreached in far flung, rural areas. AMARC's members say about Community Radio that, Firstly, Community Radio is characterised by the active participation of the community in the process of creating news, information, entertainment and culturally relevant material, with an emphasis on local issues and concerns. With training, local producers can create programmes using local voices. The community can also actively participate in the management of the station and have a say in the scheduling and content of the programmes.

Secondly, it is essentially a non-profit enterprise. In these days of highly commercialised broadcasting, the ethos of community radio remains independence and responsibility to serve the community, not the advertiser. As the community owns the station, it also maintains some responsibility in the running of the station.

Thirdly, the community designs community radio programming, to improve social conditions and the quality of its cultural life. The community itself decides what its priorities and needs are in terms of information provision.

Directorate of G. B. Pant University of Agriculture and Technology, Pantnagar has approached the Ministry of Information and Broadcasting, Government of India to establish a Community Radio System (F.M. Radio) at Pantnagar in order to broadcast knowledge based programmes as per the demand of the local people.

Farm Schools on Air

Farm School on Air Realizing the importance of radio in rural communication a programme on 'Farm School on Air' has been launched since 2001 to provide systematic radio lessons to interested farmers on different important topics of current interest in agriculture and allied subjects. So far Farm School on AIR has been broadcast on themes likely Rice Production, Sugarcane, Organic Farming, Litchi Production, Importance of Agriculture Implements, Floriculture, Integrated Insect and Pest Management and Fisheries Production, etc.

Agriculture department runs Farm Schools on All India Radio Station of state. A series of lessons prepared by experts is broadcast on a fixed day and time every week. Questions at the end of lesson are asked, feed back of listeners is ensured through reply to these questions by postcard. News items and press released relating to agricultural subjects are also sent to the All India Radio.

Farmers and Rural Magazines

Print media has also played an important role in agriculture. Many magazines like Agriculture Today, Kissan World, etc helps extension worker to be up to date with the recent advances in agriculture in the country. Other than this Seasonal zone wise as well as state level Package of Practices booklets, handouts, pamphlets, posters etc. are regularly printed to keep the extension workers/farmers abreast of the latest developments is agriculture are published in their regional languages.

Agri-Portals

Agri-portals have been established as an electronic network in collaboration with input agencies. The information on the technology package at village levels to the farmers is made possible through these portals. The portal organizes the information on agricultural practices, agricultural news, animal husbandry, agricultural machinery, aromatic and medicinal plants, agricultural credit, insurance, prices of inputs, etc. It also provides dynamic information like marketing and weather. In India many private companies have their agri-portals and farmers can access their desired information on the portals free of cost. Following are the main portals prevailing in India.

e-Choupals

Launched in June 2000, 'e-Choupal', has already become the largest initiative among all Internet-based interventions in rural India. 'e-Choupal' services today reach out to more than 4 million farmers growing a range of crops–soyabean, coffee, wheat, rice, pulses, shrimp–in over 40,000 villages through 6450 kiosks across 8 states (Madhya Pradesh, Karnataka, Andhra Pradesh, Uttar Pradesh, Maharashtra, Rajasthan, Uttarakhand and Tamil Nadu).

Taking agri-business into the new millennium, the concept was pioneered by one of India's largest exporters of agricultural commodities, Indian Tobacco Company's International Business Division (ITC- IBD). ITC's 1,200 Internet kioks in villages across 18 states allow this agri-business company to procure soya and other produce directly from farmers, eliminating the middleman and saving producers time and money.

Traditionally, choupals are community-gathering places in the village where locals meet to discuss issues and iron out their problems. In the digital age, e-choupals are gradually revolutionising the way Indian farmers do business. The system constitutes an Internet-enabled kiosk in a village, which is manned by a prominent local farmer who is familiar with computers, known as the '*Choupal Sanchalak*'. The setting up of each e-Choupal entails an investment of between Rs 1-3 lakh.

The *Sanchalak* mans the kiosk, is in touch with company representatives and guides farmers on the use of the technology. Given the levels of literacy and infrastructure limitations the sanchalak acts as the interface between the computer and the farmer. Farmers can use the kiosks to check the current market prices of their commodities, access market data, information on local and global weather and best farming practices. The entire contents of the site are accessible to registered choupal sanchalaks only.

The company has set up

- ✰ Soya-choupals in Madhya Pradesh,
- ✰ Wheat-choupals in Andhra Pradesh,
- ✰ Coffee-choupals in Karnataka and
- ✰ Aqua-choupals in Andhra Pradesh and Orissa.

Shashank Joshi a soya farmer in Mendki village, Madhya Pradesh, has a new status as a 'sanchalak' in ITC's soya e-Choupal. A computer was installed in his house and farmers often visit to access data and drop their bags of soya off at his home-office. The days of hanging around the 'mandi', waiting for the agents to examine their stock and dictate prices, are over. Prices of major 'mandis' are transparently provided on the computer screen, giving the farmer the option of selling his stock to ITC or a 'mandi' of his choice.

KIOSK: An Agriculture Helpline

KIOSK is a multilingual online question and answer forum–Almost All Questions Answered that provides online answers to questions asked by farmers and agri-professionsals over the Internet, kiosks where kiosk operators charge farmers Rs 10 per query for the use of the KIOSK. Almost all of the questions have been answered in the local language with English answers given to people posting from outside the state. Local farmers trust KIOSK service and are sensitive to the time it takes for the answers to arrive (24-48 hours on normal load). KIOSK has a multimedia portals such as video clips of improved varieties of various crops and information about the cultivation practices.

Services Provided by Kiosk

- ✰ Agriculture
- ✰ Education
- ✰ Health and Hygiene
- ✰ Employment
- ✰ Legal Services
- ✰ Veterinary Services

There are three options on the screen of the KIOSK

1. *Forums*: Crop Diseases, Animal Diseases, Market information, Farmer Scheme, education etc.

2. *Query Mailing systems*: In which farmer has to ask the question.
3. *Previous question*: Which will be answer later on.

Figure 19.1: Inauguration of KIOSK in Fatehgarh Sahib

It has a definite role to play in rural education health and agriculture. In the Indian situation Kiosk can turn into a 'Cyber Dhabha' for villager where they can access websites of their choice and get information that they can use.

Experience has shown that if the information has direct relevance to the people and has a potential to results in commercial gains, people are willing to pay for such services.

Ikisan Limited

Ikisan limited is an initiative of the Nagarjuna group of companies, established in 2000 with the objective of facilitating continued enhancement of agricultural productivity and rural prosperity, and thereby making the Indian farmers globally competitive. Initiative includes both Ikisan portal and information kiosks at the block/ village level for disseminating information. Farmers can access this portal free of cost. It provides information in Hindi and five major regional languages (Tamil, Telugu, Kannada, Marathi and Punjabi).

Ikisan maintained 21 kiosks–14 in Andhra Pradesh and 7 in Tamil Nadu. These kiosks had facilities like computer, internet connection and other infrastructure. A facilitator was also available at each kiosk for disseminating information and providing services of Ikisan. A 'member registration' system had been evolved for Ikisan at the village level with the twin aims of developing a rapport between the farmers and Ikisan and stimulating revenue generation through specific service charges. The extension officers of the company provided farm advisory service at farmers' fields, which acted as complement to the information, provided through kiosks.

Computers on Wheels (COW)

The Computers on Wheels (COW) is a novel initiative taken up since 2003 by an individual developmental entrepreneur–Ms Rajeswari Pingali and has been supported by Stanford Reuters Digital Vision Program and Digital Partners Institute, Seattle,

USA (about US$ 40,000). The fund seeks to support innovative ICT model for socio-economic development. The objective of the initiative is to enable information empowerment to citizens by providing expanded access to information and exposure through technologies in the rural areas. To provide access to information, a motorcycle is used which is equipped with a solar-powered laptop computer and facilities like Internet connectivity, printer, digital camera, and a mobile telephone. It provides doorstep services at the village level. Seven remote villages in the Mahbubnagar district of Andhra Pradesh currently receive these services and get access to information on agriculture, healthcare and a range of other rural issues. The information provider, who is the facilitator in the model, visits every village once in a week for both getting queries and providing solutions to the queries collected during the previous visit.

CARD: Computer-Aided Registration of Deeds

Land registration offices throughout Andhra Pradesh now operate computerized counters to help citizens complete registration requirements within an hour, instead of several days as earlier. The lack of transparency in property valuation under the old system resulted in a flourishing business of brokers and middlemen, leading to corruption. Antiquated procedures such as manual copying and indexing of documents, and storage in paper forms in ill-maintained back rooms have all been replaced. The cost of CARD project was funded entirely by the Andhra Pradesh government.

Video-Conferencing Technology

Video-conferencing is the new way of communication by which one can talk to other side face-to-face. It has a Microphone, Camera and TV. One can see the picture of other side on TV screen and talk to each other. It enables top management to communicate quickly and effectively sitting at multiple locations. Video-conferencing helps to reduce travel costs, improves usage of management time and speeds up decision-making. It also provides an effective way of delivering cost-efficient training to individuals without the need to constantly travel to central locations. About 250 Video-conferencing studios have been set up in India

Agricultural Resources Information System (AGRIS)

The Department of Agriculture and Cooperation is mandated to achieve agricultural growth through optimum utilization of natural resources. It is proposed to develop Agricultural Resources. Information System on various typologies in identified Districts as a pilot project. Developments of Decision Support Systems (what to grow–when and where) facilitating Agricultural Resources Management are also envisaged using the state of-the-art IT tools.

Agricultural Resources Information System Network (ARISNET)

Agricultural Research Information System Network (ARISNET) is a network of agricultural research, extension and educational institutions in India over NICNET.

A Satellite based Computer Communication Network of NIC called NICNET. It provides global access of information to scientists and facilitates them to build up their linkage among national and international research community and promotes voluntarily exchange of information between different institutions and sharing of their resources.

The ARISNET, a close user group of institutions in National Agricultural Research System of India (NARS) on NICNET, links ICAR Institutions (90) and their Regional Research Stations, Central Agricultural University, State Agricultural Universities (29) and their Colleges, Krishi Vigyan Kendras (283), Zonal Research Centres (120) and Project implementing centres with the ICAR Headquarters. ARISNET has become an integral part of agricultural research, extension and education process.

Agricultural Extension Information System Network (VISTARNET)

The Research, Education, Extension, and Training are considered as four pillars of Sustainable Agriculture. Generation and Transfer of Technology have become very crucial the world over. As a step towards making technology reach the Small Holders (Resource-Poor-Farmers), efforts are being made to establish "VISTARNET–NICNET based Agricultural Extension Information System Network," in India, linking extension functionaries at Central, State, and District level has been taken up. The required funds for implementing VISTARNET will be provided through the National Agricultural Technology Project (NATP). As a part of VISTARNET, Informatics Development for the Directorate of Extension has been taken up by NIC.

Various Web Based Dissemination Systems

Warna Wired Village Project

The Project has been covered approximately 80 villages in Maharashtra. The project has been intended to serve the needs of the farmers on different crop cultivation practices of major crops, sugarcane cultivation practices, pest and disease control, marketing information dairy and sugarcane processing information etc right upto village level by KIOSK system on co-operative basis. Network connection is maintained by "Central Hub" as the main server station, "computer booth" as information centres for the farmers and six "Information technology centres" as training units of staff, students and farmers of the village.

Information Village Project

Information village project is aimed at bringing the benefits of modern information and communication technologies to the poor families in the villages in Puducherry and value addition centre which is the hub of information net work and four information shops have been established in different villages M.S. Swaminathan Research Foundation–International Development Centre, ran the project.

Honey-Bee Knowledge Network of IIM, Ahemedabad

Honey- Bee Knowledge network of IIM, Ahemedabad has been used to augment grass-root inventors and over come language, literacy and localism. A large number of grass root inventions have been identified and documented as short multimedia presentations. Future plans include creating of database of such innovations and making them accessible via a Wide Area Network.

Akshaya Project

Launched in 2002, the Akshaya project is an initiative of the Kerala State IT Mission (KSITM) to provide equitable access to ICT throughout the state. The project involves setting up 5,000 multipurpose community technology centres across Kerala, each within 2 to 3 kilometers of every household so as to cater to the needs of 1,000–3,000 families. A variety of services like business process outsourcing, online services, vocational training, and financial services, including e-banking, have been made available to the rural population. The project piloted in the district of Malappuram is being replicated all over the state. The Intel Learn Program is a key partner in this effort.

Intel Teach Programme

Launched in India in November 2004, the Intel Learn Programme focuses on building technology literacy and 21st century skills for youth living in rural communities who have little or no access to technology. The initial programme was piloted as part of the Akshaya project in the Mallapuram district in collaboration with the Kerala IT Mission, and has been implemented through over 300 community technology centres. The programme has since expanded to other states and programs and to date has affected more than 48,000 youth across 14 states and union territories.

Conclusion

Major developments are causing a revolution in computer technology. The new information technology tools and techniques can play a great role to improve learning and extension process through visualization. The animation can explain many complex processes better than text and graphics and music can provide additional enjoyment to the study of extension material. This dream has already been realized in many sectors of science and technology and can now be realistically applied to enhance extension activities. Hopefully every extension worker and farmer would welcome the technological advancements like multimedia and enjoy the benefits while disseminating information to the information seeker in the easiest and most economic and efficient way. It will definitely accelerate development in the field of agricultural extension and ultimately farmers of the country will be able to harvest the benefits to the desired level. In future, it would be possible to link villages and training facilities by means of high-speed, high bandwidth data communication network in order to instruct and to learn through online facilities.

Using the online facilities/multimedia the training can be provided to extension workers and farmer at community level effectively. Forecasting and forewarning models can be displayed and farmers can be advised about out break of diseases and insect and other natural calamities.

Agricultural Implications of IT

Information technology provides answers to a number of questions to the farmers. For example, what are the benefits of more irrigation? Is it cost-effective to apply additional chemicals? When is the best time to sell crops or buy inputs? With improved record-keeping, more detailed cost analysis and more sophisticated marketing strategies, farmers are making better decisions and earning higher profits. The Internet is increasing communication and business opportunities within the agricultural community, which previously operated in the relative isolation of rural areas. Farmers, agricultural researchers, cooperatives, suppliers and buyers use the Internet to exchange ideas and information, as well as to conduct business with each other. Machinery, seed chemicals and other types of agricultural products can be purchased and sold online. People can search for jobs and employees.

It is to be noted that, the farmer is in no position to use IT directly. The literacy levels, language barrier as most of the application software are predominantly in English, cost of computers, poor communication infrastructure make it impossible for individual farmers, particularly small farmers to directly adopt IT. This calls for institutional effort to harness to create the necessary IT based services to farmers. But in India, one prominent problem is that most of the farmers own small holdings, this seems to be difficult. In this situation, it may be made possible by adopting the corporate farming system, which is the need of the hour with advent of new agricultural policy. By taking up corporate farming, a group of farmers can install a computer and any educated young man from that group can undergo training of how to browse the internet. He can provide the farmer's current commodity, analysis reports on world markets and trade for different commodities. Food market overviews provide valuable information about some of the most important export market. IT can help to provide the information on the likely price distribution of key commodities over the coming years. Such information helps farmers and traders make decisions on when and in what ways to market their grain. Whether, to sell at harvest or store on-farm in anticipation of higher later in the season. When combined with enterprise budget data, the information can also be used in deciding which crops to produce in the coming season. In order to encourage farmers to obtain best possible price, information on various agricultural output markets is also being provided. The objective of this activity is to provide status of price at different markets to facilitate farmer to move his produce to the market where he can expect better price. The entire exercise will not be useful unless necessary arrangements are made to ensure that the farmers utilise this facility.

The contribution of information technology in bringing down costs, increasing efficiency and improving productivity and thereby contributing to the bottom line needs no special emphasis. In the fertilizer-marketing context, IT can play a major role in efficient sales operations, checking the marketing costs, safeguarding market share and providing efficient customer services. A well conceived IT set up could endow decision makers at all levels with better reflexes to effectively respond to market conditions. IT helps producers monitor and respond to weather variability on a day-to-day basis. Solar-powered weather stations in the field can be hooked up to a farmer's computer to relay information about current air and soil temperature, precipitation, relative humidity, leaf wetness, soil moisture, day length, wind speed and solar radiation. Producers use the Internet to monitor prices quickly and as often as they like. Farmers from around the world can exchange ideas, post questions and get answers about specific topics. Thus, it is said that the importance of information technology in the field of agriculture is emerging. The challenges of cost intensive, highly technical agricultural technologies are knocking the door to the farmers, which is ignited by the globalisation. There is significant shift from agriculture supply driven to demand driven paradigm in new emerging and changing economics policy. It is viewed that future agricultural growth would be information driven. New information must reach to the ultimate user at the fastest speed to harness its potential benefits. Information like seed, water, nutrients and plants protection biologically is one of the key inputs for successful farming. Knowledge intensive and precision farming techniques will be the guiding lines for sustainable agriculture in the future.

Extension

Harnessing Information Technology for agricultural extension is going to receive high priority in future. While developing any system of IT for Agriculture technology, the farmer should be kept in focus as a player, generator and user of the knowledge. Extensive use of modern information technology should be promoted for two-way communication between scientists, extension workers and farmers to transfer technologies and information more cost effectively.

References

Sahoo, K.C., 2004. *Information Management with IT Application.* Medallion Press, Ludhiana, p. 177–178.

Ray, G.L., 2007. *Extension Communication and Management*. Kalyani Publisher, Ludhiana.

COW, 2003. *Global DM Finalist Proposal Brief,* Computers on Wheels, http://wbln0018.worldbank.org.

Devraj, K.K. Chaturvedi and Singh, Pravin Kumar, 2001. Multimedia: An Emerging Technology for Extension. *Agri. Extn. Review,* 13: 3–8.

Chapter 20

Mobile Agricultural School and Services (MASS): The Latest Technology of Extension System in Agriculture Sector

☆ ***Vijay Bharat***[1]

Introduction

India is shining. It has become the role model for the rest of the world. Credit goes to it for being the largest exporter of technical manpower in the globe, especially in the field of IT. Hyderabad has become the Cyberabad and Bangalore has become the silicon valley of India. There are many cities that are in the race of becoming modernized. But what an irony? All of them are lacking vision in some way or the other. All these technical advancement and training are guided towards the core industrial sectors only. Ask any software professional where he shall go to contribute his technical expertise; his answers would be everything except agriculture.

Agriculture-the backbone of Indian economy–has still been treated as a neglected child of IT development. It has still to taste the first bliss of IT revolution. While reducing knowledge gaps and increasing knowledge sharing is an essential step for increasing

---

1 Founder cum Managing Director, Mobile Agricultural School & Services, Ranchi.

productivity and boosting growth in rural agrarian society. Empowering farmers with relevant and timely information about their ability to withstand biotic and abiotic stresses can significantly reduce farming risks. Though in some advanced laboratories, it may have got a place and might be benefiting the agricultural sector indirectly but its firsthand exposure to the field was a distant dream. Our management institutions and vocational department of universities went hi-fi but change in the training module of the farmers was never thought of sincerely. The same old-rotten method of going inside the block cluster dilapidated hall, training the farmers through the ineffective lecture method were accruing a good expenditure to the exchequer without bringing the desired result.

But we were moving without much commitment in deep slumber. Answer of these problems was with us but being apathetic and insensible to the farmers' problem, we were not able to find it.

Solution in IT

The answer is in the form of IT itself. The need of IT in the field of agriculture definitely gets the place in several seminars but the real thing is to explore the modus operandi. In my point of view, *agri- experts are serious* but *not very sincere.* What is very recent approach to reach to the beneficiaries was to disseminate the information technique to the target group.

In one of its research conducted by Acharya N. G. Ranga Agricultural University (ANGARU) conducted a SWOT analysis and found some weaknesses of Agricultural Extension Scientists. One of them was No exposure to IT.

Development of computers and improvements in telecommunication offer farmers many new opportunities to obtain technical and economic information quickly and use it effectively for decision making. This was realized but not actualized to a satisfactory level.

Here comes MASS as a live example to settle all questions.

Story behind MASS

Vijay Bharat, an agriculture post-graduate from RAU, Pusa, Samastipur and brain behind the innovative Mobile Agricultural School and Services, served in different NGOs before launching his dream project MASS in May 2005.While rendering his services to the farmers of interior tribal region, he felt it sincerely that imparting training to the least educated society is the biggest challenge. It requires not only a good knowledge in the subject but the socio-psychological aspect of the tribal people has also to be dealt with. What he sensed that these training programmes are in the one-way mode without taking the feed back of the farmers, these training programs were designed according to the convenience of scientists and not the vice–versa. Receiving lunch packets, a gift-packet and making signatures were all interest behind the training programmes.

After much retrospection, it has been realized that farmers do not merely require a fish; rather they want to know how to fish themselves. The idea of imparting training in the best possible clinical way became the mantra. But the real challenge of transforming his thoughts into action began from here. Knocked in vain the doors of government machineries, which have always its technical obligations to initiate the things, Vijay Bharat put up his proposals before Bank of India. With their much desired support, he could purchase a bus and modeled it into a digitally equipped training hall, having a sitting capacity of 50.The bus is fitted with portable genset, laptop, computer with internet connectivity LCD projector with 4.4 ft. display screen. Henceforth, it rolled on the road to a new dawn.

Modus Operandi

If the bus and the scientists get all the limelight at training centre, IT is its behind the screen hero. And, I consider IT to be the real hero.

The mode of action of MASS is two-fold:

Development of Training Package *viz.*, CDs

It involves

Script writing—Shooting—voice-over–editing in lab—final production.

Our package development process is flexible and need-based. If necessary for a specific target audience, we go for dubbing in regional language also.

Training at the Spot

The Steps followed during training

- ☆ Showing of folk music dances to bring them into the rhythm of learning
- ☆ Displaying the CD package at the training spot
- ☆ Collecting feedback
- ☆ Sorting out the problems
- ☆ Passing out the relevant information by connecting through internet
- ☆ Field visit to solve the current problem of standing crops

The spirit of our service:

- ☆ We do not call farmers, we reach to them.
- ☆ We don't lecture, we do practicals.
- ☆ We don't ask, rather we answer.
- ☆ We don't provide what we have, we provide them what they don't have.
- ☆ We don't make them serious, we make them sincere.
- ☆ We don't administer, we simply facilitate.
- ☆ We don't make them to understand our language; rather we understand their language.

Aims and Objectives of our Training

By providing training through audio-visual medium, we keep this objective in our mind that its effect should everlast in the psyche of farmers. Also, it should motivate them to accept new things beyond the social norms. So far, MASS has developed several CD's on different packages and practices of different crops and allied subjects. Very recently three films namely, "Ketari Baba", financed by SRI, Pusa and "Soil Health" and "Biofertilzers and their uses" financed by NICHE Area project, BAU, Ranchi, have been produced under MASS production and have received huge approbation not only from the planners gallery but the beneficiaries also.

Outcome

This training programme through MASS has come up with very exciting results not only that agriculture training programme has got new outfit but also it has changed the socio-psychological framework of the local farmers. They are on the verge of shading out their inhibitions for the new things. Now they consider this training programme economically viable and content wise superb. Not only men but women are also the front runner to be benefited by this Agri-IT revolution in the shape of MASS. Sabita Rani, a resident of Kutiatu village of Namkum block, Ranchi snaps out,

"It was really a wonderful experience sitting inside a cozy bus and getting training through film. It was awesome. We wish MASS should visit us every month."

But Sabita Rani is not alone to witness the impact of this MASS mission. MASS has, in fact done a subsidiary job of uniting more than dozen women of the same village, with whom she has formed a SHG, and they are also into this new game. Shivanand Singh of Murto(Bedo Block, Ranchi) village is all happy man after saving a handsome amount during the last *kharif* season by simply saying goodbye to chemical pesticides implementing the advice of an educational film produced by MASS especially designed for this purpose only. Now, he alongwith his other villagers, is all aware of IPM which is not only cheaper but also eco-friendly.

The impact of media is everlasting; there is no question about it. By tying his knots with IT sector, Vijay Bharat, being an agrico has a vision and knows the extent and limitation of IT in training extension programme. So, MASS ends up not only with the demo of packages and practices but also have instant solution to the farmers. It carries soil-testing kit where the sample is tested on the spot. This on-farm testing and hence instant solutions have touched heart of the beneficiaries. MASS repay them by conducting follow up, through interactive programme twice in a year to the same people. Winner of Governors' first prize at Krishi Mela, pampered by many top brass think tanks *viz.* B.K.Jha, Assistant Professor,BAU and others, MASS gets all support from institutions like HARP,Palandu; BAU, Kanke,Ranchi; RAU,Pusa. MASS wishes to win the confidence of farmers and wiping out their socio-psychological inhibitions for the new. In very near future they will accept everything new, which is good for the health of the field, livestock and family. Through the easy access of the training bus, we should listen what Lakhan, a young local farmer has to say,

"We could the training and expert advice at our doorstep and this is the biggest advantage. Also, it has created a thirst in us to rear our field with the help of the latest technologies. It flushes all our worries out. Now we feel free and relaxed."

The Future Plan

MASS has an ambition to conduct courses like Diploma in Computer Applications with the help of IT professionals. Our emphasis shall be on rural and tribal children so that we can catch them young. Obviously they are our future farmers and they should be electronically literate to have an entrepreneur outlook. Once they learn to handle the key board, then ploughing hands would reap only gold.

I, think what Lakhan and other village folk say, these all would create a butterfly effect and its resonance will go worldwide. Programmes for extension, education, and communication for farmers will certainly help in propagating the message of "Sustainable Agriculture". Computer application, computer discs, press and magazines help keep the farmers up to date on latest management systems and technology development.MASS on its basis of pure experience, can assure that such programmes provide direct communication between the farmers and the agricultural counselors because the field of agriculture need two things to be fast *i.e.*, the current market rate of agricultural produces and timely help of infested crops which can be solved by proper utilization of IT.

Conclusion

MASS has given a new approach for transfer of technology in agriculture sector.

It is very fast in performing training at doorstep of the farmers and solves the problems of targeted group. Reaching at the doorstep of the farmers with well equipped training centre is a new invention in the field of Agricultural extension system.

NABARD, Ranchi has also accepted this new methodology and recommended the central Govt. to include it as a training centre all districts of Jharkhand under the scheme of Rastriya Krishi Vikash Yojna. Farmers, Implementing agencies and Policy makers appreciated our approach significantly and requested to run it on village level.

Examples of Feedback:

1. This should be exhibited in every village: S. K. Jha(SP, Chibasa,Jharkand)
2. Mobile Agricultural School and Services is very latest techonology and useful: Y.K.Singh (Member, Vidhan Sabha, Barh,Bihar)
3. I liked this programme. Farmers became happy and a hope has been arised in cultivation. My wish is to run it regularly in our locality: Kalpna Biruli(Mother-in-law of ex CM of Jharkhand *i.e.* Madhu Koda)

We have got several feedbacks of different top officials as well as downtrodden farmers. All have same version that MASS is the best option for training and taking care of overall development of the farmers.

Chapter 21

Role of Farm Literature as a Source of Latest Agricultural Information for Agricultural Development Officers of Punjab

☆ *Dharminder Singh*[1] *and D.S. Dhillon*[2]

Introduction

The information and knowledge are increasingly being seen as the new factors of agricultural production partially replacing the traditional factors of production–land, labour and capital (World Bank Report, 1998-99). The information need is the most basic of all human needs. Agricultural Information and Knowledge System (AKIS) is becoming more complex than ever. Information management has a large role to play for effective communication. The growth of information and communication technology, a process that is both a product of and a stimulus is the parallel phenomenon of globalization (Marrow, 2002).

Use of advanced information and communication technologies tends to reduce the gaps in yields of agricultural crops between research and farmers fields.

---

1 Assistant Professor (Extesion Education), KVK, Moga (Punjab).

2 Professor and Head, Department of Extension Education, PAU, Ludhiana – 141 004 (Punjab).

Effective provision of information and its precise communication at the initiation of an intervention is of utmost importance (Frewer *et al.*, 1998). In Mauritius, information and communication technologies have played a vital role in promoting the precision agriculture for improved crop production along with protection of environment.

Farming system in Punjab is undergoing dynamic transformation. Crop diversification, contract farming, integrated approaches of crop management, resource conservation technologies, organic farming, environment safety are some of the emerging issues now-a-days. Besides this, international trade is under paradigm shift and competition is becoming stiffer with the introduction of World Trade Organization (WTO) regime. The Department of Agriculture, Punjab is one of the main development departments working for the welfare of the farmers of the state. The Agricultural Development Officers (ADO's) are the grass root level functionaries of the Agriculture Department, Punjab. They are responsible for communicating the latest farm innovations, agricultural programmes and policies to the farmers. Farm literature is one of the important sources of information used by ADO's to acquaint them with the latest agricultural information.In the present study, an attempt has been made to analyse the pattern of use of farm literature by the ADO's of Punjab state

Materials and Methods

The study was conducted in Punjab state. The state is divided into six agro-climatic regions *viz.* sub mountain undulating region, undulating plain region, central plain region, western plain region, western region, and flood plain region. To ensure randomization of the study sample, one district was selected randomly from each agro-climatic region. Thus, in total six districts *viz.* Hoshiarpur, Jalandhar, Ludhiana, Faridkot, Bathinda and Ropar were selected for conducting the present study. All the Agricultural Development Officers in position from the six selected districts were taken as the respondents for the present study. Thus, the total sample size for conducting this study comprised of 170 Agricultural Development Officers.

A questionnaire was prepared for collection of data pertaining to the pattern of use of farm literature by Agricultural Development Officers. The questionnaire was pre-tested on 20 non-sampled respondents. The data were collected with the help of distributed questionnaire during monthly meetings of Agricultural Development Officers held at district headquarters in each of the six selected districts. The data were tabulated and analysed with the help of appropriate statistical tools such as frequencies, percentages, mean scores and ranks etc.

Results and Discussion

Form in which Agricultural Information is Available to Respondents

Availability of improved agricultural information in different forms may be helpful for Agricultural Development Officers in its dissemination to the farmers in a better way. On the other hand, availability of improved agricultural information in more than one form makes its understanding better for the respondents. It can be observed from the data given in Table 21.1 those popular articles in newspapers, leaflets/pamphlets and printed books/bulletins were the chief forms of farm literature in which the respondents got the farm information. Popular articles were available to all of the respondents and ranked first with mean score 1.87. More than half of the respondents (54.12 per cent) got leaflets and pamphlets frequently whereas 22.35 per cent got them sometimes. In case of printed books/bulletins, these were available frequently to 42.94 per cent of the respondents, whereas 37.65 per cent got them sometimes. Only 19.41 per cent of the respondents reported that they never received desirable books or bulletins for getting the latest farm information. Magazines and journals got fourth rank as far as their availability in different forms is concerned. In addition to these, only 10 per cent of the respondents reported that they frequently received posters and charts, whereas 51.18 per cent of them got them sometimes. Least available forms of agricultural information were found to be slides/film strips and floppies/CDs with mean scores of 0.12 and 0.19. Bhople (1985) also reported that the Agricultural Officers (AOs) and Village Extension Workers (VEWs) received the farm recommendations for their use mostly through cyclostyled messages, circular letters and printed literature.

Table 21.1: Distribution of Respondents According to the Availability of Agricultural Information in Different Forms (n =170)

Sl.No.	*Form of Information*	*Extent of Availability*			*Mean Score*	*Rank*
		Frequently	*Sometimes*	*Never*		
1.	Popular Articles in Newspapers	148(87.06)	22(12.94)	0(00.00)	1.87	I
2.	Leaflets/Pamphlets	92(54.12)	38(22.35)	40(23.53)	1.30	II
3.	Poster and Charts	17(10.00)	87(51.18)	66(38.82)	0.71	V
4.	Photographs	0(00.00)	75(44.12)	95(55.88)	0.44	VI
5.	Printed Books/Bulletins	73(42.94)	64(37.65)	33(19.41)	1.23	III
6.	Magazines/Journals	47(27.65)	68(40.00)	55(32.35)	0.95	IV
7.	Slides and Film Strips	0(00.00)	21(12.35)	149(87.65)	0.12	VIII
8.	Floppies/CDs	0(00.00)	32(18.82)	138(81.18)	0.19	VII

Figures in parentheses indicate the percentage of respondents.

Consultation of Farm Literature

A perusal of the data presented in Table 21.2 reveals that publications of Punjab Agricultural University, Ludhiana severed as the primary source of farm literature for consultation by the respondents. PAU publications *i.e., Package of Practices* for *Rabi/Kharif* crops, Progressive Farming, *Changi Kheti*, PAU Agricultural Handbook and Field Problems of Important Crops were consulted by 68.82, 32.35, 21.18, 7.06 and 4.70 per cent of the respondents, respectively. It also implies that some of the respondents consulted more than one PAU publication. Extension bulletins published by the Department of Agriculture, Punjab were consulted by 48.82 per cent of the respondents (Table 21.2).

Table 21.2: Distribution of Respondents According to Consultation of Farm Literature (n=170)

Sl.No.	Publication	Frequency*	Percentage
1.	Package of Practices for *Rabi/Kharif* crops	117	68.82
2.	Extension Bulletins Published by Deptt. of Agriculture, Punjab	83	48.82
3.	*Changi Kheti*	36	21.18
4.	Progressive Farming	55	32.35
5.	Modern Kheti	19	11.18
6.	*Kheti Duniya*	37	15.88
7.	PAU Agricultural Handbook	12	7.06
8.	Indian Farming	19	11.18
9.	Intensive Agriculture	15	8.82
10.	Science Journal	1	0.59
11.	Field Problems of Important Crops	8	4.70
12.	*Uttam Kheti*	6	3.53
13.	Safal Sunehri Kheti	10	5.88

* Multiple Responses.

As far as private publications are concerned, *Kheti Duniya* and *Modern Kheti* were consulted by good number of respondents *i.e.*, 21.76 and 11.18 per cent respectively. In case of agricultural journals, 11.18 and 8.82 per cent of the respondents consulted Indian Farming and Intensive Agriculture, respectively (Table 21.2). The findings of the study further revealed that majority of the respondents (87.06 per cent) purchased the agricultural books annually. However 12.94 per cent respondents did not purchase any book related to agriculture during year under study. Majority of the respondents purchased two to five books and only 22.94 per cent of them purchased more than five books annually. These results are in agreement with those of Kaur (2003), who in her study reported that most frequently used sources by the home scientists for generating scientific information were text books, manuals, journals, research papers and participation in workshops, conferences, seminars and discussion with subject-matter-specialists. Similarly Balasubramanyan and Menon (1977) observed that the important sources of information about high yielding varieties of paddy for the extension personnel in Tamil Nadu included State Department of Agriculture, university publications, journals, personal visits to researchers, correspondence with scientists, visit to research stations, and visit to extension personnel in the descending order.

The respondents were also asked to give their reactions about the consulted farm literature and these reactions have been presented in Table 21.3. *Package of Practices* for *Rabi/Kharif* crops were reported to be relevant and understandable by majority of them but only 57.26 per cent of the readers considered it as complete. All the readers found the extension bulletins as relevant and understandable. However,

38.55 per cent readers reported these bulletins to be incomplete. Similarly half of the readers found *Changi Kheti* as incomplete magazine for getting agricultural information. Forty per cent and 34.55 per cent of the readers reported that *Progressive Farming* lacked completeness and relevancy. *Modern Kheti* when compared with *Kheti Duniya* (both private Punjabi agricultural magazines), former was found to be more comprehensive, complete, relevant and understandable. *Indian Farming* lacked completeness and understandability while *Intensive Agriculture* lacked completeness and relevancy. Field Problems of Important Crops is a new book brought out by PAU. It contains information regarding various insect-pests, weeds and other field problems of crops along with pictures of their symptoms. Majority of the readers (62.50 per cent) complained the lack of comprehensiveness

Table 21.3: Reactions of Respondents about Farm Literature

Sl.No.	*Publication*	*Comprehensive**	*Complete**	*Relevant**	*Understandable**
1.	Package of Practices for *Rabi/Kharif* crops (n=117)	92(78.63)	67(57.26)	102(87.18)	105(89.74)
2.	Extension Bulletins Published by Deptt. of Agriculture, Punjab (n=83)	63(75.90)	51(61.45)	83(100.00)	83(100.00)
3.	*Changi Kheti* (n=36)	28(77.78)	18(50.00)	20(55.56)	26(72.22)
4.	Progressive Farming (n=55)	47(85.45)	33(60.00)	36(65.45)	42(76.36)
5.	Modern Kheti (*n=19*)	15(78.95)	14(73.68)	14(73.68)	19(100.00)
6.	*Kheti Duniya* (n=37)	24(64.86)	16(43.24)	18(48.65)	25(67.57)
7	PAU Handbook (n=12)	4(33.33)	6(50.00)	7(58.33)	11(91.67)
8.	Indian Farming (n=19)	16(84.21)	9(47.37)	13(68.42)	9(47.37)
9.	Intensive Agriculture (n=15)	11(73.33)	8(53.33)	8(53.33)	10(66.67)
10.	Science Journal (n=1)	1(100.00)	1(100.00)	1(100.00)	1(100.00)
11.	Field Problems of Important Crops (n=8)	3(37.50)	5(62.50)	5(62.50)	6(75.00)
12.	*Uttam Kheti* (n=6)	4(66.67)	4(66.67)	5(83.33)	5(83.33)
13.	Safal Sunehri Kheti (*n=10*)	5(50.00)	7(70.00)	6(60.00)	9(90.00)

Figures in parentheses indicate the percentage of respondents.

* Multiple Responses.

Problems Faced by the ADO's in the Use of Farm Literature

In the present study, an attempt was also made to know the various problems encountered by the ADO's in the use of farm literature for the dissemination of latest farm information to farmers/farm women of State.

Writing for Newspapers

Majority of respondents (51.18 per cent) reported the problem of lack of proper training to write features and success stories for newspapers whereas 27.06 per cent reported low educational level of the farmers as a problem in this regard (Table 21.4).

Lack of subject-matter knowledge and inadequate support from the senior officers were some of the other problems faced by 12.35 and 8.82 per cent of the respondents respectively. It can be further noted from the data given in Table 4 that 7.65 per cent of the respondents considered this method as ineffective. According to them everyone reads the features, but nobody acts on the advised guidelines.

Table 21.4: Distribution of Respondents According to Various Problems Faced by them in Communicating Improved Agricultural Information through the Use of Farm Literature (n=170)

Sl.No.	*Problem*	*Writing for Newspapers*		*Preparation of Farm Literature*	
		*Frequency**	*Percentage*	*Frequency**	*Percentage*
1.	Lack of training	87	51.18	53	31.18
2.	Lack of funds	-	-	128	75.29
3.	Lack of subject-matter knowledge	21	12.35	-	-
4.	Lack of opportunity	-	-	42	24.71
5.	Inadequate support from the subordinates	-	-	11	6.47
6.	Inadequate support from the senior officers	15	8.82	16	9.41
7.	Low educational level of farmers	46	27.06	46	27.06
8.	Ineffectiveness of the method	13	7.65	-	-

*: Multiple Responses.

Preparation of Farm Literature

A critical look at Table 21.4 shows that lack of funds was the major problem in the preparation of farm literature and it was faced by 75.29 per cent of the respondents. Lack of training was conceived by 31.18 per cent of the respondents while 27.06 per cent reported low educational level of farmers as a problem. As many as 24.71 per cent of the respondents reported that they did not get any opportunity for preparation of farm literature. Inadequate support from the subordinates and the senior officers was reported as problem by 6.47 and 9.41 per cent of the respondents, respectively.

References

Balasubramanyan, U.A. and Menon, K.R., 1977. Extension personnel communication linkage with researchers. *Ind. J. Ext. Edu.*, 13: 61–63.

Bhople, R.S., 1985. Communication patterns in farm innovation extension, acceptance and adoption by the farmers in Maharashtra State. *Ph.D. Thesis*, PAU, Ludhiana, India.

Frewer, L.J., Howard, C. and Shepherd, R., 1998. The influence of initial attitudes on responses to communication about genetic engineering in food production. *Agriculture and Human Values*, 15: 15–30.

Kaur, K., 2003. Communication pattern of home scientists regarding development and dissemination of scientific information. *Ph.D. Thesis*, Punjab Agricultural University, Ludhiana, India.

Morrow, K., 2002. The ICT agenda: global action plans and local solution. *LEISA,* 18: 9–10.

World Bank, 1999. World *Bank Development Report 98/99: Knowledge to Development.* Washington, DC, USA.

Chapter 22

Utilization Pattern of Communication Sources Channels Among the Rural Women

☆ *Sonika Gupta[1], J.P. Singh[2] and R.P. Singh[3]*

Introduction

Quick dissemination of agricultural and home science practices very much depends on the use and availability of different communication sources to the farmer, farmwomen, rural youth etc. The ultimate aim of information system is how best possible communication system is developed and put into practice so that the consumers of information are in the position to utilize properly at their level. The success of any program depends to a great extent on the timely availability of information sources to the' audience. The contribution of women in agriculture and home management also depends to a great extent on the prompt availability of resources and their proper utilization by them.

Communication can help, creating awareness about technology literacy, malnutrition, health, population and hunger etc. Links can be created between

---

1 Research Scholar, Department of Agricultural Extension, R.B.S. College, Bichpuri, Agra.

2 Head, Department of Agricultural Extension, R.B.S. College, Bichpuri, Agra.

3 Associate Professor, Department of Agricultural Communication and Rural Development, SVBP University of Agriculture and Technology. Modipuram, Meerut.

government agencies and people as well as people to people (Bishnoi and Dixit, 2003). Efficient transfer of innovations and their practical applications to farm and home situation is the key to economic and social development of India. Extension services in India have a network of number of large professional extension workers at national, state, block and village level. Desired effect in this direction has been achieved but to a limited extent.

'There is a wide gap between desirable/available information, improved practices and technologies and the actual knowledge level and adoption of technologies and practices relating to home and farm, there by helping them improve their standard of living and live their life happily healthily. Adequate knowledge in the field of food and nutrition, childcare, clothing and textile, family resource management, sanitation, health and hygiene and skill development among women would certainly promote their well being. Information and technological empowerment of women is the prime concern of home science and would help them to play their economic, social, educational and nurturing role more productively and effectively. Similarly, break through in the field of agriculture and animal husbandry is not possible without an effective communication support to disseminate research findings. Speedy dissemination of scientific information and technological know how to the rural community is essential for bridging the gap between the scientist and the rural women. There are many sources available in the rural areas like extension worker, local leader, radio, television, newspaper and neighbors for the purpose of agricultural communication to the rural masses (Goswami *et al.*, 2003).

The study of communication behaviour of farm women to identify the availability of infrastructure facilities would be worthwhile. The extension worker cannot expect to bring about change unless he is able to communicate effectively. He has to create a situation, where information can be effectively transmitted with greater impact on people.

It is important to know the type of communication media being used by people, the extent or frequency of their use by people and also the degree of usefulness of various channels to the rural women in the field of home science. Effective use of communication channels for dissemination of information helps to reduce the time taken in the adoption process. Knowledge about the communication channels would be helpful in developing a suitable extension strategy to uplift socio-economic status of the women.

In the recent past women's development and contribution has become a subject of global consideration. Technology exposure for empowering women in an integrated manner through active learning has been a major concern. This study is an attempt to analyse, utilization pattern of communication sources of rural women.

Methodology

The study was conducted in randomly selected ten villages, five from each panchayat samiti namely, Sewar and Kumher of Bharatpur district of Rajasthan state. From each selected village, thirty respondents under three categories namely landless, marginal and small farm families were selected randomly on the basis of their proportion in the village. Thus total number of respondents selected for the study was three hundred farm women. The data were collected through personal interview with the help of interview schedule developed for the present study. The data thus collected were analyzed in light of the objectives of the study.

Results and Discussion

Among the total respondents 44.67 per cent belonged to the middle age group of 31 to 45 years, majority of the respondents (99.67 per cent) were married, more than half of the landless respondents (52.78 per cent) were from scheduled caste and scheduled tribe while in the small and marginal farm families, most of the respondents 47.06 per cent and 44.44 per cent, respectively belonged to the other backward classes. Majority of the small (78.43 per cent) and marginal (76.98 per cent) farm families had farming as their main occupation, whereas, majority of the landless families (63.89 per cent) earned their major source of income by working as labourer, the. percentage of the nuclear families (59.67 per cent) is more than the joint families (40.33 per cent) among the total respondents. Majority of the respondents (71.00 per cent) belonged to the medium family size (5 to 8 members). Majority of the respondents (66.00 per cent) were illiterate. Most of total respondent families (79.33 per cent) had annual income upto 0.5 lakh. Majority of the respondents had poor social participation with 68.33 per cent of farm families having no membership in any organization. The maximum of the respondents (51.33 per cent) rarely visited urban areas. It was found that nearly half of the small (49.02 per cent) and marginal (49.21 per cent) respondents had medium extension contacts and majority (84.72 per cent) of landless respondents had low extension contacts. Less than 50 per cent respondents of small (40.20 per cent) and marginal (46.83 per cent) and majority, (65.27 per cent) of landless had low mass media exposure.

Majority of the respondents had kuchha cattle shed (73.33 per cent). Majority of the respondents in small (72.55 per cent), marginal (80.16 per cent) and landless (87.50 per cent) possessed 1-2 animals.

Utilization Pattern of Personal Localite Sources

The information regarding the utilization pattern of personal localite sources among the respondents is presented in Table 22.1.

Table 22.1: Distribution of Respondents According to the Utilization Pattern of Personal

Personal Localite Sources	*Mean Per cent Score*				*Rank*
	Small	*Marginal*	*Landless*	*Total*	
Progressive farmers/ farm women	63.73	58.73	52..78	59.00	IV
Village or farm leaders/ panchayat	53.92	52..78	51.39	52.83	V
Friends	97.55	98.81	97.92	98.17	II
Neighbours	98.53	99.21	99.31	99.00	I
Relatives	96.57	97.62	95.14	96.67	III
Total	82.06	81.43	79.31	81.13	

Among the personal localite sources, the most used source by the respondents is neighbours with highest mean per cent score (99.00). This was closely followed by friends (98.17 MPS) and relatives (96.67 MPS). The other lesser used sources were progressive farmers/farm women (59.00 MPS) and village or farm leaders/panchayat (52.83 MPS).

Among the small, marginal and landless respondents the neighbours, friends and relatives had mean per cent scores ranging from 95.14 per cent to 99.31 per cent. This depicts that these sources are very much used by the respondents of all the three categories. Thus, it can be generalized that personal localite sources have great positive impact among the respondents.

The overall mean percent score of all the personal localite sources was 82.06 per cent among the small, 81 *A3 per cent* among the marginal, 79.31 per cent among the landless and 81.13 per cent among the total respondents. The results depict that among the three categories of farm families there was very little variation in the total mean percent scores of personal localite sources.

It can be concluded that among the various personal localite sources; neighbours, friends and relatives are the sources utilized very frequently. The above findings are in conformity with the findings of Suriyanarayanam and Tamilselvi (2003) and Bishnoi and Dixit (2003).

Utilization Pattern of Personal Cosmopolite Sources

The information regarding the utilization pattern of personal cosmopolite sources among the respondents is presented in Table 22.2.

Table 22.2

Personal Localite Sources	*Mean Per cent Score*				*Rank*
	Small	*Marginal*	*Landless*	*Total*	
Village extension workers	54.90	59.92	55.56	57.17	III
Input dealer	50.98	48.81	46.53	49.00	VI
Personnel of co-operatives	56.86	53.57	52.78	54.50	IV
Auxi					

Among the personal cosmopolite sources Auxiliary Nurse Midwife (ANM) was the most frequently used source having mean per cent score (MPS) 63.50 per cent, followed by Anganwadi worker! sathin (62.00 MPS). Next in order was village extension workers (57.17 MPS), personnel of co-operative (54.50 MPS) and personnel of NGO! voluntary organization (54.33 MPS). Input dealers with 49.00 MPS and SMS/university scientists/State agricultural officers with 42.17 MPS were the lesser-used sources.

The highest used source was auxiliary nurse midwife among the small (65.69 MPS). Marginal (64.68 MPS) and landless (58.33 MPS) category. However, the least used source was SMS/university scientists, agricultural officers among the three categories small (41.18 MPS), marginal (46.03 MPS) and landless (36.81 MPS) respondents. The above findings are in line with Singh *et al.* (2003).

The overall mean percent score of all the personal cosmopolite sources was 55.67 among the small, 56.46 in marginal and 50.10 among the landless and 54.67 among the total respondents. The results depict that there was not much variation among the different categories. However, it was noted that the landless respondents had the least scores when compared with other categories, *i.e.*, small and marginal.

Utilization Pattern of Personal Cosmopolite Channels

The information regarding the utilization pattern of personal cosmopolite channels among the respondents is presented in Table 22.3.

Table 22.3: Distribution of Respondents According to the Utilization Pattern of Personal Cosmopolite Channels

Personal Localite Sources	*Mean Per cent Score*				*Rank*
	Small	*Marginal*	*Landless*	*Total*	
Farm and home visit	56.86	57.14	43.75	53.83	III
Demonstration	54.90	51.98	50.00	52.50	IV
Office call	0.00	0.00	0.00	0.00	VII
Tour/Field trip/field visit	25.00	30.95	24.31	27.33	VI
Group discussion/Meeting	58.82	57.94	52.78	57.00	I
Training	34.80	38.10	33.33	35.83	V
Exhibition	52.94	58.73	51.39	55.00	II
Total	40.48	42.12	36.51	40.21	

The Table 22.3 reveals that among the personal cosmopolite channels, the most used source was group discussion/meeting (57.00 MPS). This was closely followed by exhibition/farmers fair (55.00 MPS), farm and home visit (53.83 MPS) and demonstration (52.50 MPS). The least used sources were training (35.83 MPS) and tour! field trip/field visit (27.33 MPS). None of the respondents were found to he using office call (0.00 MPS). The results are in confirmation with Singh *et al.* (2003).

The mean percent scores for group discussion! meetings, exhibition/farmers fair, demonstration and farm and home visit ranged from 43.75 per cent to 58.82 per cent. Whereas, mean percent scores for training were low among the small (34.80 per cent), marginal (38.10 per cent) and landless (33.33 per cent) respondents, which indicates that more training programmes should be organized for the respondents. The least used source was tour/field trip/field visit with 25.00 per cent, 30.95 per cent and 24.31 per cent mean percent score among small, marginal and landless categories, respectively.

It is observed that the overall mean percent score of the personal cosmopolite channels was highest in the marginal category (42.12 per cent), followed by small (40.48 per cent) and landless (36.51 per cent) respondents.

Utilization Pattern of Impersonal Cosmopolite Channels

The information regarding the utilization pattern of Impersonal Cosmopolite Channels among the respondents is presented in Table 22.4.

Table 22.4: Distribution of Respondents According to the Utilization Pattern of Impersonal Cosmopolite Channels

Impersonal Cosmopolite Channel	*Mean Per cent Score*				*Rank*
	Small	*Marginal*	*Landless*	*Total*	
Television	72.06	57.94	42.36	59.00	I
Radio	45.59	40.08	50.00	44.33	II
Poster/Chart	39.22	28.97	11.81	28.33	IV
Folder/booklet/leaflet etc.	17.16	15.87	0.00	12.50	VII
Newspaper articles	19.12	14.68	0.00	12.67	VI
Farm Journal	3.92	2.78	0.00	2.50	VIII
Hoarding/banner	41.18	30.16	22.22	32.00	III
Circular letters	1.47	0.40	0.00	0.67	IX
Field board	17.65	17.46	0.00	13.33	V
Total	28.59	23.15	14.04	22.81	

Television (59.00MPS) was the most source of information. This was followed by radio (44.33 MPS), hoarding/banner (32.00 MPS) and poster/chart (28.33 MPS). The lesser-utilized sources among the impersonal cosmopolite channels were field board (13.33 MPS), newspaper article (12.67 MPS) and folder/booklet/leaflet (12.50 MPS) and farm journals (2.50 MPS). Circular letter was the least utilized by the respondents (0.67 MPS).

It was found that television had the highest score among small (72.06 per cent) followed by marginal (57.94 per cent) and landless (42.36 per cent). Whereas, the mean percent score for radio was 45.59, 40.08 and 50.00 among the small, marginal and landless respondents, respectively.

It was noted that the mean percent score was 17.16 and 15.87 for folder/booklet! leaflet, 19.12 and 14.68 for newspaper article 3:1d 3.92 and 2.78 for farm journals among the small and marginal farm categories, respectively. Among the landless respondents the mean percent scores of folder/booklet/leaflet, newspaper article and farm journals was nil. The reason behind this was higher illiteracy among the landless respondents.

It was found that the overall mean percent score of the impersonal cosmopolite channels was highest in the small farm category respondents (28.59 fv1PS), followed by marginal (23.15 MPS) and landless respondents (14.04 MPS).

Television and radio were the most used sources respondents of all the three categories. Similarly, Sangha and Kalra (1993) reported that radio and television were used to a greater extent than other mass media sources. Gupta *et al.* (2003) also observed that electronic media like television and radio were mostly used' and preferred for seeking information on agriculture and development. The findings are also in confirmation with Singh and Mishra (2002) and Goswami (2003).

Utilization Pattern of Overall Communication Channels Sources

The information regarding the utilization pattern of overall communication channels/sources among the respondents is presented in Table 22.5.

Table 22.5: Distribution of Respondents According to the Utilization Pattern of Overall Communication Channels Sources

Communication Sources Channels	*Mean Per cent Score*				*Rank*
	Small	*Marginal*	*Landless*	*Total*	
Personal localite sources	82.06	81.43	79.31	81.13	I
Personal Cosmopolite sources	55.67	56.46	50.10	54.67	II
Personal Cosmopolite channels	4U.48	42.12	36.51	40.21	III
Impersonal Cosmopolite channels	28.59	23.15	14.04	22.81	IV

The Table 22.5 makes it evident that the most used source by the respondents were personal localite sources with highest mean per cent score 81.13 followed by personal cosmopolite sources (54.67 MPS) and personal cosmopolite channels (40.21 MPS). The least used source was the impersonal cosmopolite channels (22.81 MPS)..

It was found that the personal localite sources were the most utilized source in all the three categories, *viz.*, small (82.06 MPS), marginal (81.43 MPS) and landless (79.31 MPS). The results are in line with Singh and Singh (1976) who reported that farmers mostly used the localite sources such as progressive farmers, neighbours, relatives and friends. The above findings are in conformity with Chand *et al.* (2001).

The personal cosmopolite sources (54.67 MPS) and personal cosmopolite channels (40.21 MPS) were ranked second and third, respectively. The results are in line with Singh *et al.* (2003) who reported less use of advisory letters, tours, agricultural exhibitions and demonstrations by farmers. The reason given by farmers for less use was their dependency on government/semi-government or private institutions responsible for carrying out such activities, which are not a regular feature due to certain limitations.

Impersonal cosmopolite channels had the lowest mean per cent scores in the small (28.38 MPS), marginal (23.15 MPS) and landless (14.04 MPS) categories. This indicates that respondents of landless families were using impersonal cosmopolite channels the least. Illiteracy was a major hindrance in the use of impersonal cosmopolite channels. Chole and Radlukar (1978) in their study found the same trend.

Conclusion

On the basis of the above findings the following conclusion can be drawn:

1. Majority of the respondent had Kachcha Catlle Shed and possessed 1-2 animals.
2. Neighbour, friend and relatives are the important communication sources utilized by majority of respondents under personal localite channels.
3. Landless respondents had the least score as compared of small and marginal farmers so for the utilization pattern of personal cosmopolite sources is concerned.
4. The over all mean present score of the personal cosmopolite channel was highest in case of marginal farmer as compared to landless respondents.
5. Television and Radio, were the most used source of communication among all the there categories of respondents.
6. Farm women by and large had utilized personal localized communication linkage over mass communication and institutional linkage.

References

Bishnoi, I. and Dixit, V., 2003. Sources of health communication of the women of Chandauli (Varanasi) district of Uttar Pradesh. *J. Comm. Studies*, 21(2): 70–76.

Bhople, R.S. and Thakre, S.M., 1994. Farm information sources utilisation, adoption transfer behaviour farmers. *Rural India*, 57: 153–156.

Chand, S., Dangi, K.L. and Bansal, V., 2001. Most credible information sources for disseminating improved technology of mustard among farmers, 19(2): 9–14.

Chauhan, J. and Singh, A.K., 2001. Use of communication sources by tribal dairy farmers: A study. *Indian Res. Ext. Edu.*, 1(1): 8–10.

Chole, R.R. and Radhukar, W.S., 1978. Source credibility pattern among small and big farmers. *Indian J. Ext. Edu.*,14(1&2): 50.

Gogoi, D.K., 1984. A village level analysis of communication network of agricultural information in relation to adoption of improved farm practices in developed and under developed areas of Punjab. *Ph.D. Thesis* (unpublished), PAU, Ludhiana.

Goswami, G., Dhawan, D. and Sareth, L.S., 2003. A study on sources of information used by rural women regarding agriculture and animal husbandry technology. *J. Comm. Studies*, 21(2): 66–69.

Gupta, S.P., Amardeep and Vir, K., 2003. Utilization of information sources by farmers under different production systems in Uttranchal and Uttar Pradesh. *MANAGE Extn. Res. Review*, 4(2): 70–85.

Sangha, G.S. and Kalra, R.K., 1993. Utilization of mass media sources by farmers. *Indian J. Ext. Edu.*, 29(1 &2): 139.

Singh, B., Narwal, R.S. and Malik, J.S., 2003. Communication sources used by extension personnel and farmers. *Indian J. Ext. Edu.*, 39(1 &2): 26–30.

Singh, U. and Mishra, O.P., 2002. Information sources availability and utilization pattern among farm families. *J. of Comm. Studies*, 20(1): 14–25.

Singh, U., Mishra, O.P. and Avinashilingam, N.A., 2003. Interrelationship among demographic characteristics (variables), media availability, media utilization and development message needs. *J. Comm. Studies*, 21(2): 33–40.

Suriyanarayanam, S. and Tamilselvi, 2003. Information sources utilization bahaviour of sericulture entrepreneurs. *J. Comm. Studies*, 21(2): 93–97.

Chapter 23

CCS H.A.U. Farmers Fair: An Evaluation

☆ ***Pardeep Kumar[1], A.K. Godara, Reena, Nasib Singh and S.K. Mehta***

Introduction

All agriculture universities and national research institutes organize farmers' fairs once or twice a year. These fairs have gained importance under Indian conditions as they offer inter-personal and face-to-face communication situation to the visiting farmers. These fairs are generally held when the kharif and rabi crops are ready for harvest within a month or so. The main objective is to bring the latest farm technology to the notice of farmers in a most convincing manner. It is, therefore, important to know how much effective these fairs are in communicating improved farm innovations to the visiting farmers. The present study, therefore, was an attempt to :

1. To know about the personal profile, frequency and purpose of visit, mode of transport used, new information received and items purchased by the visiting farmers.
2. To know the opinion of farmers on the usefulness of different avenues and the kind of inspirations they received on account of visiting farmers' fair.
3. To find out the degree of satisfaction and willingness of attending next fair expressed by farmers.

---

1 Department of Extension Education, CCSHAU, Hisar.

Methodology

The study was conducted in the state of Haryana at the time of a two days farmers' fair during March, 2008 held at Chaudhary Charan Singh Haryana Agricultural University, Hisar. A specially structured interview schedule consisting of mostly multiple choice questions were prepared for data collection. The data were collected at random from 100 farmers, as against about 18,000 visiting farmers after they had visited almost all the avenues, namely, stalls, exhibitions, demonstrations, experimental plots, buzz sessions, publication stalls etc. They were also free to move in the fair at their own convenience, choice and aptitude. The respondents were categorized into four categories, namely, marginal, small, medium and big based on their size of land holdings.

Results and Discussion

Profile of Respondents

District-wise distribution of respondents revealed that 77 per cent of them were from Hisar, Jind, Sirsa and Bhiwani districts. Data presented in Table 23.1 indicated that majority (72 per cent) of the respondents were found to be of young in age followed by middle aged (23 per cent). Further, majority of them were either medium (41 per cent) or big (40 per cent) farmers, educated up to matric (35 per cent), or middle (22 per cent) level. Only 22 per cent of them had no education. Cow/buffalo raring was reported as subsidiary occupation by all the respondents. Only one of them had installed an oil expeller. Forty per cent of had less than five years professional experience followed by the 35 per cent who had five to ten years of experience. Only 25 per cent of them had the experience of more than 10 years.

Table 23.1: Personal Profile of Respondents and Farming Experience

	Personal Profile	*Categories of Farmers*				
		Marginal	*Small*	*Medium*	*Big*	*Total*
(*i*)	Land holding (Acres)	2	17	41	40	100
(*ii*)	Age					
	(*a*) Young (35 yrs)	2	13	37	20	72
	(*b*) Middle (35–50 yrs)	–	4	4	15	23
	(*c*) Old (above 50 yrs)	–	–	–	5	8
	Total	2	17	41	40	100
(*iii*)	Education					
	(*a*) Illiterate	–	3	7	11	21
	(*b*) Read and write	1	–	1	–	2
	(*c*) Primary	–	5	3	5	13
	(*d*) Middle	1	2	10	9	22
	(*e*) Matric	–	7	17	11	35
	(*f*) Graduate	–	–	3	4	7
	Total	2	17	41	40	100

Contd...

Table 23.1–Contd...

	Personal Profile	Categories of Farmers				
		Marginal	*Small*	*Medium*	*Big*	*Total*
(*iv*)	Subsidiary occupation					
(*A*)	Rear Cattle & Buffalo					
	(*a*) Selling milk	–	–	1	2	3
	(*b*) Do not sell milk	2	17	40	38	97
	Total	2	17	41	40	100
(*B*)	Oil expeller	–	–	–	–	–
(*v*)	Farming experience					
	(*a*) <5 years	1	5	10	24	40
	(*b*) 5–10 years	1	6	17	11	35
	(*c*) 10–15 years	–	5	8	3	16
	(*d*) Above 15 years	–	1	6	2	9
	Total	2	17	41	40	100

Source of Information and Mode of Transport Used

Perusal of data presented in Table 23.2 indicate that highest number (57 per cent) of the respondents came to know about the fair from their friends and relatives

Table 23.2: Information Sources and Mode of Transport Used by Farmers for Reaching the Fair

	Particulars	Categories of Farmers				
		Marginal	*Small*	*Medium*	*Big*	*Total*
(i)	Information Sources					
	(*a*) Relatives and friends	1	8	27	21	57
	(*b*) Radio	1	10	17	24	52
	(*c*) VEW/ADO	–	1	6	10	17
	(*d*) Newspaper	–	1	3	6	10
	(*e*) University Extension Staff	–	3	–	3	6
	(*f*) Visit to HAU	–	–	3	1	4
	(*g*) Publicity poster/banner etc.	–	–	–	1	1
(ii)	Mode of Transport					
	(*a*) State transport	1	14	28	22	65
	(*b*) Own vehicle	–	–	2	8	10
	(*c*) Friends/relatives vehicle	1	2	9	4	16
	(*d*) Hired vehicle	–	–	2	6	8
	(*e*) Train	–	1	–	–	1
	Total	2	17	41	40	100

followed by radio (52 per cent), extension workers (17 per cent) and newspapers (10 per cent). Nearly two-third (65 per cent) of the respondents travelled by bus whereas 10 per cent were given lift by their friends and relatives.

Purpose and Frequency of Visit to Farmers' Fair

It is evident from Table 23.3 that most of the respondents visited the fair for the purchase of seeds. More than 50 per cent said that purpose of their visit was to get information about agricultural practices besides the purchase of seeds. However, one-forth of them reported that they had come for entertainment. When the respondents were asked about the frequency of their visit then eight reported that they had come for the first time. About half of the respondents said that they visit the fair every year followed by those (30 per cent) who come to the fair once in two or three years.

Table 23.3: Purpose and Frequency of Visit by Farmers for Visiting Farmer's Fair

	Particulars	*Categories of Farmers*				
		Marginal	*Small*	*Medium*	*Big*	*Total*
(*i*)	Purpose of visit					
	(*a*) To purchase seed	1	15	39	40	95
	(*b*) To get information about agril. practices/ innovations	–	7	23	24	54
	(*c*) Entertainment	1	3	16	5	25
	(*d*) To get soil and water samples tested	–	3	2	–	5
	(*e*) To participate in competitions	–	2	1	–	3
(*ii*)	Frequency of visit					
	(*a*) For the 1st time	–	2	5	1	8
	(*b*) Once in 4–5 years	1	4	6	2	13
	(*c*) Once in 2–3 years	–	2	16	12	30
	(*d*) Every year	1	9	14	25	49
	Total	2	17	41	40	100

Items Purchased and New Information Got from the Fair

All the big and more than 50 per cent of medium farmers had purchased cotton seed whereas one-forth of the total respondents purchased bajra seed (Table 23.4). Seven per cent respondents purchased rhyzobium culture. Of the farm publications only five per cent and three per cent respondents purchased *Haryana Kheti* and package of practices for *kharif* crops, respectively. On further probe 44 per cent reported that they had got new information from the fair for the first time as 30 of them said that they had come to know about the varieties of crops for the first time (Table 23.4).

Table 23.4: Items Purchased and New Information got from Fair by Farmers

	Particulars	Categories of Farmers				
		Marginal	Small	Medium	Big	Total
(i)	Items purchased					
1.	Seeds					
	(a) Cotton	–	14	22	40	76
	(b) Bajra	–	–	–	–	–
	(c) Jowar	–	5	15	7	25
	(d) Fodder	–	–	–	6	6
	(e) Bhindi	–	1	1	1	3
	(f) Chilli	–	–	1	–	1
2.	Rhizobium culture	–	–	2	5	7
3.	Books					
	(a) Haryana Kheti	–	1	2	2	5
	(b) Package of practices (Hindi)	–	–	–	3	3
(ii)	Areas of New Information					
1.	New varieties of crops	–	6	8	16	30
2.	Insect control in cotton	–	1	3	1	5
3.	Silage making	–	–	1	3	4
4.	Sunflower cultivation	–	–	2	1	3
5.	Urea treatment of fodder	–	1	1	–	2

Opinion on Usefulness of Avenues of the Fair

Data pertaining to opinion of respondents on the usefulness of the information communicated to them through different avenues of the fair have been presented in Table 23.5. A majority of the medium and big farmers opined that the material displayed in the exhibition informative stalls was useful. However, one-third and one-fifth of the farmers reported that they had not visited the exhibition stalls put up by private organizations and agricultural institutions including the host university, respectively. Likewise, nearly 50 per cent of the farmers also said that due to lack of time they could not visit the demonstration show/experimental plots and the buzz session in the fair. However, those who had seen (33 per cent) the demonstrations/experimental plots on the farm and had attended the buzz session in the fair found these avenues useful to them. Sharing of experience with fellow farmers was termed as very useful and useful by 18 per cent ad 22 per cent respondents, respectively. Yet, 60 per cent of the respondents had not shared at all their experience with others. Similarly, respondents who had not brought soil/water samples with them were found to be 95 per cent whereas 93 per cent did not visit the animal show in the fair. Here, it is worth mentioned that all the respondents had reported rearing of cattle/buffaloes as their subsidiary occupation.

Table 23.5: Farmers' Opinion about Usefulness of Different Avenues of the Fair

	Avenues of the Fair	*Very Useful*	*Useful*	*Somewhat Useful*	*Did not visit/ Participate*
I.	**Exhibition stalls**				
(a)	*Private organization*				
	Category of farmers				
	Marginal	–	1	–	1
	Small	–	9	3	5
	Medium	3	24	1	13
	Big	3	20	2	15
	Total	6	54	6	34
(b)	*CCS HAU & other sister Institutions*				
	Category of farmers				
	Marginal	–	1	–	1
	Small	–	12	3	2
	Medium	4	26	–	11
	Big	6	29	–	5
	Total	10	8	3	19
II.	**Demonstration show and experimental plots**				
	Category of farmers				
	Marginal	–	–	–	2
	Small	–	4	–	13
	Medium	2	12	13	14
	Big	3	17	–	20
	Total	5	33	13	49
III.	**Scientific information from scientists in buzz session**				
	Category of farmers				
	Marginal	–	–	–	2
	Small	–	2	1	14
	Medium	12	8	–	21
	Big	14	10	–	16
	Total	26	20	1	51
IV.	**Share of experience with fellow farmers**				
	Category of farmers				
	Marginal	–	–	–	2
	Small	3	7	–	7
	Medium	5	12	–	24
	Big	10	3	–	27
	Total	18	22	–	60

Contd...

Table 23.5–Contd...

	Avenues of the Fair	Very Useful	Useful	Somewhat Useful	Did not visit/ Participate
V.	**Soil and water testing service**				
	Category of farmers				
	Marginal	–	–	–	2
	Small	3	–	–	14
	Medium	2	–	–	39
	Big	–	–	–	40
	Total	5	–	–	95
VI.	**Visit to animal show**				
	Category of farmers				
	Marginal	–	–	–	2
	Small	–	–	–	17
	Medium	–	3	–	38
	Big	1	3	–	36
	Total	1	6	–	93

Inspiration from Farmers' Fair

It is apparent from Table 23.6 that 96 per cent of the respondents got inspiration for increasing their crop production by using improved seeds. Others received inspiration on management of insect pests and diseases (52 per cent), use of bio-fertilizers (47 per cent), purchase and use of farm and home implements (3 per cent). It is very surprising to note that the fair did not influence the farmers much on the subsidiary occupations except vegetable growing. It suggests that farmers are not being attracted to the new enterprises like mushroom growing, bee-keeping, fisheries, poultry keeping etc. However, when they were asked whether they were satisfied with their visit a majority of them reported that they were fully satisfied followed by those who were somewhat satisfied with their purpose of visiting the fair. On further probe 70 per cent said that they were willing to attend the next fair. However, the rest 30 per cent were either undecided or not willing.

Specific Suggestions Offered by Respondents

Respondents were also requested to give their specific suggestions for improving the utility of the fair. Most of the farmers (96 per cent) suggested for opening of more seed distribution counters. Two days duration of the fair was felt inadequate the 52 per cent of them suggested the duration of at least three days. More than 50 per cent respondents also suggested that there should be transport facility from fair venue to the bus stand. Discount on the sale of various agro-inputs and farm implements in the fair should be given to the farmers (72 per cent).

Table 23.6: Kind of Inspirations got by Farmers from the Fair

	Kind of Inspirations	*Categories of Respondents*				
		Marginal	*Small*	*Medium*	*Big*	*Total*
(*i*)	To increase crop production by using:					
	(*a*) Improved seeds	2	16	38	40	96
	(*b*) Management of insect–pests and disease	–	8	25	19	52
	Use of biofertilizers					
	(*c*) Recommended doses of fertilizers	–	5	24	18	47
	(*d*) Purchase and use of new farm and implements hove	–	2	1	–	3
(*ii*)	To start subsidiary occupation viz.					
	(*a*) Poultry	–	–	1	–	1
	(*b*) Dairy for selling milk	–	2	–	1	3
	(*c*) Orchard	–	–	–	4	4
	(*d*) Vegetables growing etc.	–	1	2	7	10
	(*e*) Mushroom growing	–	1	–	–	1

Conclusion

Since, majority respondents were from Hisar, Jind, Sirsa and Bhiwani districts, so the findings of the study suggest that efforts should be made by the University through its KGKs and KVKs to persuade the farmers in other districts also. As only few farmers had got information about the fair from extension personnel of the department of agriculture and the university so necessary directions should be given to the field extension staff to make publicity of the fair in the villages. Efforts should be made by the extension staff for getting the visits of farmers from different districts sponsored through different public and private organizations. It was also observed that majority of the respondents could not visit all the avenues of the fair because they had to spend many hours on the purchase of seeds due to rush at the counters. So university should open more counters for the sale of seeds. Further, the university at the outstations should make the seeds available before organizing the fair. The findings of the study has also revealed that some of the avenues like soil and water testing facility, animal show, demonstrations/experimental plots etc. were remained almost untouched by the respondents because of lack of time. Therefore, the very objective of organization of the fair for communicating improved farm practices/ innovations and also to bring to the notice of the visiting farmers all sorts of informations which help in inculcating positive attitudes towards scientific farming and subsidiary enterprises could not be achieved unless and until farmers are able to visit the fair without any tension in their mind of the time of their visit. However, the fair inspired the majority of farmers for increasing their crop production by using various agro-inputs and satisfied with their visit, as the majority of them were willing to attend the next fair.

References

Abdul, Mazid, Tantray and Khasir Mohd., 1996. Technological gap in cultivation of season. *Indian Journal of Extn. Edu.*, 32: 75–76.

Anonymous, 2003. FAO Report.

Anonymous, 2005. *Agricultural Statistics at a Glance.*

Chaparro, A., 1995. Role expectation and adoption of new farm practices. *Ph.D. Thesis*, Pennasylvania State University, University Park.

Das, N.P., Sinha, H.S.P. and Ranjan, Rajiv. *Information Gain in Kisan Mela.*

Desai, B.R., Woman, G.K. and Girase, K.A., 2000. Adoption of recommended technology for rain fed cotton NHH–44: *Maha. Journal of Extn. Edu.*, 19: 48–51.

Deswal, D.S., 1993. A study of extent of adoption of plant protection measures by *Kharif* rice growers in Kheda district of Gujarat state. (Unpublished) *Master's Thesis*, G.A.U., Anand Campus, Anand.

Dewan, P.K., 1987. A study of technological gap in adoption of large cardamom cultivation practices in north district of Sikkim state. *Master's Thesis*, (Unpublished), G.A.U., Anand Campus, Anand.

Gupta, M.P., 1981. Communication effectiveness of farmers' fair (Kisan Mela). *Indian J. Extn. Edu.*, 17 (3 and 4): 92–95.

Khalge, M.L., Jonahale, S.G. and Wakle, P.K., 1998. Knowledge and adoption of recommends package of practices for pearl millet crop by dryland farmers. *Maha. Journal of Extn. Edu.*, 17: 33–36.

Kolte, V.R., 1970. Factors related to knowledge of Package of Practices and their adoption by hybrid jowar growers. *M.Sc. (Ag,) Thesis*, (Unpublished), P.K.V., Akola, Maharastra.

Kushwaha, R.K. and Prakash, A., 1995. Adoption level of plant protection measures in different crops. In: *Souvenir and Abstracts National Symposium on respective in Eco-friendly Approaches to Plant Protection*, 7–8 September, 1995. pp. 29.

Malik, S., 2000. Technological gap in adoption of Sugarcane Production Technology in Meerut district. (Unpublished) *Ph.D. Thesis*, Department of Agriculture Extension Education, Meerut University, Meerut.

Meti, S.K. and Henchinal, S.X., 1995. Adoption of dry land technologies recommended for cotton crop. *Maha. Journal of Extn. Edu.*, 14: 55–58.

Nikhade, D.M., Bhople, R.S. and Kale, N.M., 1995. Adoption gap among small and big cotton growers. *Indian Journal of Extn. Edu.*, 31(1–4): 120–123.

Saxena, K.K., Jain, N.C. and Panday, K.K., 1990. Transfer of rainfed wheat technology and its adoption by the farmers in Malwa region. *Indian Journal of Extn. Edu.*, 26(3 and 4): 70–73.

Chapter 24

Impact of Krishi Mahotsav-2007 in TOT at Village Level in Gujarat State

☆ ***Nikulsinh M. Chauhan***[1]

Introduction

The Government of Gujarat celebrates the 'Krishi Mahotsav Programme'.The main aim was to boost up the agriculture and allied production. Under this programme, various extension activities carried out in all the villages of Gujarat state with the support of all line departments. Looking to the importance of Krishi Mahotsav-2007 the study was carried out to know the effectiveness of Krishi Mahotsav.

Objectives

The overall objective of the research study was to know the effectiveness of Krishi Mahotsav-2007. The specific objectives of study were:

1. To know the socio-economic characteristics of farmers.
2. To know the source of information about Krishi Mahotsav.
3. To know the overall perception of Krishi Mahotsav.
4. To know the opinion of farmers about time and duration of Krishi Mahotsav.
5. To know the facilities to be provided to farmers for next Krishi Mahotsav.

---

1 KVK, Mangalbharti – 391 125, Ta-Sankheda, Dist. Vadodara, Gujarat.

6. To know the opinion of farmers regarding information of Animal Husbandry given during Krishi Mahotsav.
7. To know the opinion of farmers regarding information of major crops given during Krishi Mahotsav.
8. To know the new knowledge gained 1st time by farmers during Krishi Mahotsav.
9. To find out the suggestions to overcome the problems faced by farmers.

Methodology

Krishi Vigyan Kendra, Mangal Bharti in Vadodara district under middle Gujarat, undertook the study. Total 10 villages were randomly selected in Sankheda taluka of Vadodara district namely Hareshwar, Kasumbiya, Manjarol, Orwada, Bhuriyakuwa, Aritha, Aambapura, Sundarpura, Kathmandava and Ratanpur. From each village, 10 respondents were randomly selected for the study. Thus, the total sample size of the respondents became 100. The data were collected through personal interview schedule and analyzed with simple statistics.

Results and Discussion

Data presented in Table 24.1 clearly indicate that majority of the respondents are belong to middle aged group, educated up to primary and secondary school, nuclear type of the family, big size of the family, possessed membership in one organization, 1.00 to 2.00 ha of land holding, possessed 3-4 milch animals and obtained annual income of Rs.20001-40000, respectively. The results paying attention on the facts that the awareness regarding family planning was poor in the region as well as agriculture along with animal husbandry was the main occupation with them for livelihood. The same was also reported by Chauhan *et al.* (2004).

Table 24.1: Socio-economic Characteristics of Farmers (n=100)

Sl.No.	*Socio-economic Characteristics*	*No.*	*Per cent*
A	Age		
	i) Young (18 to 35 years)	13	13
	ii) Middle (36 to 50 years)	61	61
	iii) Old (above 50 years)	26	26
B	Education		
	i) Illiterate	06	06
	ii) Primary (1 to 7 std.)	47	47
	iii) Secondary (8 to 10 std.)	32	32
	iv) Higher Secondary (11 to 12 std.)	09	09
	v) Graduate	06	06
C	Type of family		
	i) Joint	41	41
	ii) Nuclear	59	59

Contd...

Table 24.1–Contd...

Sl.No.	Socio-economic Characteristics	No.	Per cent
D	Size of family		
	i) Small (up to 2 members)	08	08
	ii) Medium (3 to 4 members)	33	33
	iii) Big (above 4 members)	59	59
E	Membership in organizations		
	i) No membership	15	15
	ii) Membership in one organization	62	62
	iii) Membership in more than one organization	20	20
	iv) Office bearer	03	03
F	Size of land holding (irrigated)		
	i) Land less	12	12
	ii) Marginal farmers (below 1.0 ha)	46	46
	iii) Small farmers (1.01 to 2.0 ha)	34	34
	iv) Medium farmers (2.01 to 4.0 ha)	06	06
	v) Big farmers (above 4.0 ha)	02	02
G	No. of milch animals		
	i) up to 2	16	16
	ii) 3 to 4	53	53
	iii) above 4	31	31
H	Annual income		
	i) up to Rs.10,000/-	20	20
	ii) Rs.10,001/- to Rs.20,000/-	19	19
	iii) Rs.20,001/- to Rs.30,000/-	29	29
	iv) Rs.30,001/- to Rs.40,000/-	20	20
	v) above Rs.40,000/-	12	12

Table 24.2: Source of Information About Krishi Mahotsav (n = 100)

Sl.No.	Source of Information	No.	Per cent
1.	Neighbour	67	67
2.	VLWs	41	41
3.	Radio	07	07
4.	Television	05	05
5.	Leaflet	12	12
6.	Poster	06	06
7.	Newspaper	16	16
8.	Krushigovidhya	24	24
9.	Krushijivan	19	19
10.	Other magazine	08	08
11.	KVK	11	11

Data in Table 24.2 revealed that the main source of information is neighbors followed by VLWs, krushigovidya, krishi jivan magazines and so on accordingly.

The overall perception of the krishirath was found medium to highly useful (84.00 per cent) as per the perception of the respondents (Table 24.3).

Table 24.3: Overall Perception of Krishi Mahotsav (n=100)

Sl.No.	*Overall Perception*	*No.*	*Per cent*
1.	Highly useful	31	31
2.	Medium useful	53	53
3.	Less useful	16	16
4.	No useful	00	00

About cent percent of the respondents opinioned that the time of the Krishi Mahotsav-2007 was timely and the same should be repeated in the next krishirath also, Table 24.4. The same fact was also reported by Patel (2008).

Table 24.4: Opinion of Farmers About Time of Krishi Mahotsav (n=100)

Sl.No.	*Opinion*	*No.*	*Per cent*
A	Opinion about time of Krishi Mahotsav-2007		
1.	Timely	97	97
2.	Earlier than expected	00	00
3.	Later than expected	03	03
B	Opinion about most appropriate time for next Krishi Mahotsav		
1.	Timely	97	97
2.	Earlier than expected	03	03
3.	Later than expected	00	00

Table 24.5: Opinion of Farmers About Duration of Krishi Mahotsav (n=100)

Sl.No.	*Opinion*	*No.*	*Per cent*
A	Opinion about duration of Krishi Mahotsav-2007		
1.	Reasonable	91	91
2.	Too long	00	00
3.	Too short	09	09
B.	Opinion about appropriate duration for next Krishi Mahotsav		
1.	30 days	91	91
2.	45 days	05	05
3.	60 days	04	04

Great majority the respondents opinioned that the duration of the krishirath was reasonable *i.e.*, 30 days and same will be repeated in the next year, too. Table 24.5 clearly indicates the same.

The data in the Table 24. 6 clearly indicate that the facilities expected by the farmers at krishirath are demonstration and distribution of the seeds, it may be due to the principle of the extension education seeing is believing and learning by doing as well as availability of the good quality seeds by the authority has been understood by the farmers.

Table 24.6: Facilities to be Provided to Farmers for Next Krishi Mahotsav (n=100)

Sl.No.	Facilities	No.	Per cent
1.	Demonstration	16	16
2.	Distribution of inputs		
	(a) Seeds	15	15
	(b) Fertilizers	09	09
	(c) Pesticides	19	19
	(d) Plant protection equipments	11	11

Table 24.7: Opinion of Farmers Regarding Information of Animal Husbandry given During Krishi Mahotsav (n=100)

Sl.No.	Item	Not Given (%)	1st Time Known Correct Information (%)	Usefulness (%)	
				Useful (%)	Most Useful
1.	Animal breeds	32	68	60	40
2.	Primary treatment of animal diseases	88	12	83	17
3.	How to increase milk productions in milch animals	10	90	00	100
4.	Need of milk of calf	48	52	100	00
5.	What to do if the respiration system is not working after birth of a calf	100	00	00	00
6.	When to cut naval cord after the birth of a calf	100	00	00	00
7.	Treatment to be given after cutting the naval cord	100	00	00	00
8.	Time of colostrums to be given to a calf after birth	08	92	87	13
9.	Reason of giving colostrums	43	57	89	11
10.	What to do if your buffalo/cow feels difficulty in calving	100	00	00	00
11.	Need of salt to be given to a new born calf	46	54	94	06

Contd...

Table 24.7–Contd...

Sl.No.	Item	Not Given (%)	1st Time Known Correct Information (%)	Usefulness (%)	
				Useful (%)	Most Useful
12.	Important fodder crop	35	65	68	32
13.	Age of a calf for consuming fodder	85	15	100	00
14.	The dehorning to a calf	100	00	00	00
15.	Balanced feed	12	88	00	100
16.	Concentrated CP to be given to milch animal	43	57	84	16
17.	Everyday need of green fodder for a milch animal	25	75	23	77
18.	How often fodder be given to a milch animal during a day?	25	75	20	80
19.	Need of dry fodder to a milch animal during a day	30	70	21	79
20.	Method of giving dry fodder	68	32	94	06
21.	Importance of chopped green + any fodder in mixed form	20	80	75	25
23.	Need of concentrate to a advanced pregnant animal (after seven months pregnancy)	100	00	00	00
24.	Need of mineral mixture to milking animal everyday	10	90	00	100
25.	Need of mineral mixture to pregnant animal everyday	100	00	00	00
26.	Need of water to be given to a milch animal everyday	10	90	00	100
27.	Which agricultural produce is convenient/suitable for bed in the shed?	100	00	00	00
28.	Importance to clean the udder before milking	23	77	09	91
29.	Contagious diseases of animals	15	85	94	6
30.	Vaccination schedule in buffalo/cow	30	70	14	86
31.	Need to vaccinate	28	72	00	100
32.	Age of calf to give foot and mouth vaccine	28	72	17	83
33.	Name of medicine for dysentery disease for a calf	33	67	91	09
34.	Name of month of the vaccine for foot and mouth disease	28	72	74	26
35.	Month the vaccine for HS disease	30	70	79	21
36.	Month of the vaccine for BQ	100	00	00	00

Contd...

Table 24.7–Contd...

Sl.No.	Item	Not Given (%)	1st Time Known Correct Information (%)	Usefulness (%)	
				Useful (%)	Most Useful
37.	Frequency to give vaccine for BQ disease	100	00	00	00
38.	Frequency to give vaccine for Foot and Mouth and HS disease	100	00	00	00
39.	Symptoms of a buffalo/cow being in heat/estrus	55	45	100	00
40.	Time for conceiving of buffalo/ cow after heat	100	00	00	00
41.	When should the buffalo/cow be served after calving ?	58	42	95	05
42.	Artificial insemination	48	52	52	48
43.	Advantages of artificial insemination?	48	52	42	58
44.	What to do if buffalo/cow is not conceived by more than 3 inseminations	100	00	00	00
45.	No. of days buffalo/cow normally repeats heat cycle	100	00	00	00
46.	Average period of gestation in buffalo/cow	100	00	00	00
47.	Optimum dry period in a buffalo/cow	100	00	00	00
48.	Period of a milch animal to give milk after parturition	100	00	00	00
49.	When to examine for pregnancy diagnosis after service	57	43	65	35
50.	Age of buffalo/cow/crossbred heifer for breeding	100	00	00	00
51.	Milking process	50	50	100	00
52.	Pace for milking a buffalo/cow?	45	55	100	00
53.	Frequency to milk in a day for 10 to 15 litres of milk producing animal	49	51	100	00
54.	Perfect time of milking	30	70	29	71
55.	Best method of milking	25	75	00	100
56.	Ideal space for a milking cow/buffalo?	100	00	00	00
57.	Ideal roof material for animal shed	100	00	00	00
58.	Light and air inside the shed	100	00	00	00
59.	Ideal floor material for animal shed	100	00	00	00
60.	Sources of contamination to milk	100	00	00	00
61.	Infection in human of TB/Anthrax/ Brucellosis/JD from animals	100	00	00	00
62.	Concept of comfortable housing	100	00	00	00
63.	Age of heifer to be calved	100	00	00	00

The majority of the farmers gained the new knowledge 1st time during Krishi Mahotsav related with Animal Breeds, How to increase milk productions in milch animals, Time of colostrums to be given to a calf after birth as well as so many other animal husbandry practices (Table 24.7). They also first time got information regarding agriculture and AAU activities, KVK activities and Kisan Credit Card in relation to agricultural enterprises (Table 24.8). The results are in the same line of Chauhan and Thakor (2005).

Table 24.8: Opinion of Farmers Regarding Cotton Crop Information (n=100)

Sl.No.	*Item*	*Not Given (%)*	*1st Time Known Correct Information (%)*	*Usefulness (%)*	
				Useful (%)	*Most Useful*
Agricultural information					
(A)	**Nursery management**				
1.	Sources of seed	09	91	34	66
2.	Suitable high yielding variety for the area	11	89	13	87
3.	Rate of seeds	100	00	00	00
4.	Land preparation	63	37	100	00
5.	Soil treatment methods	88	12	100	00
6.	Place of availability of soil treatment inputs	91	09	100	00
7.	Seed rate	19	81	79	21
8.	Price of soil treatment inputs	100	00	00	00
(B)	**Post nursery phase**				
9.	Sowing time	40	60	80	20
10.	Depth of sowing	38	62	83	17
11.	Method of sowing	38	62	50	50
12.	Spacing	12	88	78	22
13.	Seed treatment inputs	19	81	74	26
14.	Gap filling	22	78	79	21
15.	Price of fertilizers	100	00	00	00
16.	Place of availability of fertilizers	100	00	00	00
17.	Name of advantageous chemical fertilizers	45	55	100	00
18.	Method and time of fertilizer application	28	72	20	80
19.	Nutrient requirements of crop	31	69	72	28
20.	Calculating the doze of chemical fertilizers	73	27	85	15
21.	Deficiency symptoms of major plant nutrients	87	13	100	00
22.	Bio-fertilizers	10	90	10	90

Contd...

Table 24.8–Contd...

Sl.No.	Item	Not Given (%)	1st Time Known Correct Information (%)	Usefulness (%)	
				Useful (%)	Most Useful
23.	Making organic matter from farm waste	79	21	76	24
24.	Organic manures	14	86	93	07
25.	Chemical weed control	50	50	58	42
26.	Price of weedicides	100	00	00	00
27.	Place of availability of weedicides	55	45	88	12
28.	Trade name of weedicides	60	40	100	00
29.	Hand weeding	35	65	93	07
30.	Schedule for irrigation	30	70	64	36
31.	Critical stages of irrigation	32	68	83	17
32.	How to save crop during shortage of water	52	48	68	32
33.	Fertilizers management during irrigation	46	54	81	19
34.	Method of irrigation	40	60	63	37
35.	Insect management	25	75	25	75
36.	Diseases management	25	75	20	80
37.	Price of insecticides and pesticides	100	00	00	00
38.	Integrated pest management	29	71	16	84
39.	Biological control of pests	77	23	56	44
40.	Useful insects	70	30	26	74
41.	Method of preparing solution of insecticides/pesticides	25	75	33	67
42.	Trade name of insecticides/pesticides	59	41	100	00
43.	Place of availability of insecticides and pesticides	50	50	40	60
44.	Proper time of harvest	27	73	23	77
45.	How to store production	90	10	100	00
46.	Care after harvesting at farm level	91	09	100	00
47.	Care during harvesting	82	18	100	00
(C)	**Marketing of the products**				
48.	Market price	100	00	00	00
49.	Quality parameters that affects price	100	00	00	00
50.	Time of market inflow	100	00	00	00
51.	Place of marketing	57	43	51	49
52.	Marketing procedure	89	11	100	00
53.	Facilities available at market	90	10	100	00

Contd...

Table 24.8–Contd...

Sl.No.	Item	Not Given (%)	1st Time Known Correct Information (%)	Usefulness (%)	
				Useful (%)	Most Useful
54.	Value addition	60	40	100	00
55.	Export marketing	93	07	00	100
(D)	**Related information**				
56.	Weather forecast	55	45	100	00
57.	Crop related government policies	100	00	00	00
58.	Credit/loan facilities for crop cultivation	100	0	00	00
59.	Insurance of crop	65	35	42	58
60.	Subsidies for crop cultivation	100	00	00	00

As per the opinion of the farmers they got much useful information regarding cotton cultivation during krishirath in the Vadodara district (Table 24.8). It is well understood fact that the cotton is the main crop of the region and farmers are more enthusiastic to get the information regarding cotton cultivation. The same was also observed in case of the pigeon pea cultivation (Table 24.9). The main problem of the farmers regarding both of the crops are related to the market and expected prices of the same. The main problems of the respondents regarding maize cultivation is also related to the pricing policies, as the maize crop cultivation is increasing day by day in the Narmada command areas of the district (Table 24.10). Chauhan and Thakor (2005) and Chauhan and Chauhan (2006), has also reported the same results. The world is shifting very rapidly.The technologies of today becomes outdated for tomorrow and agriculture is not an exception of the same. Farmers are anxious and become more desirous regarding the world scenario of the agriculture for acquainting with continuous changing world. This may be the probable reason for this opinion of the farmers. Chauhan (2007) has also reported the same output.

Table 24.9: Opinion of Farmers Regarding Pigeon Pea Crop Information (n=100)

Sl.No.	Item	Not Given (%)	1st Time Known Correct Information (%)	Usefulness (%)	
				Useful (%)	Most Useful
Agricultural information					
(A)	**Nursery management**				
1.	Sources of seed	13	87	24	76
2.	Suitable high yielding varieties for the area	13	87	10	90
3.	Rate of seeds	100	00	00	00
4.	Seed rate	32	68	76	24

Contd...

Table 24.9–Contd...

Sl.No.	Item	Not Given (%)	1st Time Known Correct Information (%)	Usefulness (%)	
				Useful (%)	Most Useful
(B)	**Post nursery phase**				
5.	Sowing time	39	61	83	17
6.	Depth of sowing	100	00	00	00
7.	Method of sowing	39	61	77	23
8.	Spacing	18	82	89	11
9.	Seed treatment inputs	37	63	68	32
10.	Gap filling	42	58	94	06
11.	Price of fertilizers	100	00	00	00
12.	Place of availability of fertilizers	83	17	100	00
13.	Name of advantageous chemical fertilizers	68	32	100	00
14.	Method and time of fertilizer application	25	75	70	30
15.	Nutrient requirements of crop	29	71	92	08
16.	Calculating the doze of chemical fertilizer	100	00	00	00
17.	Deficiency symptoms of major plant nutrients	93	07	100	00
18.	Bio-fertilizers	12	88	17	83
19.	Making organic matter from farm waste	100	00	00	00
20.	Organic manures	18	82	84	16
21.	Chemical weed control	60	40	100	00
22.	Price of weedicides	100	00	00	00
23.	Place of availability of weedicides	60	40	68	32
24.	Trade name of weedicides	69	31	100	00
25.	Hand weeding	35	65	87	13
26.	Schedule for irrigation	37	63	100	00
27.	Critical stages of irrigation	45	55	100	00
28.	How to save crop during shortage of water	100	00	00	00
29.	Fertilizer management during irrigation	54	46	100	00
30.	Method of irrigation	58	42	80	20
31.	Insect management	27	73	67	33
32.	Diseases management	25	75	64	36
33.	Price of insecticides and pesticides	100	00	00	00
34.	Integrated pest management	33	67	47	53

Contd...

Table 24.9–Contd...

Sl.No.	Item	Not Given (%)	1st Time Known Correct Information (%)	Usefulness (%)	
				Useful (%)	Most Useful
35.	Biological control of pests	79	21	100	00
36.	Useful insects	84	16	100	00
37.	Method of preparing solution of insecticides/pesticides	41	59	81	19
38.	Trade name of insecticides/pesticides	60	40	87	13
39.	Place of availability of insecticides and pesticides	63	37	78	22
40.	Proper time of harvest	47	53	100	00
41.	How to store production	78	22	100	00
42.	Care after harvesting at farm level	83	17	100	00
43.	Care during harvesting	83	17	100	00
(C)	**Marketing of the products**				
44.	Market price	100	00	00	00
45.	Quality parameters that affects price	100	00	00	00
46.	Time of market inflow	100	00	00	00
47.	Place of marketing	86	14	100	00
48.	Marketing procedure	100	00	00	00
49.	Facilities available at market	100	00	00	00
50.	Value addition	87	13	100	00
51.	Export marketing	100	00	00	00
(D)	**Related information**				
52.	Weather forecast	73	27	100	00
53.	Crop related government policies	100	00	00	00
54.	Credit/loan facilities for crop cultivation	100	00	00	00
55.	Insurance of crop	80	20	70	30
56.	Subsidies for crop cultivation	100	00	00	00

The majority of the farmers gained the new knowledge 1st time during Krishi Mahotsav were AAU activities, KVK activities and Kisan Credit Card. The probable reason for that may be the district contains 12 blocks and it is quiet difficult for all line departments for reaching the unreached all the time (Table 24.11). The ratio of the VLW: FARMERS is 1000 and in some pockets it is more than that, too.

Table 24.10: Opinion of Farmers Regarding Maize Crop Information (n=100)

Sl.No.	Item	Not Given (%)	1st Time Known Correct Information (%)	Usefulness (%)	
				Useful (%)	Most Useful
Agricultural information					
(A)	**Nursery management**				
1.	Sources of seed	11	89	13	87
2.	Suitable high yielding varieties for the area	15	85	07	93
3.	Rate of seeds	100	00	00	00
4.	Land preparation	69	31	100	00
5.	Soil treatment methods	100	00	00	00
6.	Place of availability of soil treatment inputs	100	00	00	00
7.	Seed rate	28	72	82	18
8.	Price of soil treatment inputs	100	00	00	00
(B)	**Post nursery phase**				
9.	Sowing time	41	59	71	29
10.	Depth of sowing	39	61	91	09
11.	Method of sowing	39	61	67	33
12.	Spacing	18	82	69	31
13.	Seed treatment inputs	25	75	38	62
14.	Gap filling	30	70	74	26
15.	Price of fertilizers	100	00	00	00
16.	Place of availability of fertilizers	78	22	81	19
17.	Name of advantageous chemical fertilizers	65	35	100	00
18.	Method and time of fertilizer application	21	79	46	54
19.	Nutrient requirements of crop	27	73	84	16
20.	Calculating the doze of chemical fertilizer	100	00	00	00
21.	Deficiency symptoms of major plant nutrients	89	11	100	00
22.	Bio-fertilizers	12	88	19	81
23.	Making organic matter from farm waste	82	18	88	12
24.	Organic manures	16	84	84	16
25.	Chemical weed control	59	41	78	22
26.	Price of weedicides	100	00	00	00
27.	Place of availability of weedicides	60	40	75	25
28.	Trade name of weedicides	68	32	100	00
29.	Hand weeding	35	65	84	16

Contd...

Table 24.10–Contd...

Sl.No.	Item	Not Given (%)	1st Time Known Correct Information (%)	Usefulness (%)	
				Useful (%)	Most Useful
30.	Schedule for irrigation	34	66	65	35
31.	Critical stages of irrigation	41	59	71	29
32.	How to save crop during shortage of water	63	37	67	33
33.	Fertilizers management during irrigation	48	52	92	08
34.	Method of irrigation	61	39	100	00
35.	Insect management	30	70	37	63
36.	Diseases management	28	72	33	67
37.	Price of insecticides and pesticides	100	00	00	00
38.	Integrated pest management	32	68	72	28
39.	Biological control of pests	82	18	100	00
40.	Useful insects	86	14	100	00
41.	Method of preparing solution of insecticides/pesticides	39	61	90	10
42.	Trade name of insecticides/pesticides	62	38	100	00
43.	Place of availability of insecticides and pesticides	61	39	92	08
44.	Proper time of harvest	30	70	80	20
45.	Ideal thrasher for thrashings	48	52	100	00
46.	How to store production	74	26	84	16
47.	Care after harvesting at farm level	100	00	00	00
48.	Care during harvesting	81	19	100	00
(C)	**Marketing of the products**				
49.	Market price	100	00	00	00
50.	Quality parameters that affects price	86	14	57	43
51.	Time of market inflow	100	00	00	00
52.	Place of marketing	89	11	18	82
53.	Marketing procedure	100	00	00	00
54.	Facilities available at market	100	00	00	00
55.	Value addition	18	82	54	46
56.	Export marketing	100	00	00	00
(D)	**Related information**				
57.	Weather forecast	67	33	100	00
58.	Crop related government policies	100	00	00	00
59.	Credit/loan facilities for crop cultivation	100	00	00	00
60.	Insurance of crop	70	30	56	44
61.	Subsidies for crop cultivation	100	00	00	00

Table 24.11: New Knowledge Gained 1st Time by Farmers During Krishi Mahotsav (n=100)

Sl.No.	*Item*	*Not Given (%)*	*1st Time Known Correct Information (%)*	*Usefulness (%)*	
				Useful (%)	*Most Useful*
1.	Research stations of crop	64	36	81	19
2.	Soil Health Card	79	21	24	76
3.	AAU activities	17	83	82	18
4.	KVK activities	43	57	63	37
5.	Possibilities of introducing new crops in the area	100	–	–	–
6.	WTO perspectives	100	–	–	–
7.	Kisan Credit Card	51	49	43	57
8.	Drip irrigation	56	44	100	–
9.	Sprinkler irrigation	100	–	–	–
10.	Low cost/high cost implements	80	20	80	20
11.	Home management	83	17	100	–

Table 24.12: Suggestions Received from the Farmers to Overcome the Problems (n=100)

Sl.No.	*Problems*	*Suggestions*	*No.*	*Per cent*
1.	Inadequate visit of VLWs	✰ Regular visit of VLWs at village level.	26	26
2.	Distribution of kits	✰ Kits should be distributed among all farmers	40	40
		✰ Kits should be given as per requirement of farmers	26	26
		✰ Good quality of kit should be given	12	12
3.	Lack of awareness regarding information	✰ Information about Krishi Mahotsav should be given in advance by VLWs	32	32
	about Krishi Mahotsav in advance	✰ Information about Krishi Mahotsav should be given in advance by Sarpanch/Talati etc.	10	10
		✰ Information about Krishi Mahotsav should be given in advance by newspaper	13	13
4.	Lack of time for discussion	✰ Sufficient time should be given for discussion	08	08
		✰ Minimum 3 hrs should be spared for discussion in each village	04	04
5.	Lack of proper guidance	✰ Proper guidance should be given	29	29
6.	Lack of practical knowledge	✰ Demonstration should be arranged for practical knowledge	16	16
7.	Lack of electricity power for agriculture	✰ Electricity power should be supplied at proper time	11	11
8.	Low market price of agro products	✰ Proper price of agro products should be provided	19	19

The main problems faced by the farmers were distribution of kits, Inadequate visit of VLWs, Lack of awareness regarding information about Krishi Mahotsav in advance and Lack of proper guidance were the most important problems faced by the farmers. The most important suggestions given by farmers to overcome the problems were kits should be distributed among all farmers as per requirement, regular visit of VLWs at village level, information about Krishi Mahotsav should be given in advance by VLWs and proper guidance regarding the same should be given well in advance for its effective and efficient utilization by the farmers (Table 24.12).

Conclusion

1. Majority of the farmers had middle age educated up to primary and secondary, belongs to small and marginal group, lived in nuclear type of family, had big family size, possessed 3 to 4 milch animals, bearing membership in one organization and income up to Rs. 30,000/- per anum.
2. Neighbor, VLWs and magazines were found the important sources of information about Krishi Mahotsav.
3. The overall perception of farmers regarding Krishi Mahotsav was medium to highly useful for the further improvement in their agriculture.
4. The great majority of the farmers opined that time and duration of Krishi Mahotsav was appropriate in effective transfer of technologies in agriculture at village level.
5. The majority of the farmers had opined that the information given in Krishi Mahotsav about Animal Husbandry and major crops of the region were useful.
6. The majority of the farmers gained the new knowledge 1st time during Krishi Mahotsav related with AAU activities, KVK activities and Kisan Credit Card.
7. Distribution of kits, Inadequate visit of VLWs, Lack of awareness regarding information about Krishi Mahotsav in advance and Lack of proper guidance were the most important problems faced by the farmers. It is the real feedback from the farmers for further improvement of the same in future.
8. The most important suggestions given by farmers to overcome the problems were kits should be distributed among all farmers as per requirement, regular visit of VLWs at village level, information about Krishi Mahotsav should be given in advance by VLWs and proper guidance should be given. It should be taken in to consideration for a grand success of the programme in Gujarat itself as well as in other states in the golden era of ICTs in agriculture for next phase of Green revolution on sustainable basis.

References

Chauhan, N.M. and Thakor, R.F., 2005. Information Seeking Behaviour of Rice Growers. *Gujarat Journal of Extension Education (GJEE)*, (15 December 2004): 25–28.

Chauhan, N.M. and Chauhan, N.B., 2006. Farmer's opinion about use of internet technology in agriculture at village level. In: *Paper presented in National Seminar on Information and Communication Technology: Opportunities and Challenges for Revitalizing Extension System,* December 27-29, at Navsari Agricultural University, Navsari, Gujarat held by Indian Society of Extension Education.

Chauhan, N.M., 2007. Information needs of the rice growers for sustainable agricultural development. In: *Paper presented at Seminar on Extension Strategies for Agricultural Development*, Organized by Society of Extension Education, Gujarat, at NAU, Navsari on 25th February.

Patel, R.B., 2008. Gujarat Krishi Mahotsav: A demand driven Extension approach for poverty eradication of Gujarat State. In: *Paper Presented in International Seminar on Strategies for Improving Livelihood Security of Rural Poor* at Goa during 24-27 September, pp. 123.

Chapter 25

Quantification of Indigenous Knowledge on Dairying

☆ *A. Saha*[1], *Swapan B. Mukhopadhya*[2] *and D. Das Gupta*[3]

Introduction

Indigenous knowledge is an immensely valuable resource that provides humankind with insights on how communities have interacted with their changing environment. Indigenous knowledge gained global recognition through the United Nation Conference on Environment and Development (UNCED) in 1992, and documents such as the World Conservation Strategy of International Union for the Conservation of Nature and Natural Resources (IUCNR, 1980) and Brundtland Commission's Our Common Future, World Commission on Environment. Indigenous Knowledge (IK) is local knowledge–knowledge that is unique to a given culture of society. IK contrasts with the international institutions and private firms. It is logical that we first ground ourselves in the knowledge that we already have. But in the mad race for modernization, we seem to have mortgaged our traditional wisdom. The technologies developed through integration of indigenous knowledge will (*i*) provide diversified technological options, which enable farmers to choose from using their own decision-making systems; (*ii*) originate from the farmers' own knowledge and

---

1 Lecturer.

2 Professor & Head.

3 Professor, Department of Agricultural Extension, Agricultural Economics & Agricultural Statistics, Pally Sikha Bhavana (Institute of Agriculture), Visva Bharati, Sriniketan – 731 236.

(*iii*) use diversified sources in active participation of research–minded farmers (Rajshekaran, 1993). Indigenous knowledge (IK) refers to the unique, traditional, local knowledge existing within and developed around the specific conditions of women and men indigenous to a particular geographic area (Grenier, 1998). Some of the synonyms to the term are "traditional knowledge", "folk knowledge", "local knowledge" and "wisdom of the elders". These synonyms refer to the local origin and promotion by a community (Seeland, 2000). Indigenous practices were evolved through constant observation of animals, understanding of day-to-day behavioral pattern; trial and error and refining the techniques based on experiences made during application. These experiences have been handed over from one generation to the other and many of the practices are time-tested, environment friendly, cost-effective, readily available involving minimum risk and location specific. The use of IK has an ethical dimension by encouraging transparency and accountability. Validation of IK is a logical step to qualify and quantify the effectiveness of the practices. Suitable modifications of the local practices, through research and development will help to develop appropriate and acceptable technologies that are more suited to our farming situations. Considering this, an attempt was being made to validate IK in livestock health care system through farmers' assessment.

Methodology

The study were conducted in Birbhum district of West Bengal. Snow-ball sampling technique was applied to identify the key informants. Participant observation and unstructured interview methods were adopted to document indigenous knowledge. Documentation is the conversion of traditional/indigenous knowledge information processed by communities into written documents, drawings or audio recordings. The main aim of such documentation is to ensure that information is not lost and also to protect communities by showing that such information is prior art. Participant observation is an omnibus field strategy in that it simultaneously combines document analysis, interviewing of respondents and informants, direct participation and observation and introspection. The purpose of unstructured interactions is to elucidate relevant information pertaining to indigenous practices that are documented during participant observation (Parasar, B, 1994).

A total of 7 IK commonly used in three different health care practices, *viz.* FMD, Wound Healing and Diarrhoea were selected based on the extent of their use throughout the region under study for assessment/validation by QuIK(Quantification of Indigenous Knowledge) method. QuIK provides a systematic method of assessing this knowledge and may be especially useful in diverse environments with multiple niches since these have their own unique research needs, which formal research cannot possibly access (De Villiers, 1996; Arya, 2003; Padaria *et al.*, 2003; De. *et al*, 2004). Prior experience in using PRA (Participatory Rural Appraisal) matrix ranking methods combined with the use of basic statistics led to the idea that the two could be linked, and hence to the development of QuIK. QuIK develops the PRA tool matrix ranking/rating to assess farmers' knowledge. Parameters for the matrix were derived from interactive farmer, researcher discussion. Five QuIk matrices were developed for five different health care practices. The IK were validated by the farmers in terms of extent of cure, cost effectiveness, ease in use, recovery time and resource availability. Five matrices of decision parameters with respect to use of practices for control of these diseases in dairy animals were administered to 15 key informants individually. Key informants or experienced farmers were identified through informal inquiry in the village. Ranking for the relative performance of the practices for each parameter was done with scoring pattern of 0-10 with maximum score of 10 for each cell in the matrix. Mean scores of farmers' response for each IK against the identified parameter were tabulated. There after, based on the average score calculated for each of the IK, ranking was done by giving the first rank to highest average score, then second rank to the next higher average score and so on and so forth. Kruskall–Wallis or H test was used to determine whether the IK and modern veterinary drugs selected for each matrix were equally effective.

Findings and Discussion

Seven common IK were selected for validation through QuIK method.

IK considered for *QuIK matrix–1* were

IK_1 = Treating FMD with *neem* leaves- *Neem* leaves were boiled in water, after cooling the green water was used to wash the infected hooves.

IK_2= Walking of animal into mud to control FMD- Animals were allowed to walk into mud, so that a thick layer of mud was formed and covered the wounds of feet.

IK_3= Treating animals with *tarpin* oil and camphor for FMD- After washing the wounds in feet or mouth with luke warm water, drying of infected portion was done, then mixture of tarpin oil (12 ml) and camphor (3 g) was applied to the infected feet.

Table 25.1: QuIK Matrix–1 for Assessing Performance of IK to Control FMD Against Different Parameters
Max. marks =10; No. of respondent, N = 15

Sl.No.	*Parameters*	*Practices*			
		IK_1	*IK_2*	*IK_3*	*MVD*
1.	Extent of cure	1.67	1.33	3.0	4.0
2.	Cost of effectiveness	2.00	4.67	2.33	1.0
3.	Ease in use	1.33	5.33	1.50	2.0
4.	Recovery time	1.67	1.33	1.67	4.67
5.	Resource availability	2.67	3.67	2.33	1.33
	Average score	1.87	3.27	2.17	2.60
	Ranks	IV	I	III	II

IK_1: Treating FMD with *neem* leaves; IK_2: Walking of animal into mud to control FMD; IK_3: Treating animals with *tarpin* oil and camphor for FMD; MVD: Modern veterinary drug.

Table 25.2: The Results of Kruskall–Wallis or H Test for Assessing Performance of IK to Control FMD

H–Value		*Conclusion*
Calculated Value	*Table Value*	
1.097	χ^2 0.5 with 3df = 7.815	Practices were equally effective

IK considered for *QuIK matrix–2* were

IK_1 = Application of warm coconut oil containing garlic extract to open wound- Coconut oil was heated for few times with garlic, them luke warm oil was applied in open wound.

IK_2= Treating open wound with lime, garlic and turmeric- Paste of garlic and turmeric was prepared and lime was added to it, then this mixture was applied in open wound after that it was covered with *akand* leaves.

Table 25.3: QuIk Matrix–2 for Assessing Performance of IK to Treat Open Wound Against Different Parameters Max. marks =10; No. of respondent, N = 15

Sl.No.	*Parameters*	*Practices*		
		IK_1	*IK_2*	*MVD*
1.	Extent of cure	1.53	3.20	5.93
2.	Cost of effectiveness	4.53	3.47	2.00
3.	Ease in use	3.47	3.53	3.00
4.	Recovery time	1.53	3.53	4.93
5.	Resource availability	3.93	4.53	1.53
	Average score	2.99	3.65	3.478
	Ranks	III	I	II

IK_1: Application of warm coconut oil containing garlic extract to open wound; IK_2: Treating open wound with lime, garlic and turmeric; MVD: Modern veterinary drug.

Table 25.4: The Results of Kruskall–Wallis or H Test for Assessing Performance of IK to Treat Open Wound

H–Value		*Conclusion*
Calculated Value	*Table Value*	
0.506	χ^2 0.5 with 2 df = 5.991	All the practices were equally effective

IK considered for *QuIK matrix–3* were

IK_1 = Feeding of rice husk with banana against diarrhoea- Farmers practiced feeding of rice husk with banana thrice a day to control diarrhoea. It is followed for 2-3 days.

IK_2= Feeding of whey to control diarrhoea

Mean scores of farmers' response for each IK against the identified parameters were presented in five matrices. As per ranking by dairy farmers (N=15) in matrix–1, IK_2 ie. walking of animal into mud to control FMD got 1st rank. Modern veterinary medicine had been awarded 2nd rank followed by IK_3 and IK_1 respectively. Matrix–2 revealed that IK_2 *i.e.* treating open wound with lime, garlic and turmeric was in the 1st position followed by IK_1 and MVD. It was obvious from matrix–3, IK_2 *i.e.*, feeding of whey to control Diarrhoea got the 1st rank followed by IK_1 and MVD. It was evident from the result of Kruskall–Walis or H test for assessing performance of the above-mentioned IK for three matrices that the hypothesis was separated as the calculated value was less than the table value and it was concluded that all the practices were equally effective.

Table 25.5: Qulk Matrix–3 for Assessing Performance of IK to Control Diarrhoea Against Different Parameters Max marks =10 No. of respondents; N=15

Sl.No.	*Parameters*	*Practices*		
		IK_1	*IK_2*	*MVD*
1.	Extent of cure	2.73	2.47	5.00
2.	Cost of effectiveness	4.07	4.47	1.47
3.	Ease in use	3.47	4.53	2.00
4.	Recovery time	3.00	2.47	4.53
5.	Resource availability	3.53	4.47	2.00
	Average score	3.36	3.68	3.00
	Ranks	II	I	III

IK_1: Feeding of rice husk with banana against diarrhoea; IK_2: Feeding of whey to control diarrhoea; MVD: Modern veterinary drug.

Table 25.6: The Results of Kruskall–Wallis or H Test for Assessing Performance of IK to Control Diarrhoea

H–Value		*Conclusion*
Calculated Value	*Table Value*	
- 1.552	χ^2 0.5 with 2 df = 5.991	All the practices were equally effective

Conclusion

Many indigenous veterinary beliefs and practices persist, given that the benefits of centralized, high-tech driven and top-down approaches to animal health care have reached only a minority of the world's stock raisers and farmers. Till now, as little as 15-20 per cent of the livestock population in developing countries has enjoyed any regular and affordable access to modern veterinary medicine. High–tech health care is beyond the reach of the dairy farmers, who cannot afford them in view of their raising cost thereof coupled with their own economic constraints and other problems like side effects associated with the conventional animal health care system. Thus leads to an increasing reliance on ethno veterinary medicine, which continues to have strong roots amongst the local community. Dairy farmers believe that this kind of traditional way of treating animal diseases by employing the knowledge passed on orally from generation to generation provides a permanent and safe cure for this commonly occurring disease. Farmers' perception provides important insights for evolving practical strategies in programmes of future development intervention in this field of study.

References

Arya, H.P.S., 2003. Methodologies for validation of indigenous technical knowledge (ITK), Training course on indigenous knowledge system for Agriculture, December 9-29. Centre of Advance Studies, Division of Agricultural Extension, IARI, New Delhi.

De, A., Arya, H.P.S., Tudu, B. and Goswami, A., 2004. Indigenous technical knowledge in animal husbandry. *Livestock Research for Rural Development,* 16(8).

De Villiers, A.K., 1996. Quantification of indigenous knowledge: a rapid method for assessing crop performance without field trial. *Agriculture Research and Extension Network,* ODA, UK.

Grenier, L., 1998. *Working with Indigenous Knowledge: A Guide for Researchers.* IDRC, Ottawa.

Padaria, R.N., Singh, B. and Kumar, S., 2003. Indigenous knowledge and intellectual property rights. *Training Manual on IKS for Agriculture,* IARI, New Delhi.

Parasar, B., 1994. Indigenous agricultural knowledge of tribals in Eastern Ghat High Land Zone of Orissa. *Ph.D. Thesis,* Department of Extension Education, Institute of Agricultural Sciences, BHU, Varanasi.

Rajshekaran, B., 1993. A framework for incorporating indigenous knowledge system into agricultural research, intension and NGOs for sustainable agricultural development. Studies on technologies and social change, No. 21. Technologies and social change programme, IOWA, State University, Ames, IOWA, 5001, USA

Seeland, Klaus, 2000. What is indigenous knowledge and why does it matter today? In: Seeland, Klaus and Franz Schmithusen (Ed.) *Man in the Forest,* DK Printworld, New Delhi.

Chapter 26

Construction and Standardization of Management Ability Scale for Mango Orchard Enterprise

☆ *N.B. Jadav*[1], *P.R. Kanani*[2] *and P.S. Gorfad*[3]

Introduction

All the enterprises are interested in increasing the productivity. The agriculture being an enterprise is not an exception to this. The farmers as the manager of the enterprise are except to bring about maximum profit with available resources irrespective of the economic, social, cultural, physical and technological environment. The farmers manage a production system to get a return from it, consciously or unconsciously. Management is the process essential to accomplish enterprise goal and objectives. In general, we may say management as the process of working with one through individuals and groups and other resources such as equipment capital and technology. Managing is always concerted with productivity, which implies the effectiveness and efficiency of individuals. Management ability in present study has been defined as the degree of ability to which an individual acquires and adopts effective factors in an enterprise to reach higher level of performance.

---

1 Subject Matter Specialist, Krishi Vigyan Kendra, JAU, Jamnagar.

2 Professor, Department of Extension Education, JAU, Junagadh

3. Agricultural Officer, Krishi Vigyan Kendra, Millet Research Station, JAU, Jamnagar.

Belshaw (1974) rightly stated that today farming enterprise is becoming more complex and complicated and therefore, management is a key to face this problem. Agriculture is considered as an important enterprise in India; so, it is worthwhile to study the related factor responsible for the productivity of the commodities involved in agriculture. Orchard growing enterprise also requires the skill and knowledge pertaining to the latest technological know how in the context of management aspect. There are several factors affecting to the productivity of orchard crops. Such as knowledge of scientific cultivation, planning, organizing, supervising, budgeting, coordinating activities, communication, controlling, decision-making etc. Number of studies have made an attempt to study management ability in a systematic and scientific way at operational level. Therefore, in the present study attempt has been made to develop and standardize and objective scale to ensure management ability of mango orchard growers.

Development of Scale to Measure Management Ability of Mango Orchard

Identification of Indicator

In order to identify the basic components of management ability, a good number of indicator and sub indicator pertaining to mango orchard grower were collected through review of literature, correspondence with experts and discussion with extension management specialists. A total number of 11 main indicators and 105 sub indicators were selected tentatively as possible indicators and sub–indicator of management ability of mango orchard grower. These 11 main indicators along with their sub-indicator were mailed to panel of judges from the field of agricultural extension development administration. Out of 100 judges, only 64 judges were responded. Finally, 60 judges were considered by rejecting four judges' responses due to incomplete responses. Thus, 60 judges were requested to indicate whether each of the indicator sent to them were relevant and suitable for inclusion in the scale and to rank the relevant items for their inclusion in the scale.

Relevancy of Scale Indicator

The responses received from the judges supported the relevancy of ten main indicator and 82 sub- indicators out of 105 sub indicators. The indicators that received more than 75 per cent response were considered as relevant for the inclusion in the scale. These main indicators are knowledge, planning organizing, supervising, budgeting coordinating, communication, control, decision making human relationship. The same result was obtained by Chari and Nandpukar, (1987). The details are furnished in Table 26.1.

Consistency of the Indicator

In order to know the consistency in rating the various items of the scale by different judges, *viz.* experts in the field of extension/administration and farmers for each main items, co-efficient of concordance 'W' as suggested by Kandel (1948) was calculated. The prefect agreement and the least agreement among the judgment of the judges are know when the value of 'W' is one and zero. The chi-square was also computed to test the significance of the co-efficient of concordance 'W'.

Table 26.1: Relevancy of Main Indicators of the Scale

Sl.No.	*Item*	*Relevant*	
		Number	*Percentage*
1.	Knowledge	60	100.00
2.	Planning	60	100.00
3.	Organizing the activities	53	88.33
4.	Supervising	54	90.00
5.	Budgeting	56	93.33
6.	Coordinating the activities	52	86.67
7.	Communication	49	81.67
8.	Controlling the activities	50	83.33
9.	Decision making	58	96.67
10.	Human relationship	48	80.00

It is conspicuous that there was a perfect agreement in rating the scale items by the experts.

Obtaining Scale Value

In order to obtain the scale value for the 10 main and sub indicator of the scale, a set of 60 judges requested to give rank for these main and sub indicator. The centile position 'P' based on the method suggested by Guilford (1954) was computed. The 'C' values, 'Rj' values and finally scale values *i.e.*, 'Rc' value were worked out by using the following formula:

$$Rc = 2.357\,Rj - 7.0$$

Whereas, for sub indicator weightage were given according to rank received for each indicator in the scale.

Development of Sub-Indicator for Each Scale Indicator

It was necessary to list out sub-indicator under each scale item to help in administering the scale and to have objectives assessment of the scale item.

Calculating the Management Ability Index

In order to calculate the management ability index for each respondent, each scale item was first rated for the maximum of the particular main component, then, the actual score obtained was multiplied by scale value, and management ability index was calculated by using the following formula.

$$MAI = \frac{\sum \text{Score obtained for the component} \times \text{Scale value of component}}{\sum \text{Score obtained for the component} \times \text{Scale value of component}} \times 100$$

Standardization of Scale to Measure Management Ability of Mango Orchard

Calculating the Reliability and Validity of the Developed Scale

Content Validity

According to Kerlinger (1976), the content validity is representativeness of sampling adequacy, of the content, the substance, the matter and the topics of measuring instrument. The content validity was worked out by two ways. Firstly, the indicators selected for inclusion in the scale were based on extensive and exhaustive literature and secondly, the opinions of panel of 60 judges were obtained to know whether the indicators were relevant for inclusion in the scale. More than 75 per cent judges endorsed proposed indicators as relevant, indicating the scale was valid.

Criterion Validity

A criterion may be an objective measure of performance or quality (Garrett, 1985). It was established by correlating the management ability scores obtained from 20 non-sampled respondents with their respective experience as a mango grower. The coefficient of correlation (0.376) between the two variables was found to be significant at 0.01 level of significance indicated that the scale was valid.

Reliability of the Scale

The split-half method was used to achieve the reliability of the scale. The scale was administered to 20 non-sampled farmers. The coefficient of correlation was used for the two sets of score. The reliability coefficient of correlation obtained (0.685) was significant at 0.01 level of significance indicated that the scale was reliable.

It can be inferred from the above findings that all the ten indicators and 82 sub-indicators were found relevant for inclusion in the scale. The results of the content and criterion validity tests were indicated that the developed scale was valid. The split-half reliability coefficient stated that the developed scale was reliable.

Relative Importance of Indicators of the Scale

In order to know the relative importance of different main indicators of management ability scale, the scale value of each main indicator were compared with the mean score assigned by the respondents. Two methods were adopted for ascertaining the correlation. Firstly between the two sets of score by applying Pearson's product moment coefficient of correlation and secondly, between the ranks of the two sets of the score by applying Spearman's rank correlation. The data are presented in Table 26.2.

The value of the Pearson's product moment Correlation Coefficient (0.731) and Spearman's rank correlation coefficient (0.795) were significant at 0.01 level of significance, indicating positive and significant correlation between the two observations. Thus, it can be inferred that all the indicators were important and hence, none of the main indicators was eliminated. However, their importance was varied as the scale value of each main indicator was different.

Table 26.2: Scale Value of Different Indicators of the Management Ability Scale (n = 60)

Sl.No.	Indicator	Scale Value	Rank	Means Score	Rank
1.	Knowledge of scientific cultivation	8.70	I	82.84	I
2.	Planning	8.35	II	74.75	IX
3.	Organizing the activities	3.75	VI	78.21	VII
4.	Supervising	5.84	V	78.74	V
5.	Budgeting	6.19	IV	76.57	VIII
6.	Co-ordinating the activities	3.24	VII	80.01	III
7.	Communication	0.53	X	78.28	VI
8.	Controlling the activities	3.52	VIII	78.89	IV
9.	Decision making	6.42	III	80.37	II
10.	Human relationship	1.20	IX	72.38	X

It can be observed from Table 26.2 that the highest scale value was obtained by knowledge of scientific cultivation followed by planning, decision-making and budgeting. The indicators *viz.* supervising, organizing the activities, coordinating the activities, controlling the activities, human relationship and communication were next in their scale values in descending order.

It can be concluded that the first rank was assigned by the judges to knowledge of scientific cultivation indicator followed by planning, decision making, budgeting and supervising. For the proper management of any mango orchard, adequate knowledge regarding scientific cultivation of mango practices is the prime requirement. Knowledge about current technology is essential to maximize return from an enterprise. It is saying that education is the panacea for the life and knowledge is the determinant factor for the education. Planning helps the farmer in proper allocation of resources and better organization of the farm activities. It enables the farmers to achieve the desired goals. Rationality in decision-making helps the farmer to clearly set up the goals in farming. Budgeting involves the allotment of the fund to the various activities and credit is an important input for production, which helps in obtaining full benefits of the technology. These may be the reasons for giving the higher ranking to these indicators by the judges.

Organizing and coordinating were given sixth and seventh rank respectively. Organizing is a process of identifying and grouping the work to be performed. Co-ordination is essential to get the production inputs timely and in adequate quantities. Mango growers used these tools for wise counseling, monitoring, evaluation and promoting teamwork.

The indicators viz; controlling, human relationship and communication were next important ones in descending order. Better communication obtains current and profitable information about modern farm technology keeps the farmer up to date about scientific cultivation. The ability to communicate effectively with other is a management skill. It helps the farmers to receive and disseminate profitable information relating to farm enterprise. Communication skill also enables the farmer to keep good contact with the agricultural support system. Communicating includes an effective and proper selection of ways and means of communication for achieving effective management ability. At the same time human relationship used for discharging their duties with a consideration of labour morale, treat them as social beings, which positively influence on management ability.

Implications

This finding of this study has both theoretical and practical implications. Theoretically, the findings of the present study add to the knowledge about the management role in orchard farming and practically it is useful in measuring the of any orchard growers with due modification.

The main indicator *viz.*, knowledge, planning, organizing activities, supervising, budgeting, coordinating, communication, controlling, decision making and human relationship will considered while working with orchard growers in farming enterprise. To maximize overall profit of any orchard, the orchard growers should have management skills particularly in planning, organizing the activities, supervising and decision-making.

References

Belshaw, D., 1974. Improving management procedures for agricultural development. In: *Paper Presented at International Seminar on Change in Agriculture Reading,* U.K.

Chari, A.P. and Nandapurkar, G.G., 1987. A scale to measure management ability of farmers. *Maha. Jour. of Extn. Edn.*, 6: 163–168.

Garrett, H.E., 1985. *Statistics in Psychology and Education.* Valals, Feter and Simons Pvt. Ltd., Bombay, p. 27–65.

Guilford, J.P., 1954. *Psychometric Methods*. Tata McGraw Hill Pub. Co., Bombay.

Kendall, M.G., 1948. *Rank Correction Methods*. Griffan, London.

Kerlinger, F.N., 1976. *Foundations of Behavioural Research*. Surjeet Publications, Delhi, pp. 198–204 and 442–478.

Chapter 27

A Scale to Measure Attitude Towards Increased Retirement Age of College Teachers

☆ *G.L. Bagdi*[1] *and Anupama Shah*[2]

Introduction

According to the University Grant Commission (UGC) directive, the retirement age of college teachers has been increased to sixty-two (62) years. Still some states have kept the retirement age as fifty-eight (58) years. It is true that the retirement of any official or staff of any organization depends on the age of a person. During young age, people have more energy to work. It is believed that after the age of 40 to 45 years the working efficiency goes decreasing. The working efficiency of the staff in an organization also depends on his/her health condition. Every phenomenon in the universe has merits or demerits. It is also true that the senior teachers have vast experience of their duties. Increasing the retirement age of college teachers or officials of any organizations is beneficial to the old persons, who are going to be retired. The older teachers can occupy themselves and of course can also earn money in old age.

---

1 Senior Scientist, (Agricultural Extension), CSWC R&T, RC, Vasad, Anand, Gujarat.

2 Professor & Dean, Department of Extension and Communication, Faculty of Family and Community Science, The Maharaja Sayajirao University of Baroda, Vadodara.

On the other side, the increased retirement age may not be good for young generation, who is waiting for recruitment for job in different organizations. The retirement age may also depend on organizational economic management. For example, a person retires at the age of sixty-two or sixty-five years then the organization has to bear the higher expenses to give salary in old age. But when an official retires early and the organization recruits new official on the same post, it requires comparatively less money to give salary to new staff. Therefore, the retirement age of college teachers may depend on organizational condition and need also. Keeping these points in mind, it is imperative to measure the attitude of people towards increased retirement age of college teachers by developing an attitude scale. A study was carried out during 1999-2000 to develop an attitude scale towards increased retirement age of college teachers.

Material and Methods

The method of equal-appearing intervals, as originally described by Thurstone and Chave (1929) has been used in this study for developing the present attitude scale to measure attitude towards increased retirement age of college teachers. Allen L. Edwards (1957) describes the same method of equal-appearing intervals. The following procedure was adopted in constructing the attitude measurement scale.

Collection of Attitude Statements

The items making up an attitude scale are called statements. A statement may be defined as anything that is said about a psychological object. The first step in the construction of an attitude scale is collection of items, that is, statements that will represent in a particular test the universe of interest. Statements covering all aspects of attitude towards increased retirement age of college teachers were made with the help of available literature, subject matter specialists and discussion with scientists. As such ninety (90) statements were prepared.

Editing of Statements

Edward (1957) wrote in his book that, Wong (1932), Thurstone and Chave (1929), Likert (1932), Bird (1940) and Edwards and Kilpatrick (1948) have suggested various informal criteria for editing statements to be used in the construction of attitude scales. Therefore, all the statements were edited by applying the fourteen informal criteria and guidelines suggested by above scientists. As a result out of the ninety statements collected initially, the eighty-three (83) generalized attitude statements relevant to increased retirement age of college teachers were selected (Appendix–I).

Judges' Rating of Attitude Statements

All the 83 statements regarding attitude towards increased retirement age of college teachers were given to a group of fifty (50) judges for their rating of unfavourableness and favourableness on the each statement. The judges selected for the study comprised of extension specialists, teachers, professors and subject matter specialists who were engaged in teaching and research from various universities and Central Soil and Water Conservation Research and Training Institute, Dehradun. The judges were requested to rate each statement on an eleven-point continuum scale. The favourableness of the statements varied from one to eleven points at equal intervals. The point six was considered as neutral. Seven to eleven points were considered favourable and points one to five were considered unfavourable. The points 1st and 11th were extremely unfavourable and extremely favourable respectively on this eleven-point continuum.

Out of the total fifty judges, thirty judges returned their statements after rating and their responses were considered for calculation of the scale values and Q-values of the attitude statements.

Calculation of Scale and Q-Values

The scale-values and Q-values for all the 83 statements were calculated with the following formulae (Edward, 1957).

Calculation of Scale Value

$$S = L + (0.50 - \Sigma\, p_{b/}p_{w}) \times i$$

where,

S: The median or scale value of the statement

L: The lower limit of the interval in which the medium falls.

Σp_b: The sum of the proportions below the interval in which the median falls.

p_w: The proportions within the interval in which the median falls.

i: The width of the interval and is assumed to be equal to 1.0.

Calculation of Q Value

Thurstone and Chave used the interquartile range or Q as a measure of the variation of the distribution of judgements for a given statement. The interquartile range contains the middle 50 per cent of the judgements. To determine the value of Q we need to find two other point measures, the 75th centile and 25th centile. The 25th centile can be obtained from the following formulae:

$$C_{25} = L + (0.25 - \Sigma\, p_{b/}p_{w}) \times i$$

where,

C_{25}: The 25th centile.

L: The lower limit of the interval in which the 25th centile falls.

Σp_b: The sum of the proportions below the interval in which the 25th centile falls.

p_w: The proportions within the interval in which the 25th centile falls.

i: The width of the interval and is assumed to be equal to 1.0.

The 75th centile will be given by as below:

$$C_{75} = L + (0.75 - \Sigma\, p_{b/}p_{w}) \times i$$

where,

C_{75}: The 75th centile.

L: The lower limit of the interval in which the 75th centile falls.

Σp_b: The sum of the proportions below the interval in which the 75th centile falls.

p_w: The proportions within the interval in which the 75th centile falls.

i: The width of the interval and is assumed to be equal to 1.0.

Then the interquartile range or Q will be given by taking the difference between C_{25} and C_{75}. Thus,

$$Q = C_{75} - C_{25}$$

The interquartile range is a measure of the spread of the middle 50 per cent of the judgement. When there is good agreement among the judges in judging the degree of favourableness and unfavourableness of a statement, Q will be small. A large Q value, indicates disagreement among the judges. Thurstone and Chave regard large Q values primarily as an indication that a statement is ambiguous. If a choice is to be made among several statements for constructing the final scale with approximately the same scale values, preference is given to the one with the lowest Q value, that is, to the one believed to be least ambiguous.

Final Selection of Attitude Statements

Based on the scale value and Q value, the statements were finally selected to constitute the attitude scale taking into account the following criteria.

1. The statements should have smaller Q values as far as possible.
2. The scale value of the statements should be fairly distributed throughout the continuum.
3. There should be statements indicating both favourable and unfavourable attitudes.

Based on these criteria, the statements for constructing the scale to measure attitude towards increased retirement age of college teachers were finally selected. Twenty four (24) statements, having Q value of three and less than three were selected.

Results and Discussion

Final Format of the Scale and Scoring Procedure

The attitude statements finally selected have been randomly arranged in the final format of the scale. There were two columns in the final format of the attitude scale for responses of respondents as agree or disagree to the statements. For quantifying the data, each response of "agree" will be assigned scale value of the statement as a score and disagree response will be assigned no score. The final format is given in Table 27.1.

Reliability of the Scale

The reliability of the scale was analyzed with the help of test–retest method. The scale was administered twice to thirty (30) college teachers at an interval of one month. The respondents were randomly selected from the different institutions. The total score of a person on all the items was divided by the total number of the items to determine score of an individual on the scale (Shah and Gupta, 1993). The correlation coefficient was calculated and the r-value 0.898 was determined, which is highly significant at 1 per cent level of probability. It indicates that the scale is reliable and has the degree of dependability on the scale for measuring the attitude of teachers towards increased retirement age of college teachers. It established that the scale is a reliable instrument for the measurement of attitude towards increased retirement age of college teachers.

Table 27.1: Scale to Measure Attitude Towards Increased Retirement Age of College Teachers

Sl.No.	*Attitude Statement*	*Scale Value*	*Agree*	*Disagree*
1.	The older teachers have more experience of teaching	9.94		
2.	The older teachers may give correct instructions to their young colleagues to carry out the work	8.83		
3.	The older teachers have more knowledge about their job	8.00		
4.	The older teachers are trustworthier than the younger teachers	6.3		
5.	I think that the older teachers are bad tempered on the job	3.3		
6.	The older teachers are absent more often than younger teachers	4.0		
7.	By increasing the age of retirement, the problem of unemployment for young generation will be increased	4.0		
8.	I believe that as the age increases the efficiency to do the work decreases	4.0		
9.	The age of the teachers has no effect on the mistakes during the job	6.0		
10.	The quality of teaching has no relevance with the age of teacher	6.0		
11.	The older teachers have practical knowledge about the teaching	9.8		
12.	The older teachers are unduly conscious about their self-respect	5.7		
13.	The older teachers try to maintain their good behaviour with the staff colleagues	7.0		
14.	The younger teachers get opportunity to learn good skill of teaching by the company of older teachers	8.9		
15.	The younger teachers are very enthusiastic in doing their duties	5.0		
16.	The younger teachers have more creativity in teaching	2.0		
17.	Organizational climate is better with young teachers due to their supportive attitude with the colleagues	6.0		
18.	The younger teachers are often persuasive in nature during teaching	6.0		
19.	The younger teachers assist the older teachers in their duty/works	5.0		

Contd...

Table 27.1–Contd...

Sl.No.	*Attitude Statement*	*Scale Value*	*Agree*	*Disagree*
20.	The older teachers have knowledge of administrative rules and regulations	9.6		
21.	The older teachers have weak audio and visual capabilities	4.0		
22.	The older teachers are capable to control students in difficult situations	8.6		
23.	The age of teachers has less relevance with learning ability of teachers	6.0		
24.	The teachers should retire at 55 years of age, because the universities and colleges may face financial crisis	3.6		

Validity of the Scale

"A scale possesses validity when it actually measures that it claims to measure", Goode and Hatt (1981). The validity of the scale was examined with the help of content validity method of determining how well the contents of the scale represented the subject matter under the study. The contents of the attitude scale were derived from various literature and discussion with teachers and subject matter specialists. All the edited statements of the scale were passed by the different teachers and scientists. Hence, the present developed scale satisfied the content validity.

Conclusion

The scale to measure attitude towards increased retirement age of college teachers or staff of any organization was developed. It will be useful to researchers to measure the attitude towards increased retirement age of college teachers, scientists and staff of any organization.

Acknowledgements

The authors are grateful to all the teachers and expert judges, who sent their responses for developing the attitude scale and their valuable suggestions and constant encouragement. The help of Smt. D.S. Macwan, Stenographer, CSWCRTI, Research Centre, Vasad, for typing the manuscript is also thankfully acknowledged.

References

Edward, A.L., 1957. *Techniques of Attitude Scale Construction.* Appleton Century Crafts, Inc., New York, pp. 83–119.

Goode, W.J. and Hatt, P.K., 1981. *Methods in Social Research.* McGraw-Hill Book Co., Singapore.

Shah, Anupama and Gupta, A., 1993. *Measurement Techniques for Affective Behaviour.* Department of Home Science Education and Extension, Faculty of Home Science, M. S. University of Baroda.

Thurstone, L. L. and Chave, E.J., 1929. *The Measurement of Attitude.* University Chicago Press, Chicago.

Appendix–I
Attitude Statements Regarding Increased Retirement Age Showing Scale Value and Q Value

Sl.No.	*Attitude Statement*	*Scale Value*	*Q Value*
1.	Mostly older teachers make less mistakes in their duties	8.83	3.54
2.	The older teachers are more careful on their job	8.0	4.83
3.	The younger teachers are somewhat careless about their duties	3.83	3.84
4.	*The older teachers have more experience of teaching	9.94	0.84
5.	*The older teachers may give correct instructions to their young colleagues to carry out the work	8.83	2.94
6.	The older teachers complete their work with high quality	7.5	4.1
7.	The older teachers are co-operative to the staff on the job	7.5	3.17
8.	*The older teachers have more knowledge about their job	8	3
9.	*The older teachers are more trustworthy than the younger teachers	6.3	3
10.	The older teachers are more conscious about the reputation of the institute or college	9.16	4
11.	I think that the older teachers are more responsible for their duties	8	4.07
12.	The older teachers with good health should be retained in teaching	9	4.05
13.	The poor eye-sight in old age effect on the teaching work therefore the retirement age should be decreased	6.16	6.16
14.	The older teachers are susceptible to different diseases so that they should retired early	5.83	6.83
15.	*I think that the older teachers are bad tempered on the job	3.3	2.7
16.	*The older teachers are absent more often than younger teachers	4	3
17.	I think the older teachers are harder to train for new techniques in the job	4	5.7
18.	*By increasing the age of retirement, the problem of unemployment for young generation will be increased	4	2.7
19.	As the older teachers are to be paid more salary, they should retire early	6.2	6
20.	The young teachers act smartly on the job than the old teachers	5	5.8
21.	The older teachers usually find it difficult to keep up with the speed needed in the modern age	6	3.1
22.	*I believe that as the age increases the efficiency to do the work decreases	4	3
23.	I think the older teachers are more rigid to change their views to accept the innovations	6	7
24.	The mostly old teachers are money minded.	2	4.6
25.	*The age of the teachers has no effect on the mistakes during the job	6	2
26.	The older teachers are attentive to their work	7	3.83

Contd...

Appendix–I–Contd...

Sl.No.	Attitude Statement	Scale Value	Q Value
27.	The younger teachers have interest to get the new experience about their job	9.1	5.07
28.	The young teachers may be interested towards sports and entertainment during the working time	8	4
29.	The older teachers have responsibility of their family so they have disturbed mind	5	5.1
30.	*The quality of teaching has no relevance with the age of teacher	6	2.3
31.	The younger teachers may have some hesitation during teaching	6	3.7
32.	All age group teachers are friendly with the students	6.1	3.2
33.	The older teachers are kind towards their colleagues and students	6	3.5
34.	The young teachers are interested to know about innovative information related to teaching	8.1	4.5
35.	The young teachers want to complete their teaching work within time	6.3	3.75
36.	*The older teachers have practical knowledge about the teaching	9.8	2.6
37.	*The older teachers are unduly conscious about their self-respect	5.7	3
38.	*The older teachers try to maintain their good behaviour with the staff colleagues	7	3
39.	The age of the teachers has less effect to carry out innovative researches in teaching	6	3.4
40.	I think the older teachers can write more articles about education	8.7	4.8
41.	The older teachers remain more busy with meetings and conferences rather than teaching	6	5.1
42.	By reducing the retirement age we have chances of induction of different new personnel in education	8.8	4.1
43.	The older teachers motivate their younger colleagues to improve their performance in teaching	7	3.9
44.	*The younger teachers get opportunity to learn good skill of teaching by the company of older teachers	8.9	2.8
45.	I think that old teachers encourage their younger colleagues to conduct innovative researches in education	8.6	3.2
46.	The older teachers seem to be happier on the job.	7	3.4
47.	The older teachers are also dependable to younger colleagues to carry out of their some duties	5.1	3.1
48.	The older teachers get upset easily.	6	5
49.	The majority of older teachers have ideas in right direction regarding to achieve aims of the university or institution	7	3.6
50.	The older teachers are kind to solve the problems of other colleagues and students	8	3.1
51.	I think majority of older teachers has good communication skill	7	3.9
52.	The mostly younger teachers are taking initiative to accept new challenges	6	4

Contd...

Appendix–I–Contd...

Sl.No.	*Attitude Statement*	*Scale Value*	*Q Value*
53.	*The younger teachers are very enthusiastic in doing their duties	5	3
54.	I believe that the mostly older teachers are taking good decisions or judgments during their job	7.1	3.7
55.	I think the older teachers have more managerial skills in the management of the college or institution	8	3.6
56.	Some of young teachers are having risk taking ability.	8.2	5.3
57.	I believe that mostly older teachers are having more patience in difficult times	8.8	3.8
58.	*The young teachers have more creativity in teaching	2	3
59.	The majority of younger age teachers are alert on the job	7.1	4.8
60.	The majority of younger age teachers have little self confidence to carry out their duties	6	3.8
61.	I feel that the younger teachers are often high energetic during their duties	7.7	4
62.	*Organizational climate is better with young teachers due to their supportive attitude with the colleagues	6	2.9
63.	*The younger teachers are often persuasive in nature during teaching	6	2.8
64.	I believe that the older teachers may have ability to lead the other colleague teachers	8.7	3.6
65.	I think the older teachers motivate their students to carry out the work	9	3.1
66.	The older teachers may have somewhat good speaking and writing skills	8.7	3.8
67.	*The younger teachers assist the older teachers in their duty works	5	3
68.	I believe that by reducing the retirement age the new young generation will get opportunity in teaching	8	5
69.	I think that the some older age teachers use their old or traditional ways of teaching	5.7	4.7
70.	I believe that the older teachers are more stereotyped in teaching	6.1	5
71.	I believe that the majority of young teachers are more punctual in office	5.9	3.8
72.	*The older teachers have knowledge of administrative rules and regulations	9.6	2.2
73.	I think most of the older teachers have past teaching experience to improve upon in future	9	3.7
74.	I feel that the older teachers have lots of influence to others on the duty	5.7	4.33
75.	*The older teachers have weak audio and visual capabilities	4	3
76.	*The older age teachers are capable to control students in difficult situations	8.6	3
77.	The age of the teachers has no effect on the teaching ability.	6.1	3.8
78.	The older teachers work according to the rules and regulations.	7	4

Contd...

Appendix–I–Contd...

Sl.No.	*Attitude Statement*	*Scale Value*	*Q Value*
79.	If the teachers retire at the age of 58 years, the entry of young teachers increases	9	4.6
80.	*The age of teachers has less relevance with learning ability of teachers	6	3
81.	Even the universities borrow money; the teachers should retired up to 65 years of age	6	7.6
82.	*The teachers should retired in 55 years of age, because university may face financial crisis	3.6	2.5
83.	The retirement of the teachers should be based on their teaching performance	7	6.1

* The statements selected for construction of final attitude scale.

Chapter 28

Assessment of Adoption Gap of Potato Production Technology of Potato Growers and their Adoption Level

☆ ***R.P. Singh*[1]*, Shiv Kumar, Shakuntala Gupta, Virendra Kumar and Amit Kumar*[2]**

Introduction

Potato is one of the most important vegetable crops in India. Its popularly known as the king of vegetables. The vegetable basket is incomplete without Potato in India. The world's Production of Potato is 311.4 million tones form 19.22 Million hectare of land with an average productivity of 16.20 tones per hectare during the year 2003 (F.A.O. report, 2003).

At present India rank 4th in area 3rd in production and 10th in productivity in the world (Luthra *et al.,* 2004). India is producing 23.27 million tones of Potato form an area of 1.28 million hectares with an average yield of 18.113 tones per hectare (Source-Agriculture Statitistics at a Glane 2005).

---

1 Associate Director, Extension, K.V.K., Hastinapur, Meerut.

2 SMS, Extension, K.V.K., Hastinapur, Meerut.

During the last 4 decades our country has progressed significantly by developing high yielding varieties of Potato, improved Agro-Techniques and other Production Techniques suitable for different agro climatic zones in country. But there are wide gap between the available technologies and their adoption. Keeping in view the present study was under taken to know the adoption in Potato Production Technology of Potato growers and their adoption level.

Methodology

The study was conducted in Meerut district of Uttar Pradesh during 2004-05. The district comprises of 14 blocks out of them three blocks were selected based on cropped area and the crop productivity. Five villages from each block and eight potato growers were randomly selected from each village. Thus, the total sample size was of 120. The Primary data were collected through personal interview with the help of structural schedule developed and secondary data were collected form FAO Report–2003 and Agricultural statistics at a Glance 2005. The data were analyzed and find out the mean values and per centage.

Results and Discussion

The data present in Table 28.1 indicates, that the highest adoption gap was found in plant protection (72.10 per cent) in case of marginal potato growers, followed by 52.00, 49.92, 46.03, 46.02, 44, 43.80, 43.52, 43.20, 42.97 and 34.28 per cent adoption gap regarding to method of sowing, improved varieties, use of improved implements, time of fertilizer and manures application, harvesting of crop, time of sowing, storage and transportation, manures and fertilizers, seed rate, water management and selection of soil respectively.

Table 28.1

Sl.No.	Potato Production Technology	Score Assigned	Marginal Potato Growers	Small Potato Growers	Medium Potato Growers	Large Potato Growers
1.	Selection of Soil	6	34.28	27.83	18.14	10.86
2.	Improved Varieties	8	52.00	42.65	38.20	30.08
3.	Seed Rate	6	43.52	33.18	32.68	24.20
4.	Time of sowing	6	44.00	34.30	29.07	24.20
5.	Methods of sowing	8	50.63	43.60	39.80	26.92
6.	Manures and Fertilizers	14	43.80	38.10	25.08	17.02
7.	Time of fertilizers and Manures application	6	46.03	35.01	34.90	29.90
8.	Water management	6	43.20	36.03	37.07	28.20
9.	Plant Protection	12	72.10	66.45	57.06	52.06
10.	Use of improved implements	6	49.92	38.01	36.03	28.98
11.	Harvesting of crop	6	46.02	36.06	34.00	29.98
12.	Storage and Transportation	8	42.97	38.01	32.90	26.10
	Total and Average	92	47.08	37.57	33.80	26.80

In case of small potato growers the highest adoption gap was found in plant protection (66.45 per cent), followed by 42.65, 38.10, 38.01, 36.06, 35.01, 34.30, 33.18 and 27.83 per cent adoption gaps regarding to improved varieties, method of sowing, use of improved implements, storage and transportation, manures and fertilizers, water management, harvesting of crop, seed rate, time of fertilizers and manures application, time of sowing and selection of soil respectively.

In case of medium potato growers again the highest adoption gap was found in plant protection (57.06 per cent), followed by 39.80, 38.20,37.07,36.03,34.90,32.00 and 18.10 per cent adoption gaps regarding to improved varieties, method of sowing, water management, use of improved implements, time of fertilizers and manures application, harvesting of crop, storage and transportation, seed rate, time of sowing, manures and fertilizers and selection of soil respectively.

While the highest adoption gap of large potato growers was found (52.06 per cent) plant protection, followed by 39.80,38.20, 37.07, 36.03, 34.00, 32.90,32.68, 35.08, and 17.02 per cent adoption gaps regarding to improved varieties, time of fertilizers and manures application, harvesting of crop, use of improved implements, method of sowing, water management, storage and transportation, time of sowing, seed rate, manures and fertilizers and selection of soil respectively.

The average adoption gaps of marginal potato growers were found highest 47.08 per cent and lowest 26.80 per cent of large potato growers. The adoption gap of small and medium potato growers were 37.57 and 33.80 per cent respectively.

The data presented in Table 28.2, indicates that the highest adoption level was found in selection of soil (89.14 per cent) in case of large potato growers, followed by 81.86, 80.98, 75.080, 75.08, 73.90, 73.08, 71.08, 71.02, 69.20 per cent adoption level with respect to the manure and fertilizers, seed rate, time of sowing, storage and transportation, methods of sowing, water management, used of improved implements, time of manures and fertilizers application, harvesting of crop, improved varieties and plant protection measures respectively.

In case of medium potato growers the highest adoption level was found in selection of soil (81.86 per cent), followed by 74.92, 70.93, 68.93, 67.10, 63.97, 63.97, 60.20 and 42.92 per cent adoption level were regarding to manure and fertilizers, time of sowing, seed rate, storage and transportation, harvesting of crops, use of improved implements, water management, method of sowing, improved varieties and plant protection measures respectively.

In case of small potato growers the highest adoption level was found in selection of soil (72.17 per cent), followed by 66.82, 65.70, 64.90, 69.97, 63.94, 61.99, 57.35, 56.40, and 33.55 per cent adoption level were regarding to manure and fertilizers, time of sowing, seed rate, storage and transportation, harvesting of crops, use of improved implements, water management, method of sowing, improved varieties and plant protection measures respectively.

While the highest adoption level of marginal potato growers was found (65.32 per cent) plant protection, followed by 58.98,57.03,56.48,56.20,56.00,53.97,52.08,49.37 and 27.90 per cent adoption gaps regarding to improved varieties, time of fertilizers and manures application, harvesting of crop, use of improved implements, method of sowing, water management, storage and transportation, time of sowing, seed rate, manures and fertilizers and selection of soil respectively.

Table 28.2

Sl.No.	Potato Production Technology	Score Assigned	Marginal Potato Growers	Small Potato Growers	Medium Potato Growers	Large Potato Growers
1.	Selection of Soil	6	65.72	72.17	81.86	89.14
2.	Improved Varieties	8	47.40	57.35	61.80	69.20
3.	Seed Rate	6	56.48	66.82	67.32	75.80
4.	Time of sowing	6	56.00	65.70	70.93	75.80
5.	Methods of sowing	8	49.37	56.40	60.20	73.08
6.	Manures and fertilizers	14	56.20	61.90	74.92	82.98
7.	Time of fertilizers and manures application	6	53.97	64.99	65.10	70.10
8.	Water management	6	56.80	63.97	68.93	71.80
9.	Plant Protection	12	27.90	33.55	42.94	47.94
10.	Use of improved implements	6	52.08	61.99	63.97	71.02
11.	Harvesting of crop	6	53.98	63.94	66.00	70.02
12.	Storage and Transportation	8	57.03	61.99	67.10	73.90
	Total and Average	92	52.92	62.43	66.20	73.20

The average adoption gaps of large potato growers were found highest 73.20 per cent and lowest 52.93 per cent of large potato growers. The adoption gap of small and medium potato growers were 62.43 and 66.20 per cent respectively.

Conclusion

It may be concluded that highest adoption level was found in selection of soil of all categories potato growers and the lowest in plant protection measures of all categories potato growers. The average adoption levels of large potato growers were found highest and lowest of marginal potato growers. The highest mean value was found in manures and fertilizers of all categories potato growers and the lowest in selection of soil of all categories potato growers. The average adoption gap of marginal potato growers was found highest and lowest of large potato growers. The investigators refer to provide training, farm advisory services and demonstrations to the potato growers for increasing the quality production of potato for human resource development

Chapter 29

Evaluation of Watershed Management Programme in Bundelkhand Region: A Case Study

☆ *Om Prakash and V.S. Katiyar*[1]

Introduction

The region of Bundelkhand (24°–26°30′ N latitude and 78°10′–80°30′ E longitude) is bounded by the river Yamuna in the north and Vindhyan Plateau in the south. It comprises six districts of Uttar Pradesh, *viz.* Hamirpur, Jalaun, Banda, Jhansi, Lalitpur and Mahoba and six districts of Madhya Pradesh, *viz.*, Datia, Damoh, Chhatarpur, Panna, Sagar and Tikamgarh. Out of total 7.04 m. ha. geographical area of the region, 3.01 m. ha. falls in the State of Uttar Pradesh (Gupta, 1981) and remained 4.01 m. ha (Annon., 1986) in the State of Madhya Pradesh. The small area comes under Lahar tehsil of Bhind district of Madhya Pradesh, respectively. Three fourth of the area depends on rainfall for its agriculture. *Rabi* is the main cropping season in the region. Appropriate and adequate soil conservation measures, like, gully plugging, check dam, peripheral bunding, terracing, leveling, contour bunding and water harvesting structures etc. have been adopted in ravined/gullied watersheds like, Sheetalpur, Hamirpur, Tejpura, Jhansi and Randhar, Jalaun which had considerably reduced the run-off and soil-loss in Bundelkhand region. Similarly, conservation practices,

---

1 Central Soil & Water Conservation Research and Training Institute, Research Centre, Datia (M.P.).

like contour furrows, stone bunding, stone diking and plantation of suitable trees on humps and grasses on bunds not only enhanced the productivity but also reduced sedimentation in Gharawa and Khariya Nala watersheds of the region (Hazra and Singh, 1992). In the present paper, the impact of ORP Sheetalpur watershed, Hamirpur has been emphasized.

Methodology

The ORP (Operational Research Project) watershed is located at village Sheetalpur in Hamirpur district of Uttar Pradesh. The Hamirpur district of Bundelkhand region is one of the most backward districts of Uttar Pradesh, is located at 25°26′ N latitude and 79°5′ E longitude. The watershed comprising the villages of Sheetalpur and Kusmera in Hamirpur district and is located at Sheetalpur village which is 6 kms from Hamirpur on Kalpi road. The watershed drains into Rohin nala which is a tributary of river Betwa. River Betwa meets rive Yamuna after travelling about 8 kms from the watershed. The total area of the watershed is 571 ha., topography is marked by mild to steep slope with shallow and medium ravines. An area of 134 ha is under ravine system is totally devoid of vegetation. The soils are divided into two groups, *viz.*, red (upland ones) and black (lowland) on the basis of colour and topography.

The watershed falls in semi-arid climatic zone with extremes of temperature in summer and winter. Rainfall is erratic and uneven with normal average rainfall of 854 mm varying between as low as 435 mm as high as 1200 mm. the climatic water balance of Hamirpur district indicated water deficit of 1196.4 mm. However, water surplus of 370.4 mm during active monsoon months shows the possibility of water harvesting. The watershed comprises land capability classes II, III, IV, VI and VII out of which VI and VII classes land are unsuitable for agriculture. There is no class I land in the watershed. Out of total land in the watershed 90 ha is under class II, 243 ha class III, 91 ha class IV, 124 ha class VI and 10 ha land comes under class VII, respectively. A major portion of the soil belongs to class III and VI. Total human population in the watershed who were directly receiving the benefits in both the villages, namely, Sheetalpur and Kusmera is 1266. There were total 257 families in the watershed of which marginal (50), small (68), medium (83) and large (56). The extensive wastelands in the form of ravines, rocky wastes and scrub-lands were the results of continued soil degradation and biotic interference.

Soil and Water Conservation Measures/Works

Mechanical Measures

The various soil and water conservation engineering measures were adopted in the watershed for the development of land and water resources (Table 29.1). The table indicates that about 200 ha land was developed through leveling and terracing with suitable water disposal structures. Contour bunding was carried out on 41.42 ha land at a vertical interval of 0.9 m. Another 190.2 ha land was treated with gully plug (check dam) structures. A pacca drop inlet structure was also provided in the gully plugs. A private tube well boring of 60 m deep was also completed in the watershed.

Table 29.1: Impact of Mechanical Measures

Particular	*Target*	*Achievement*
Leveling and terracing (ha)	230.0	199.9
Contour bunding (ha)	41.1	41.4
Gylly plug (check dam) structures (ha)	183.0	190.2
Tubewells (no.)	5	1+1*

*: Private tubewell.

Agronomical Measures

Introduction of High Yielding Crop Varieties

Different high yielding crop varieties were introduced in the watershed for increasing the production of cereals, pulses and oil seeds to meet the requirements of growing population (Table 29.2).

Table 29.2: High Yielding Varieties of Different Crops Introduced in the Watershed

Crops	*Varieties Introduced*
Arhar	T-21
Jowar	CSH-5
Til	T-4
Wheat	WH-147 HD-2204 RR-21
Gram	JG-315 K-850 L-550

For the development of dry land agriculture, various cultural practices like, date of sowing, method of sowing, seed rate, sowing across the slope, application of fertilizers, plant protection, intercropping, timely weeding and inter-culture operations to conserve moisture in the soil were also demonstrated to the farmers of watershed.

Vegetative Measures

Under horticulture programme budding on *desi* ber plants was done by using improved varieties, like, banarasi-karka, apple and amravati bud stocks. The entire forest land in the watershed was afforested with different tree species for the protection and production of watershed area. The important species were *Prosopis juliflora, Acacia nitotica, Acacia leycocephala, Azadirachta indica, Dendrocalamus strictus, Dalbergia sissoo, Albizia lebbek* and *Bombax ceiba* etc. The seeds of trees were shown on staggered contour trenches and cattle proof trenches, however, the seedlings of different species were planted in ring pits (45 × 45 × 45 cm^3 size).

Results and Discussion

Groundwater Table

After the implementation of the programme in the area, there was a considerable improvement in the ground water level. Even in the severe drought year (1987) the ground water table remained in the depth range of 20-26 m. and an area of about 100 ha land has been brought under tube well irrigation. A private tube well boring of 60 m deep was also completed in the watershed, which has brought some more area under irrigation. The irrigated area has increased from 23 ha during 1983-84 to 43 ha during 1987-88.

Increased Crop Productivity

Out of total 571 ha area of the watershed, 463 ha was cropped during the year 1984. However, after carried out the land development programme, the cropped area was progressively increased about 550 ha in a year with the production level of more than 6000 q. It is remarkable to reveal that the year 1987 was the drought year when cropping intensity in the district Hamirpur was not more than 50 per cent but in the watershed 123 per cent area was cropped during kharif season (Table 29.3). Thus, it is clearly indicated that by appropriate conservation measures, moisture conservation is better managed during drought condition to sustain the productivity.

Table 29.3: Total Cropped Area and Production of Field Crops in the Watershed

Year	*Cropped Area (ha)*			*Production (q)*		
	Kharif	*Rabi*	*Total*	*Kharif*	*Rabi*	*Total*
1984	187	276	463	2058	2429	4487
1985	202	291	293	2004	2452	4456
1986	255	298	553	2759	3959	6688
1987	230	316	546	2398	4220	6618

Consumption of Fertilizers

The data in Table 29.4 revealed that the use of fertilizers during pre-project period was almost negligible (2.6 kg/ha) in the watershed. As a result of different soil and water conservation measures, better soil moisture was assured in the root zone of the crops prompting the farmers to use fertilizers for increasing their crop yields. The total consumption of fertilizers in 3 years (1985, 1986 and 1987) was increased by about 36 per cent.

Table 29.4: Consumption of Fertilizers in the Watershed

Year	*Kharif*		*Rabi*		*Total*
	Rate kg/ha	*Total Used*	*Rate kg/ha*	*Total Used*	
1985	2.64	533	14.98	4359	4892
1986	4.06	1035	16.50	4917	5952
1987	4.50	1017	18.45	5830	6847

Impact on Livestock Improvement

Due to increase in cropped area and productivity of agricultural commodities, availability of feed and fodder also enhanced in the watershed. As a result, the stall-feeding was introduced and the people started to keep greater number of milch animals (buffaloes and cows) bullocks and goats (Table 29.5). Further, the table showed that the position of livestock has increased from 955 during 1982–83 to 1791 during 1990–99.

Table 29.5: Livestock Position in the Watershed Before and After the Project

Livestock	*Number*		*Per cent Increase*
	Before (1983)	*After (1991)*	
Buffaloes	95	275	189
Cows	187	264	41
Bullocks	172	430	150
Goats	257	742	188
Sheeps	191	80	(-)58
Horse/Mare	9	Nil	-
Pigs	44	Nil	-
Total	955	1791	-

Impact of Demonstrations on Crop Productivity

Through the adoption of different soil and water conservation measures, yields can be improved on sustainable basis. On the developed land, scientifically improved agro-techniques can further increase the productivity to a greater extent, which is remarkably demonstrated in Table 29.6. The table indicates that the yield of rainfed crops, like gram + mustard increased up to 46 per cent in demonstrated field over non-demonstrated field followed by gram alone 37 per cent and mustard alone 22 per cent respectively. Whereas in case of irrigated crops like, mustard, wheat and wheat + mustard the yields increased up to 50, 49 and 24 per cent, respectively in demonstrated area as compared to non-demonstrated area. Thus, the above results indicate that improved package of practices not only increased the overall yield by about 40-50 per cent but also attracted the farmers of watershed for continuous adoption of improved farm technologies. Further, the table showed that even on non-demonstrated fields, the yield of most of the crops was increased manifold as compared to pre-project yields. This is because of the fact that the farmers were being exposed to latest farm technologies through various demonstrations conducted at their fields. In addition to this, different extension programmes, like, field visits, kisan gosthis and farmers fairs etc. were also organized to expose the farmers for adoption of recommended agro-techniques in the region.

Table 29.6: Impact of Improved Agro-techniques on Crop Yields in the Watershed

Crop/Cropping System	*Crop Yield*			
	Pre-project (1982-83) (q/ha)	*Post-project*		*Per cent Increase over Non-demons*
		Under Demons. (q/ha.)	*Non-Demons. (q/ha)*	
Rainfed				
Gram (6 yr.av)	6.30	8.46	6.91	37
Mustard (4 yr.av)	2.20	12.16	9.95	22
Gram+Mustard(4 yr.av)*	–	12.47	8.52	46
Irrigated				
Wheat (6 yr.av)	7.60	22.42	15.02	49
Mustard (5 yr.av)	NA	14.49	9.63	50
Wheat+Mustard (4 yr.av)*	NA	28.24	22.75	24

*: Prevailing local prices were taken into account while calculating main crop equivalent yield.

Conclusion

Thus, the study revealed that the integrated watershed development programme has been successfully implemented in the region for sustainable development and increased productivity due to adoption of soil and water conservation measures and improved agro-techniques to make it an environmentally feasible, economically viable and socially acceptable programme.

References

Anonymous, 1986. A report of Bundelkhand Authority, Sagar, M.P.

Bhushan, L.S., Sharma, A.K., Singh, B.L. and Yadav, R.C., 1997. *Bulletin on Management of Ravine Watershed, Sheetalpur, Hamirpur, Bundelkhand Region*. Central Soil and Water Conservation Research and Training Institute, Research Centre, Agra.

Sastry, G. and Patnaik, U.S., 1990. *Bulletin on Watershed Management*, ICAR, New Delhi.

Gupta, R.N., 1981. Soil and Water Management in Bundelkhand region of U.P. In: *Paper Presented at National Seminar of Rainfed Agriculture and Forestry*, Jhansi, December 17–19, pp. 1–14.

Hazra, C.R. and Singh, D.P., 1992. A Silvipastoral approach for revegetation of barren hills and wastelands: An experience of Gaharawa watershed in Bundelkhand. *J. Soil Water Conservation*, India, 36: 105–114.

Chapter 30

Impact of IPM Team Work on Diffusion and Adoption of IPM Practices in Pigeonpea Crop for Sustainable Yield in North Karnataka

☆ ***S.K. Meti***[1]

Introduction

Gulbarga district in Karnataka state is called one of the pulse bowl and bearing exporting of pure quality *dal* to Burma and other countries. The farmers are cultivating pigeon pea as sole crop since many years in Gulbarga district. As it is evidence by several survey report and researcher that, the sole crop of pigeon pea cultivation invites more pest and disease attack. This is because of every year cultivating the same crop in the same land. For this crop, the major impediments are pod borer *Helicoverpa armigera* (Huborer) which affects the crop production and productivity. In the last decade three severe out break of this pest were noticed and the latest incidence was during 1997. Due to the severe pest incidence caused on average pod borer damage was 90 to 100 per cent of the crop during 1992-93 and 1997-98 and several farmers' suicidal deaths were reported due to crop loss and no expected return. Further, farmers are commonly spraying more than 12 times of pesticides on

---

1 Department of Agricultural Extention Education, College of Agriculture, Raichur, U.A.S., Dharwad, Karnataka.

un-limited basis without proper knowledge. The indiscriminate use of pesticide on the crop increases cost of cultivation of farmers, which leads for suicide and environmental pollution.

Keeping the incidence and event of 1997-98 the University of Agricultural Sciences, Dharwad has taken a lead in exploiting IPM technology through team work for the benefit of pigeon pea growers. The impact of IPM and teamwork can be judged by its effectiveness on the farmers and farm yield. The team work consists of University Scientist and State Department of Agriculture at district level and taluka level.

The teamwork main job is to conduct rapid growing servey twice in a week from the flower initiation period in pigeon pea to harvesting of pigeon pea. Based on the pest incidence (ETL) level, the appropriate control measures through mass media like radio, television, pamphlets, dangura etc. were suggested for effective diffusion of IPM message with in short time. The farmers are being cultivated long duration variety (150–170 days) ICPL 87119. Further, the team has to conduct IPM demonstration in farmers field, supervision of demonstration, advise the farmers in the field on the spot. The dissemination of IPM information were made among the pigeon pea growers by coducting training and pre-seasonal training at off the campus and on the campus. Keeping this teamwork efforts on dissemination of IPM technologies, the study was designed to analyze the following factors.

Objectives

1. To study the impact of team work and IPM technologies on knowledge, adoption, yield and cost of cultivation among pigeon pea growers.
2. To know the opinion about IPM team work efforts and awareness of information by the farmers.

Methodology

Integrated pest management (IPM) is one of the effective control measures to combat the pod borer damage in pigeon pea crop, which was developed by Agriculture Research Station at Gulbarga. The study was designed to critical analysis of impact of IPM practices and IPM teamwork effectiveness on dissemination of IPM technological knowledge to protect the pod borer damages on pigeon pea crop. In Gulbarga district, three talukas were selected for the study namely, Gulbarga, Aland and Sawalagi based on the highest cultivable area of pigeon pea crop. In each taluka 2 villages were selected and from each village 15 IPM farmers were selected. The total size of the sample for the study was 60 IPM and 60 non IPM farmers comprising of small, medium and large farmers. The information was gathered in pre-tested interview schedule and analyzed the information by appling appropriate statistical device. The analyzed information were presented in tabular form.

Results and Discussion

It could be observed from the Table 30.1 that, the majority (70.00 per cent) of the farmers opined more satisfactory about IPM practices and team work in management of Pigeon pea pod borer at Gulbarga District.

Table 30.1: Opinion of Respondents Regarding IPM Team Work on Rapid Growing Survey (n=60)

Sl.No.	*Category*	*Frequency*	*Percentage*
1.	Most satisfactory	12	20
2.	More satisfactory	42	70
3.	Less satisfactory	06	10

This might be due to the frequently visiting of team experts to different villages during rapid growing survey and provided the information to needy farmers.

The result in Table 30.2 depicted that, the majority (75 per cent) of the respondents were opined that, the IPM practices information utilized in time. This might be due to timely dissemination of information through mass media like TV, newspapers, radios, Pamphlets, Bulletins etc.

Table 30.2: Opinion of Respondents Regarding Utilization of Information (n=60)

Sl.No.	*Category*	*Frequency*	*Percentage*
1.	Aware of practice	45	75.00
2.	Adopted the practice	42	70.00
3.	Aware but not adopted	15	25.00

Table 30.3: Adoption of IPM Practices in Pigeon Pea Crop by the Pigeon Pea Growers (N=60)

Sl.No.	*IPM Practices*	*Frequency*	*Percentage*	*Rank*
1.	Fall ploughing to expose pupae to hot sun and natural enemies	57	95.00	1
2.	Placing of pheromone traps@5 per ha. to attract male moths and to monitor the pest	56	93.33	2
3.	Erecting bird perches @ 20 per ha.	55	91.66	3
4.	First spray with ovicides like Thiodiocarb 75 WP @ 0.6g/l or profenofos 50 EC 2ml/l or Methomyl 12.5L @2ml/l of water	50	83.33	4
5.	Second spray with 5 per cent neem seed kernel extract or 2ml/l of commercial neem formulations	48	80.00	5
6.	Third spray with Ha NPV @ 250 LE/ha LE/ha	45	75.00	6
7.	Fourth spray with quinplphos 25EC 2ml/l or Chloropyriphos 20EC @2.5ml/l or Endosulphan 35 EC @ 2ml/l of water	38	63.33	7
8.	Fifth spray wuth cypermethrin 10EC or Alphamethrin 5EC @ 5ml/of water	35	58.33	8
9.	In case of scarcity of water dust endosulfan 4 per cent followed by quinolphos 1.5 per cent and fenvalrate 0.4 per cent @ 25 kg/ha	30	50.00	9

It could be seen from the Table 30.3 that, out of nine IPM practices the fall plough to expose pupae to hot sun and natural calamities were adopted by (95.00) of the respondents followed by the placing of pheromone traps@5 per ha. to attract male moths (93.33 per cent) erecting bird perch at the rate of 5/ha (91.66 per cent) and first spray with ovicidal chemical spray like prophenophos 50 EC 2 ml/liter (83.33 per cent) of the farmers were adopted. Further, less than 80 per cent of the respondents adopted second spray with 5 per cent of neem seed kernel extract. Whereas, third spray with HNPV @ 250 LE/ha (75.00 per cent) and fourth spary with qunolphos 25 EC 2ml/lit. (63.33 per cent). Followed by the fifth spray cypermethrin 10 EC or Alphamethrin 5EC @5ml/lit. of water (58.33 per cent) and dusting of chemicals (50.00 per cent). The over all results shows that out of nine practices six practices were adopted by more than 75 per cent of respondents.

It was observed from the Table 30.4 that the average pod damage was 7.99 per cent in IPM demonstration field as compared to 18.69 per cent in non IPM farmers field.

Table 30.4: Pod Damage and Yield Data of IPM and Non-IPM Farmers in Demonstration Plot

Sl.No.	*Demonstration Plots (Area in acres)*	*IPM Farmers*				*Non-IPM Farmers*			
		No. of Pods/Plant		*Total Pods*	*Per cent of Pods Damages*	*No. of Pods/Plant*		*Total Pods*	*Per cent of Pods Damages*
		Good	*Damaged*			*Good*	*Damaged*		
1.	5	203.5	8.50	212.0	4.01	102.454	22.68	125.13	18.13
2.	2	107.6	17.50	125.10	13.98	116.80	20.43	137.23	14.88
3.	1	103.0	11.45	114.45	10.01	114.00	25.43	139.43	18.23
4.	2	134.8	4.70	139.50	3.37	92.0	22.00	114.00	19.29
5.	5	85.70	7.20	92.90	7.75	86.0	23.00	109.00	21.10
6.	5	98.60	12.48	111.08	11.24	106.45	25.00	127.45	19.61
7.	5	109.6	7.30	116.90	8.24	93.15	23.00	116.15	19.80
8.	3	94.36	12.30	106.66	11.53	92.00	21.00	113.0	18.54
	Average	117.1	10.18	127.32	7.99	100.35	22.81	122.67	18.69

Further it was observed from the result that, the economics of pigeon pea crop production by the IPM and Non IPM farmers. The average gross income of IPM farmer was around Rs. 8,084/acre whereas non IPM farmers was Rs. 5,010/acre. The net income from IPM field was Rs. 6,780/acre compare to non-IPM-farmers was Rs.1,926/acre. It is quite interesting to note that, the average cost of cultivation of IPM farmers was around Rs. 2,310/acre whereas high cost of cultivation from non IPM farmers was is Rs. 3,084/acre.

Table 30.5: The Economics of Pigeon Pea Crop Production of IPM and Non-IPM Farmers

	IPM				*Non IPM*				
Demonstration Fields	*IPM Plot Yield (Qtls)*	*IPM Cost of Cultivation (Rs)*	*Gross Income*	*Net Profit*	*Non IPM Plot Yield (Qtls) per Acre*	*Non IPM cost of Cultivation*	*Gross Income*	*Net Profit*	*Per cent Income Increase Over Non IPM*
1	6.50	2100	9,750	7,650	3.00	2623	4,500	1,877	6.07
2	5.50	1925	8,250	6,325	3.50	2500	5,250	2,750	3.66
3	6.0	1555	9,000	7,445	2.50	3000	3,750	750	12.43
4	6.30	2500	9,450	6,950	4.00	3200	6,000	2,800	2.75
5	5.75	3000	8,625	5,625	3.00	3500	4,500	1,000	11.25
6	6.0	2700	9,000	6,300	3.50	3150	5,250	2,100	5.86
7	6.50	2300	9,750	7,450	2.75	3200	4,125	925	12.92
8	6.0	2450	9,000	6,550	4.50	3500	6,750	3,250	2.25
Average	6.06	2310	8,084	6,780	3.34	3084	5,010	1,926	–
Total	48.48	18,480	64,675	54,245	26.72	24,672	40,080	15,408	–

Rs. 1500 per q. market price.

Conclusion

It is clear from the result that, the IPM practices, team work efforts and participatory approach of farmers significantly contributed to get higher yield with less cost of cultivation in pigeon pea crop. Hence, there is a need for build up of local level IPM teams on coordinated manner by which we can share and care of farmer interest on pigeon pea crop pest management. Further, the district and taluka level advisory service has to be strengthened for frequently visiting to interior and remote corner of villages for farmers field visit and advice in time. The media has to play vital role in consultancy with the team of experts and message has to be treated and channelized through locally available channel in local language for community approach as a whole.

References

Ahmed, Z., 2000. Integrated pest management in Pakistan. *The News*, August 1.

Chandrakandan, K., 1994. Oilseed technology transfer package through mass campaigns, video teaching, demonstration and small group approach. IDRC sponsored research project report. TNAU, Coimbatore.

Patil, S.L., 1998. A study on extent of adoption of recommended cultivation practices of DCH-32 cotton by farmers and its economics in Malaprabha command Area of Dharwad district. *M.Sc. (Agri.) Thesis*, University of Agricultural Sciences, Dharwad.

Chaudhry, S.A., 1981. Plant protection. In: *Paper Presented at 2nd Agricultural Research Conference*. October 24-27, PARC, Islamabad.

Sumathi, P. and Alagasan, V., 1998. Awareness and adoption of IPM practices among groundnut growers. *Journal of Extension Education*, 9(2): 2011–2015.

Singh, H. and Patil, H.N., 1990. Knowledge level and socio-economic characteristics of contract and non contract farmers. *Maharashtra Journal of Extension Education*, 9: 178–180.

Chapter 31

Impact Study of Sulphur Nutrition on Productivity of Soybean in Vertisols of M.P

☆ *A.K. Dixit, D.S. Tomar and Arvind Saxena*[1]

Introduction

Modern agriculture has become highly efficient and has contributed greatly to national economic growth. Madhya Pradesh is generally called "The Soybean State" of India contributing more than 70 per cent in terms of area and 64 per cent of the total production (Joshi, 2003). The soil type is pre-dominantly black-cotton soils, which are low in nitrogen, low to medium in phosphorus, and medium to high in potassium content with alkaline in nature. The area has been identified under >70 per cent deficient in Sulphur. Sulphur deficiency is increasing throughout the world as a result of decreased use of Sulphur (S)–containing and high analysis fertilizers, intensive cultivation, increased irrigation, use of high-yielding varieties, and use of S-free fungicides. The magnitude of deficiency widely differs according to soil types and crop management (Malik, 1999).The major crop of the *kharif* season is soybean occupying an area of 4, 16,000 ha which amounts to 94 per cent of the total *kharif* season. In the wake of acute injudicious use of integrated nutrient management system in Vertisols of Ujjain, Sulphur can play a vital role in augmenting the production and productivity of soybean. It is necessary to study the relationship between the increase in production and the factors involved in such production. Krishi Vigyan Kendra,

---

1 Programme Coordinator, Subject Matter Specialist, Krishi Vigyan Kendra, RVSKVV, Ujjain.

where in farming families are being taught through modern technology step by step through learning by doing with scientists of KVK, is playing significant role.

Methodology

Krishi Vigyan Kendra-Ujjain is engaged with the motto of transformation of technology to the grass root level to boost up the agricultural production by restoring the inherent fertility status of soil. The district is located at 20^0 43 to 23^0 36 latitude and longitude of 75^{0}-76^{0}:30 with altitude of 527 meters above mean sea level. It falls under X^{th}- Agro-climatic Zone, *i.e.* Malwa Plateau and Agro-ecological region-V^{th}. The study was carried out in Ujjain district under two ecological situation and five plots of the farmers of Madhya Pradesh to assess the technology as per the needs of farmers. KVK has a selected technology module for assessment: INMS *i.e.* [RDF= 20:60:30 + 20 kg S/ha] as recommended by AICRP on soybean, 2007.On the basis of PRA conducted, the most prominent problem emphasized for causing the low yield (Average 10.40 q/ha) by the farming community and ranked first, was the imbalance use of fertilizers including no use of micronutrients and secondary nutrients like Sulphur and Zinc. The ranking of problem under PRA was based on adopted village, area wise distribution, importance of enterprises, seriousness of problem and number of affected families. Keeping in view the seriousness of problem, afore said OFT was conducted in five villages adopting full package of practice using variety JS-335 which covers over 80 per cent area to achieve higher production as well as restore the inherent fertility status of the soil. Sulphur deficiency in soils (>70 per cent in area) was diagnosed in soil tests.

Result and Discussion:

The data in Table 31.1(a) shows that the plant height, effective number of nodules per plant and dry matter accumulation was significantly more at 45 DAS and 75 DAS in treated plots as compared to the control. The increase in dry matter accumulation at two crucial stages was higher by 28 and 22.7 per cent over control which amounts to 0.068 g/plant/day and 0.08 g/plant/day, additional dry matter accumulation.

Table 31.1(a): Effect of Sulphur on Crop Growth and Yield Attributing Characters

Treatment Sulphur kg/ha	*Plant Height (cm)*		*Effective Nodules/ Plant*		*Dry Matter Accumulation at (g)*	
	45DAS	*75DAS*	*45DAS*	*75DAS*	*45DAS*	*75DAS*
0	57.5	90.4	47	68	10.75	26.46
20	69.4	96.8	62	83	13.82	32.7
CD (P=0.05)	3.96	3.82	3.54	6.82	0.98	2.57

This was also supported by the yield attributing characters (Table 31.1b) where in the number of pods/plant and 100 seed weight was higher by 77 and 5.22 per cent, respectively. The cumulative effect of all the characters elaborated above finally contributed towards higher yield in sulphur treated plots ie; 19.1q/ha as compared to 13.90q/ha in control plots. The above results are in conformity with the findings of Malik and Khurshid, 2007.

Table 31.1(b): Effect of Sulphur on Crop Growth HI and WUE

Treatment Sulphur kg/ha	*No. of Pods/ Plant*	*Yield (q/ha)*	*Stover Yield kg/ha*	*Harvest Index (Per cent)*	*100 Seed Weight (g)*	*WUE (kg grain/cm of Water Used)*	*Sulphur Use Efficiency (kg grain/kg S Used)*
0	31	13.90	48.40	34.1	11.50	17.07	–
20	55	19.10	54.7	38.27	12.10	23.46	26
CD (P=0.05)	8.50	0.67	6.06	–	–	–	–

The figure on metrological parameters further indicates that during the entire crop growth period the total rainfall received was 814 m.m distributed mainly between the seed emergence stage to seed development. This led to the better utilization of the inputs used particularly fertilizers as a result of which the sulphur use efficiency and the WUE in treated plots accounted to 26 kg grain/kg S and 23.46 kg grain/cm of water, Ghosh *et al.*, 2003. The total increase in WUE owing to better moisture condition over control was 37.43 per cent.

Data in Table 31.3 reveal that the yield of soybean increased by 37 per cent with the application of 20 kg Sulphur/ha, giving an overall net profit of Rs 10150 over the farmers practice and thus the incremental B: C ratio was 1:3.26, which indicates the economic feasibility and viability of the technology under assessment.

Table 31.3: Analysis of B : C Ratio in the Technology Assessed

Technology Assessed/ Refined	*Production per unit*	*Net Return (Profit) in Rs./Unit over Farmers Practice*	*Incremental B:C Ratio*
Farmer's practice	1390	10150	1:3.26
Technology assessed RDF (20:60:30)+ 20 kg sulphur	1910		

Horizontal Spread

Looking into the benefits of the technology farmers rapidly adopted in larger areas which occupied 15 per cent area of during the current kharif season-2008 with the application additional sulphur containing fertilizers amounting to 28447 MT SSP, 15 MT gypsum 15 MT ammonium sulphate based on the sales data of the district. Data in Table 31.4, reveals the direct impact of the technology in terms of the additional production, total increase in income, amount of employment generated to produce the additional production and the contribution of the technology in terms of total revenue generated. This reflects towards the economic sustainability of the technology adopted by the farmers.

Table 31.4: Horizontal Spread of Technology and Projected Increase in Production, Employment and Income Generation

Year	*Area*	*Productivity*	*Total Production*	*15 Per cent Area of District*	*Production in 15 per cent Area*
2007-08	425000	11.5	4887500	63750	733125
2008-09*	428350	12.2	5225870	64253	783880.5

Increase in Production	*Per cent Increase in Production*	*Total Increase in Income (Lacs)*	*Employment in Man Days Generated*	*Extra Revenue Earned in Employment (Lacs)*
		0.0	0	0
50755.5	6.92	659.8	457632	37

Applicability

Soybean being the most economic crop of Malwa Plateau zone during *kharif* season has changed the economic scenario of the farmers over the last 10 years, but previous gains have not sustained today, because of increasing cost of production and yield stagnation whereas the genetic potential of varieties is more than 2.5 t/ha. Keeping this in view and the suitable technology available for yield enhancement, farmers realized that the importance of sulphur utilization in soybean resulting in excellent yield increment as well as the better economic crop returns.

Sustainability

Almost the entire area is occupied under soybean cultivation with improved varieties having high production potential. Due to continuous soybean-wheat cropping sequence and injudicious use of fertilizer the soil health has deteriorated.On the basis of Soil Testing data the area has been categorized under severe deficiency zone *i.e.,* 70 per cent. After conduction of On-Farm Trials farmers were fully satisfied and agreed with the importance of sulphur application in soybean. Hence this technology is bound to be sustainable owing to its better performance.

Equitability

The result of sulphur application in soybean @ 20 kg/ha assessed in different location in the similar agro-ecological situations has showed that yield increments were almost equal in all plots with variation of 2.25 per cent in minimum and maximum yield. Hence, the result has characteristics of equitability.

References

Ghosh, P.K., Bandopadhyay, K.K., Tripathi, A.K., Hati, K.M., Mandal, K.G. and Mishra, A.K., 2003. Effect of integrated management of farmyard manure, phosphocompost, poultry manure and inorganic fertilizers for rainfed sorghum (Sorghum bicolor) in Vertisols of central India. *Indian Journal of Agronomy,* 48(1): 48–52.

Joshi, O.P., 2003. Future perspectives of soybean in India. *Soybean Research,* (1): 29–42.

Malik, R.S., 1999. Crops in India need sulphur. *Indian Farming,* pp. 11–13.

Malik, T.H. and Khurshid A. Sofi, 2007. Effect of *Rhizobium* and different levels of sulphur on growth, nodulation, yield and oil content of soybean *Glycine max* (L). *Environment and Ecology,* 25S(2): 325–326.

Chapter 32

Impact of Watershed Management on Land Use Pattern and Productivity Level of Different Crops of Jamui District (Bihar)

☆ ***R.B. Prasad, Dilip Kumar, R.K. Sinha and V.K. Singh*[1]**

Introduction

Majority of people in India live in rural area and depend on agriculture for their livelihood. They spend a high proportion of their income on food. Population density is continued to increase and land available for expansion of agriculture is becoming scarce. So scientists and policymakers have laid stress on the need to increase food production substantially to keep with growing population. During early phase of planned development efforts were made to achieve self-sufficiency in food grain through assured irrigated agriculture. The objective of overall food availability in the country has been achieved no doubt but during the development process proper attention was not paid toward rainfed agriculture which resulted in shape decline in the annual production of food grain during drought years when rainfed crops suffer.

---

1 ARI, Mithapur Farm, Patna , Rajendra Agricultural University, Bihar.

Rainfed farming system includes 148 million of land in our country, so scientific development of these area is much essential. Scanty rainfall on one hand and high intensity of rainfall on the other hand are the foothills. Water harvesting is a big problem in this area as the entire area is having high and steep slope with tendency of soil erosion during monsoon season.

Agriculture department started soil conservation work mainly in these areas which has given increase not only in cultivable area but also in productivity. Table 32.1 indicates that area under paddy in sample villages have increased from 25 per cent in Muratarn village to 280 per cent in Mahugai village. The total area of paddy has increased from 416 ha. To 869 ha. which is clear cut indication that water shed practices has given a positive impact on increase of area under paddy cultivation. Further Table 32.1 also indicates that there is increase of area of Maize + Arhar from 11.25 per cent in sample village of Pratappur to 142.8 per cent in Khashara village with over all increase in area under maize + arhar from 158 ha. to 316 ha that is 100 per cent increase of area within span of 15 years. But there is decrease in Ragi crops up to 30 per cent though it is not significant. Due to positive impact on watershed management practices limited area is covered with vegetable crop. Not only this in Rabi season people have started wheat crop and the area under that has increased from 119 to 259 ha. Similar increasing trend has also been observed in case of Kulthi and Moong.

Productivity Level of Major Crops

Since maize, arhar ragi and paddy are major crop of the sample areas under watershed management, so the productivity of these crops was compared before and after the implementation of watershed and soil conservation programme. The result of which are presented in Table 32.2. It could be seen from the table that there is a significant increase in the productivity level of different crops over a period of 15 years. The average yield of paddy increased by 5.56 q/ha and similar increase was also found in case of maize + arhar. When we look the Table 32.2 carefully it becomes clear that increase in paddy yield was found 25 per cent in sample Pratappur village and 64.28 per cent in case of Mahugai village. It is interesting to note that though there is decrease in the area of Ragi crops but it has an increase trend in their yield due to adoption of better agronomical practices and use of drought tolerance varieties of ragi crops.

Therefore integrated watershed area development approach is persuaded. Rain fed and marginal and small farmers are primarily practicing dry land agriculture. So this section of society should be given paramount importance. Any effects of watershed development project should reflect in the component like adoption of agricultural technology by farmers on the large scale, higher productivity level of the crops, change in land use pattern, generating more income from the watershed area and ultimately improving overall socio economic status (SES) of the farmers.

Considering this Government of India launched watershed project in the country with the help of different state government. Consequently Government of Bihar, Department of Agriculture and soil Conservation is operating different watershed management for development of the people of rainfed and dry areas. Agriculture

Table 32.1: Change in Areas of Different Crops Before and After Implementation of Watershed in Sample Village (area in ha)

Village	*Paddy*		*Maize+Arahar*		*Ragi*		*Vegetable*		*Wheat*		*Kulthi*		*Moong*	
	A	*B*	*A*	*B*	*A*	*B*	*A*	*B*	*A*	*B*	*A*	*B*	*A*	*B*
Dhond	15	40 (166)	25	40 (60)	10	8.9 (–11%)	2	12	16	24 (50%)	–	–	–	–
Pratappur	30	40 (333)	8	9 (11.25%)	12	8.4 (–30%)	–	–	40	40 (–)	–	–	1	4
Keluadih	8	1250 (%)	4	650 (%)	15	120 (–20%)	1	3	–	–	0	1	–	–
Khiria	25	35 (40)	15	25 (66.6)	15	10.5 (–30.6)	1	5	–	–	4	10	–	–
Murataru	8	10 (25%)	4	6 (50)	3	2.0 (–25%)	–	–	–	–	–	–	–	–
Mahugahi	140	532 (280)	32	60 (87.5%)	–	–	–	–	23	75 (226%)	–	–	–	–
Kharahara	180	200 (11.1)	70	170 (142.8)	10	7.0 (–30%)	20	30	40	120 (200%)	10	20	20	20
Total	416	869	158	316	65	48.8	24	50	119	259	14	31	21	24
Diff	453		158		(–) 16.2		26		140		17		3	
% Increase	108		100		24.92		108.3		117.6		121.42		12.28	

A: Yield Before 1980; B: Yield After 2006.

Figure in parenthesis indicates changes in percentage.

Table 32.2: Effect of Management on Productivity of Different Crop in the Sample Village (q/ha)

Village	*Paddy*		*Maize+Arahar*		*Ragi*		*Wheat*		*Kulthi*		*Moong*	
	A	*B*	*A*	*B*	*A*	*B*	*A*	*B*	*A*	*B*	*A*	*B*
Dhond	20	25(25%)	10	15(50%)	4	6(50%)	10	14(40%)	–	–	–	
Pratappur	16	20(25%)	15	20(33.3%)	3	5.5(75%)	12	15(25%)	–	–	2.5	
Keluadih	12	18(50%)	7	10(42.55%)	4	5.5(37.5%)	–	–	0	3	–	
Khiria	17	22(29.41%)	10	14(40%)	3	5(66.6%)	–	–	2.5	3.5	–	
Murataru	11	15(36.36%)	8	12(50%)	2	3.5(75%)	–	–	–	–	–	
Mahugahi	14	25(64.28%)	12	20(66.66%)	–	–	16	20(25%)	–	–	–	
Kharahara	16	20(25%)	8	18(125%)	2	5(150%)	15	22(46.6%)	3	4	4	6
Total	106	145	70	109	18	30.5	53	91	5.5	10.5	6.5	10

A: Yield Before 1980; B: Yield After 2006.

Figure in parenthesis indicates changes in percentage.

department has spent a lot of money during last 15 years on these projects. Now, Government of Bihar is interested to know the impact of these schemes on the people by knowing the increase in productivity of different crops of the area change in the land use pattern and improvement in this SES. This work was assigned to Agricultural Research Institute, Patna to know the impact of this work on the farmers of the area in term of increase in productivity level, enhance in land use pattern and improvement in SES. Thus this study was undertaken with the following objectives:

1. To study the change in land use pattern for different crops in last 15 years.
2. To know the productivity level of different crops of the watershed area.
3. To study impact of soil conservation and watershed management on the S.E.S. of the farmers.

Methodology

The study was conducted in seven villages of Lower Kuil watershed in Jamui District of Bihar State. From each of the identified village a list of beneficiaries having land in watershed was prepared. Than from the list of beneficiaries data were collected by using participatory Rural Technique (PRA). The benchmark data was obtained from the secondary data available with the office of the watershed development (Soil Conservation Office). Jamui. The ex-post facto research design was employed in the study and data was analyzed. The land use pattern and productivity level of major crops, which are grown in project area, before and after implementations of the project were considered in the present study.

Findings And Discussion

Dry land agriculture is not only restricting the expansion of the cultivable land but it also limited the increase in productivity of Watershed Area of Lower Kuil of Jamui District of Bihar which is a hilly track reddish as well as having sandy soil. Few patches are having heavy texture. In case of wheat the increase was found 9.50 qt/ha in the project area. The increases in the productivity level of different crops were clearly suggest the adoption of watershed management practices by the farmers. These practices helped in not only in water harvesting but also it protected the top soil fertility which enabled the crop to survive during dry spell, thus avoiding possible crop failure. Further more, enhanced technical guidance to the farmers inforce them to take up timely agricultural practices, using drought prone varities which helped farmers in increasing crop yield. This shows clear-cut positive impact of watershed management practices on the land use pattern and productivity.

Conclusion

The lower Kuil watershed project is one of the model watersheds being operated by the Department of Agriculture and department of soil conservation where State Agricultural University as well as Krishi Vigyan Kendra has lent their technical support and technical guidance. It can be concluded that with the help of Agriculture Department and Department of Soil Conservation considerable change in land use pattern and increase in area with increase in productivity level of different crops shows the impact of government effort in increasing the S.E.S. of the farmers of those area. In general, these programmes proved to be beneficial. Hence, such programme can be taken up extensively with adequate financial Assistance in order to achieve watershed objectives.

Chapter 33

Impact Analysis of Apiculture Training Programme: An Evaluatory Study of Samastipur District in Bihar

☆ ***Anuradha Ranjan Kumari*[1], *Meera Singh*[2] *and Sudhir Das*[3]**

Apiculture on scientific lines is an ideal activity for the development of women in rural areas. This opens up opportunities and incentives for developing entrepreneurship among rural women. It does not require any sophisticated equipments and the technology rather is simple and within easy grasp of illiterate rural women. Women can easily perform all that operations involved on apiculture. Training is vital and essential to enhance motivation, create confidence and inculcate efficiency among them. It is an important educational tool which can be effectively used to improve, refresh or upto date the farm women's and major factor that is required to increase the agro-based production apart from availability of input in development of these women trough training. Ultimately, it will enhance capabilities, improve the competency and help in getting assumed income and self-employment

---

1 Research Associate (H.Sc.), KVK, Sheohar, R.A.U., Pusa

2 Dean, College of Home Science, R.A.U., Pusa.

3 Programme Co-ordinator, KVK, Sheohar, R.A.U., Pusa.

from the enterprise. However, training programmes will be effective and purposeful only when they are based on felt needs of rural women. Keeping this in view, the present study was, therefore undertaken to know the impact of training on different aspects of Apiculture technology with the following objectives.

1. To study the level of knowledge gained by women through apiculture technology.
2. To measure the extent of adoption of recommended apiculture technology.
3. To study the relationship between independent variables with dependent variables regarding acquisition of knowledge, skill and adoption of apiculture technology.

Methodology

Seven villages from Pusa Block of Samstipur district were selected purposively for study, because maximum women were trained from these villages. From this block, 50 trained and 50 untrained women were selected, thus, total 100 women were selected for the study. The fourteen independent variables and three dependent variables *i.e.* knowledge, skill and adoption of apiculture technology were undertaken during the study. The data were obtained through personal interview with the help of interview schedule. Analysis of data was resorted to frequency count and t-value. The correlation and multiple regression analysis were also done to know the relationship of independent variables with knowledge, skill and adoption of apiculture technology.

Results and Discussion

Data presented in Table 33.1 indicate that out of 50 trained women, 96.0 per cent had high knowledge, only 4.0 per cent had medium knowledge and none of these fell under low level of knowledge about apiculture technology, but in case of untrained women maximum 98.0 per cent women had medium knowledge followed by only 2.0 per cent had low knowledge. It was also revealed that both the categories of women *i.e.*, trained and untrained differed widely from each other in respect of level of knowledge about apiculture technology. The mean knowledge score of trained and untrained beekeepers were 117.34 and 70.30 respectively. The computed t-value (47.4370) is highly significant.

In case of skill, the data indicate that highest percent (98.0 per cent) had high level of skill followed by only 2.0 per cent had medium level of skill apiculture technology. The finding further indicated that trained respondents were more prone in attaining more skill of bee management technology as compared to untrained beekeepers women. The mean skill score of trained and untrained beekeepers were 102.42 and 60.36 respectively. The computed t-value (35.9099) is highly significant.

In case of adoption of apiculture technology, the trained respondents who were adopted at high level were 88.0 percent followed by 12.00 per cent had medium level adoption. There were no respondent who have low level of adoption in trained groups.

Table 33.1: Distribution of Respondents According to Extent of Knowledge, Skill and Adoption of Apiculture Technology

Sl.No.	Variables	Categories	Trained (n=50)		Untrained (n=50)	
			F	%	F	%
1.	Knowledge	Low	0	0	1	2
		Medium	2	4	49	98
		High	48	96	0	0
		Mean	117.34		70.3	
		t-value	47.437**			
2.	Skill	Low	0	0	1	2
		Medium	1	2	49	98
		High	49	98	0	0
		Mean	102.42		60.36	
		t-value	35.909**			
3.	Adoption	Low	0	0	10	20
		Medium	6	12	38	76
		High	44	88	2	4
		Mean	31.16		14.6	
		t-value	27.848**			

The mean score of trained and untrained women were 31.16 and 14.60 respectively. The t-value 27.8487 is highly significantly. It is obvious from this table that trained beekeepers had significantly higher level of adoption than those of untrained beekeepers. Apiculture training had increased the professional skill of trained women and they were more equipped with latest technical know how of honey production, they might had adopted more recommended technology in their apiary as result of which the production of honey bee colony was also significantly better than of untrained women.

Relationship of Independent Variables with Knowledge, Skill and Adoption of Apiculture Technology

It is clear from Table 33.2 that the variables caste, education, family education, annual income, cosmopoliteness and economic motivation were found to be positively and significantly correlated with level of knowledge, skill and adoption at 1 percent level of significance of trained beekeepers. Age was negatively but significantly correlated with level of knowledge, skill and adoption. The remaining seven variables were found to be non significant whereas in case of untrained beekeepers the correlation co-efficient analysis revealed that the variables *viz.* caste, education, family education, annual income, cosmopoliteness and economic motivation were positively and significantly correlated with level of Knowledge, Skill and Adoption at 5 per cent level of significance. Age was negatively but significantly correlated with level of Knowledge, Skill and Adoption of apiculture technology.

Table 33.2: Relationship Between Selected Independent Variables and Knowledge, Skill and Adoption of Apiculture Technology by the Rural Women

Sl.No.	Independent Variables	Correlation Co-efficient 'r' Value					
		Knowledge		Skill		Adoption	
		Trained	Untrained	Trained	Untrained	Trained	Untrained
1.	Age	-0.4344**	0.3259*	-0.4002**	-0.3240*	-0.3063**	-0.3299*
2.	Caste	0.7147**	0.3038*	0.6065**	0.3010*	0.6686**	0.2849*
3.	Personal Education	0.3691**	0.3221*	0.4352**	0.3182*	0.4087**	0.3408*
4.	Family Education	0.4315**	0.2838*	0.5188**	0.2816*	0.4987**	0.3209*
5.	Occupation	0.0179	0.0646	-0.0305	0.0610	0.0287	0.0367
6.	Family size	0.2187	-0.1219	0.2130	-0.1246	0.0484	-0.0860
7.	Family type	0.1863	-0.0283	0.0703	-0.0281	0.2399	-0.0223
8.	Size of land holding	-0.0579	0.0847	-0.0611	0.0897	-0.0393	0.0991
9.	Annual income	0.4338**	0.3517	0.2813**	0.3371	0.3713	0.3493*
10.	Type of house	0.0670	0.0706	0.0037	0.0678	0.1794	0.0714
11.	Household material possession	-0.0745	0.0978	-0.1307	0.1028	-0.0363	0.1024
12.	Social participation	-0.1969	-0.1233	-0.0539	-0.1370	-0.1446	-0.0796
13.	Compositeness	0.4247**	0.3327*	0.4740**	0.3361*	0.5720**	0.3340*
14.	Economic Motivation	0.4607**	0.3233*	0.4536**	0.3122*	0.3745**	0.3199*

**: Significant at 0.01 level; *: Significant at 0.05 level.

Date presented in Table 33.3 revealed that the coefficient of determination (R^2) for 14 selected variables jointly explained 86.56 per cent variation in the knowledge. It has been further revealed that the regression co-efficient was significant only in case of three variables *viz.*, caste, family size and economic motivation. This shows that these three variables had significant influence on the knowledge of women towards apiculture technology of trained women.

In case of untrained women the coefficient of determination (R^2) from 14 variables was jointly predicted to the extent of 61.95 per cent on the knowledge.

Table 33.3 also revealed that on skill of strength of coefficient of determination (R^2) of 14 selected independent variables of trained women was 84.53 per cent. It further revealed significant coefficient in case of only three variables namely caste, family education and economic motivation. In case of untrained women the coefficient at determination (R^2) was found 61.12 per cent on the skill.

In case of adoption, the data show that coefficient of determination (R^2) for 14 selected variables jointly explained 83.21 per cent of trained women. It was also revealed that the regression coefficient was significant in case of three variables *viz.* caste, family education and family type were positively and significantly contributed in the prediction of adoption. The contribution of remaining eleven variables were

Table 33.3: Multiple Regression Analysis Between Selected Independent Variables and Knowledge, Skill and Adoption of Apiculture Technology by the Rural Women

Sl.No.	Independent Variables	Knowledge				Skill				Adoption			
		Trained (n=50)		Untrained (n=50)		Trained (n=50)		Untrained (n=50)		Trained (n=50)		Untrained (n=50)	
		b	t	b	t	b	t	b	t	b	t	b	t
1.	Age	–0.1104	1.21	–0.2421	–1.61	–0.1252	–0.92	0.2043	–1.59	0.0699	–0.79	–0.0513	–1.58
2.	Caste	3.8211	4.72**	0.6923	0.40	3.5922	2.97*	0.6149	0.42	2.0172	2.57*	0.0513	0.34
3.	Personal Education	–0.6502	–1.10	–0.4432	–0.20	0.1295	0.15	0.4272	–0.32	–0.6057	–1.06	–0.0258	–0.08
4.	Family Education	1.0690	1.60	1.3544	0.72	3.2535	3.25*	1.0944	0.68	1.5093	2.32*	0.4486	1.11
5.	Occupation	0.7874	1.12	–0.5688	–0.44	1.6436	1.54	–0.4638	–0.42	0.1395	0.26	–0.2154	0.76
6.	Family size	1.8823	2.31*	–1.7380	–0.69	2.4389	2.00	–1.6025	–0.74	–0.3582	–0.45	–0.1461	–0.27
7.	Family type	0.60000	0.85	1.0526	0.41	0.0400	0.38	0.9570	0.43	1.5691	2.30*	0.1089	0.19
8.	Size of land holding	–0.8914	–1.55	–1.0984	–0.56	–1.0907	–1.27	–0.8720	–0.52	–0.3212	–0.58	–0.2104	–0.50
9.	Annual income	–0.5662	–0.70	2.7995	1.52	–1.7759	–1.47	2.2327	1.42	–0.3765	0.48	0.3046	1.52
10.	Type of house	0.4915	0.87	0.2464	0.16	–0.1938	–0.23	0.1669	0.14	0.8661	1.50	0.0237	0.08
11.	Household material possession	–0.6741	–0.53	–0.6804	–0.32	-3.7629	-1.96	–0.4153	–0.23	-2.1003	–1.68	–0.2256	–0.49
12.	Social participation	–0.6613	–0.59	-3.8177	–1.10	1.3410	0.81	-3.4914	–0.17	0.4476	0.41	–0.5136	0.68
13.	Casmopoliteness	–0.0449	–0.53	0.2802	0.61	0.0083	0.04	0.2607	0.66	0.2913	1.98	0.0638	0.66
14.	Economic motivation	0.7501	3.26*	0.6120	1.49	1.1284	3.28*	0.4934	1.40	0.3958	1.77	0.1299	1.46
		R2=0.8656		**R2=0.6195**		**R2=0.8453**		**R2=0.6112**		**R2=0.8321**		**R2=0.6202**	

*: Significant at 0.05 level; **: Significant at 0.01 level.

non-significant However in case of untrained, out of fourteen variables under study none of the variables contributed significantly towards level of knowledge, skill and adoption of scientific apiculture technology.

Conclusion

The above findings clearly indicated that the trained beekeepers possess higher level of knowledge, skill and adoption of apiculture technology. Apiculture training has been successful in producing a significant impact on the participants.

References

Chahal, B.S., 1995. *Success of Beekeeping in Punjab.* Beekeeping School on All India Radio Pub., Deptt of Entomology, P.A.U., Ludhiana, p. 60–64.

Das, P. and Sharma, J.K., 1998. Impact of training on knowledge and perception of rural youth about scientific beekeeping. *Journal of Ext. Edn.*, 9(2): 1957–1963.

Goyal, R.A., 1999. Apiculture gainful employment for rural people, *Kurukshetra,* 47(10): 44–48.

Katariya, J.S. and Singh, A.P., 1987. Factors associated with gain in knowledge through farmers training, *Maha. J. Ext. Edn.*, 6: 45–50.

Mahipal and Prasad, M.S., 1997. Impact of training programmes on extension personnel. *Maha J. of Ext. Edn.*, 16: 111–115.

Singh, R., 2000. Impact of training on adoption behaviour of beekeepers. A study of Muzaffarpur districts. Unpublished *M.Sc. Thesis,* Deptt. of Ext. Edn. RAU., Pusa, Bihar.

Shrivastava, K.K. and Lakhera, M.L., 2003. Impact analysis of mushroom training programme. *Maha. J. of Ext. Edn.*, 22(2): 22–27.

Shrivastava, K.K., Bareth, L.S. and Sarkar, J.D., 1996. Impact of farmers training centre on rural women. *Maha. J. Ext. Edu.,* 15: 141–145.

Verme, J.P., 2003. Effectiveness of training programme for farmers. *Kuruskshetra,* 51(3): 30–37.

Chapter 34

Impact Assessment of Potato Technologies in Indo-Gangetic Alluvial Plains of Bihar

☆ *Barsati Lal*[1], *Anil Kumar*, *T.K. Sinha*[2] *and N.K. Pandey*[3]

Introduction

Potato is an important crop of Bihar. Sustainability of the region for potato cultivation was realized long back with this background. Central Potato Research Institute was established at Patna in 1949. The average productivity of the state has reached to 19,0 t/ha during recent years. This increase in productivity is the results of latest potato production technologies evolved by the Central Potato Research Institute (CPRI) as well as the training and development work being done by the Institute with close collaboration with the state government and farming groups. Keeping in view, CPRI has developed several high yielding short, medium and long duration potato varieties for the agro- climatic conditions of the state. It also includes the promoshing varieties like Chipsona–1, 2 and 3. Besides this balanced fertilizer application, Integrated pest management practices, Seed Plot Technique (SPT) for quality seed production, forecasting of late blight disease, package of practices for better potato seed production were also evolved by the Institute.The present study was conducted with the objective to evaluate the some of the latest potato production technologies which were transferred among the farmers of the state.

---

1 Scientist, S.S, CPRS, Patna.

2 Technical Officer , CPRS, Patna.

3 Senior Scientist and Head, Division of Social Sciences, CPRI, Shimla, H.P.

Methodology

The study was conducted in purposively selected villages of Kurkuri, Dhanaut and Newraganj of Phulwarisharif and Bihta blocks of Patna district of Bihar. Demostration were laidout on 54 farmers' field, selected after identification of technological gap in the areas of; use of high yielding variety, seed preparation, plan ting operations, nutrient management, intercultural operations, irrigation management, insect/pest and disease management and harvesting and post harvest management.. Demonstrations were laid out on the use of high yielding variety of potato, balanced fertilizer application, cutworm management, late blight management and true potato seedling raising. During the implementation phase, adopted farmers were also supported by training, literature, field visits, Radio/T.V. talk etc.

Findings and Discussion

Impact of Improved Seed of High Yielding Variety of Potato

Under the intervention-improved seed of high yielding variety of potato *i.e.* Kufri Ashoka, Kufri Pukhraj, Kufri Sutluj, and Kufri Ananad were planted in 200 m^2 area on each farmer's field. After maturity of the crop, yield was recorded. Finding indicated that potato variety Kufri Ashoka fetched maximum yield (242.16q/ha) followed by Kufri Sutluj (240.00q/ha) against the farmers variety "C-1" (140-170q/ha). There was also a significant increase in tuber yield (48.77 per cent) recorded in high yielding potato varieties (Table 34.1). The farmers were also get benefited by a net return of Rs. 37600/ha to Rs. 52000/ha by adopting the interventions.

Table 34.1: Yield and Economics of Interventions on Farmers Field

Sl.No.	*Interventions*	*Av. Yield (q/ha)*		*Yield Adv. (Per cent)*	*Net Return Rs/ha*	*B:C Ratio*
		Farmers Plot	*Demo. Plot*			
A.	**High Yielding potato variety**					
1.	Kufri Ashoka (n=10)	162.66	242.16	48.87	50972	1.87
2.	Kufri Sutluj (n=1)	170.00	240.00	41.17	52000	1.92
3.	Kufri Pukhraj (n=2)	165.00	224.00	35.75	44800	1.80
4.	Kufri Anand (n=1)	140.00	208.00	48.57	37600	1.67
B.	**Balanced fertilizer application (n=19)**					
1.		208	280.88	72.88	57352	2.04
C.	**Late blight management (n=16)**					
1.		267	338.66	71.66	80464	2.46
D.	**TPS seedling raising (n=2)**	140.00	128.00	-8.57	14200	1.38

Impact of Balanced Fertilizer Application

Farmers of the area were applying imbalanced doses of chemical fertilizers. Therefore, under this intervention, balanced fertilizer application trial using nitrogen, phosphorous and potash was applied @ 150;60:100 kg/ha respectively. Half dose of nitrogen and full doses of phosphorous and potash was applied at the time of planting. Rest dose of nitrogen was given at the time of earthing-up (35 days of planting). Perusal of data presented in Table 34.1 clearly showed that there was 72.88 per cent increase in tuber yield of balanced fertilized potato crop as against the farmers practice (80:40:00). Further, farmers were also obtained a net return of Rs. 57352/ha due to the adoption of the practice (Table 34.1).

Impact of Interventions on Late Blight Management

Late blight, a dreaded disease of potato causing havoc in the area when it founds a favourable weather conditions. Upto 40-80 per cent crop damaged occurs by the disease in the area. Farmers of the area are indiscriminately using fungicides to protect their potato crop against the disease. To overcome the problem, interventions were laidout on applying need based spraying schedule. Under the intervention, two spray of fungicide Dithane-M-45 was applied @ 2.00 kg/ha at 10-12 days intervals. The first spray was applied on the crop between 15-20 December followed by second spray at 10-12 days of interval. Similarly, third spray was done by applying Ridomil M.Z. @ 2.00 kg/ha. Data were recorded at the time of maturity of the crop. Results indicated that there was a significant increase in crop yield (72.66) over the farmers practice (Table 34.1). Further, by adopting the intervention farmers were also got a net return of Rs. 80,464/ha with a high B: C ratio of 2.46.

Impact of True Potato Seedling Crop

Keeping in view, the scarcity of cold storage facilities in the state True Potato seed (TPS) technology have a vast potential for wider adoption among the farmers. Therefore, under the project nursery of TPS was raised at Central potato research station, Patna and the seedling were transplanted (at a distance of 6–10 cm plant to plant and 50 cm from row to row) in 200 m^2 area on the farmers field. Balanced doses of nitrogen, phosphorous and potash were applied @ 150: 60: 100 kg/ha. respectively. After the harvesting of the crop data on yield and economics was recorded. Data presented in Table 34.1 clearly showed that by adopting the technology, farmers had obtained a net return of Rs. 14200/ha as against the farmers practice. Though the yield was comparatively low (128.00q/ha) as against the farmers practice (140.00q/ha) but considering the seed cost, TPS technology was found economical on the farmers field. Hence, have a good potential for its adoption among the farmers in those areas where cold storage facilities are in scanty.

Conclusions

To conclude, the potato technologies have very good impact in enhancement of yield and economics of the potato farmers and proved beneficial for them as well as cost effective as compared to the local varieties and methods being adopted by the farmers.

Chapter 35

Constraints in Adoption of Technologies by Farmers of Watershed Areas in Different Agro-Ecological Regions of India

☆ *Om Prakash*[1] *and Y.V.R. Reddy*[2]

Introduction

Watershed Development Programme has been carried out on large scale with an annual allocation of funds to the tune of about Rs.1000 crores in India with an objective of developing and stabilizing natural resources in addition to improving productivity of crops, employment opportunities to rural people, income generation and standard of living of rural people in arid/semi-arid regions in India. During the past three decades, many crop production technologies were evolved and spread to farmers' fields. Since 1983-84, technologies of dryland crops were also transferred to farmers' fields effectively on watershed basis, as technological options and choices available to dryland farmers are many and varied and it may be a problem tofarmers to choose suitable package of technologies suiting to their fields. Though it is difficult

---

1 Senior Scientist, CSWCRTI, Research Centre, Datia – 475 661, M.P.

2 Central Research Institute for Dryland Agriculture (ICAR), Hyderabad.

to enlist any single component of technology that would bring out rapid and dramatic transformation, yet it is possible by adoption of a series of improved components of technologies to minimize instability in yield which is a common characteristic of dryland farming. The improved production technologies in dryland consist of high monetary (fertilizers, plant protection chemicals etc.), low monetary (seed, weeding etc.) and non-monetary (timeliness and precision) inputs. Thus a dryland farmer is in a better position to choose varieties, crops, crop mix and agronomic practices to suit the local environmental conditions. Recommendations on soil and water conservation, alternate land use systems, soil and crop management are also available to farmers to make efficient use of resources and environment. The dryland technologies in relation to crop production at Research Stations are to be transferred to farmers' fields through effective extension agencies. Farmers' management adequacy and resource facilities that farmer could draw upon ultimately decide the success or failure of a viable improved crop production technology which is determined by a number of socio-economic parameters and constraints. Their attempt has been made in this study to know the levels of adoption of dryland technology in different parts of India and reasons for non-adoption.

Materials and Methods

The four components of improved crop technology *viz.* improved seed, application of fertilizers, plant protection chemicals and weeding were considered for this study in different Watershed Development Programme areas vis-à-vis non-watershed areas (adjacent to watershed area) under different agro-eco system regions. The data were collected from 50 farmers in watershed villages and 50 farmers in non-watershed villages during 2001-2002. Detailed information were collected and tabular analysis method was followed in analysis of data. The study focuses on adoption levels of four components of dryland crop technologies covering 3700 farmers from 37 locations representing different agro-ecoregions and also covered an area of about 7250 hectares.

Results and Discussion

Thirty seven locations from where the farmers covered were fairly well served by Extension Agencies in watershed villages but in non-watershed villages, farmers learnt from watershed farmers/Agricultural Officers/magazines/seed and pesticide companies etc. Thus the farmers in watershed villages are exposed to crop production technologies in addition to soil and water conservation technologies whereas farmers in non-watershed villages were not exposed directly but they observed farmers practicing in watershed village in addition to learning through Agriculture Officers, magazines, and advertisements etc., which have indirect influence on the farmer in non-watershed villages. Though there were number of crops at each location and also crops varied based on soil type, environment, topography of soil, rainfall and its distribution etc. from location to location, the average adoption of technologies by farmers at each location was summarized in this paper to have an overall view of adoption of improved dryland technologies in different parts of the dryland tracts in the country. Though the data pertained to agriculture year 2000-2001 in watershed and non-watershed villages, even data were collected after withdrawal of watershed programme. The watershed programme was carried out for 5 years and then withdrawn. The stage of watershed programme withdrawal may be any time during 1990 to 2000 at the locations. Thus there was a scope to learn from watershed area farmers by non-watershed village farmers due to prolonged time also.

The reason given by farmers for non-adoption of technologies were found to be more or less same and crops over locations under two different situations.

Adoption of Improved Dryland Technologies

The adoption levels of improved seed, fertilizer, plant protection resources and improved weeding are presented in Table 35.1.

Table 35.1: Percentage of Farmers Adopted Under Different Crop Production Technologies Under Two Situations at Different Locations in India

S.No.	State	Watershed Location	Watershed Development Programme Villages				Non-watershed Villages			
			*S	F	P	W	S	F	P	W
			NWDPRA Watersheds							
1.	Andhra Pradesh	Chittoor	12	48	6	98	8	42	4	86
2.	Assam	Jorhat	42	68	18	88	36	52	12	66
3.	Bihar	Ranchi	16	82	22	72	10	75	16	70
4.	Chattisgarh	Raipur	45	90	42	80	40	86	30	76
5.	Gujarat	Banaskantha	68	86	36	96	50	82	24	88
6.	Karnataka	Chitradurga	20	60	16	100	12	38	8	92
7.	Madya Pradesh	Indore	90	92	70	80	86	90	64	76
8.	Maharashtra	Solapur	92	84	52	92	82	76	38	82
9.	Orissa	Pulbhani	18	36	4	66	8	30	4	60
10.	Punjab	Relmajra	76	92	36	98	74	90	30	96
11.	Tamil Nadu	Madurai	50	56	24	82	30	40	20	70
12.	Uttar Pradesh	Agra	28	78	68	96	26	76	66	94
13.	Uttar Pradesh	Mirzapur	18	42	18	78	14	36	10	74
			MRD Watersheds							
14.	Andhra Pradesh	Medak	22	62	16	92	10	52	6	78
15.	Karnataka	Gollahally	36	78	48	98	30	72	40	92
16.	Maharashtra	Kolhapur	26	68	26	76	20	62	16	72
17.	Orissa	Koraput	20	58	22	78	18	56	18	70
18.	Punjab	Hoshiarpur	40	86	76	92	36	82	70	88
19.	Rajasthan	Bhilwara	28	66	18	82	22	62	8	80
20.	Tamil Nadu	Coimbatore	22	62	12	76	16	54	6	74

Contd...

Table 35.1–Contd...

S.No.	*State*	*Watershed Location*	*Watershed Development Programme Villages*				*Non-watershed Villages*			
			**S*	*F*	*P*	*W*	*S*	*F*	*P*	*W*
			ICAR Watersheds							
21.	Haryana	Bhiwani	30	86	22	88	22	78	16	82
22.	Karnataka	Bellary	16	70	16	86	8	62	6	80
23.	Karnataka	Bijapur	18	66	12	82	10	60	4	78
24.	Karnataka	Kolar	10	52	10	88	8	48	8	82
25.	Maharashtra	A'nagar	38	72	22	94	26	60	10	80
26.	Rajasthan	Kota	26	68	12	86	18	52	8	82
27.	Uttar Pradesh	Jhansi	32	76	28	88	20	60	18	84
		International Agency supported Watersheds								
28.	Haryana	Adilabad	24	66	18	92	12	52	8	82
29.	Karnataka	Maheswaram	28	70	22	86	26	66	18	84
30.	Karnataka	Junagadh	60	72	60	98	48	72	28	88
31.	Karnataka	Tiptur	32	68	18	88	16	54	10	80
32.	Tamil Nadu	Ramanathapuram	6	38	4	60	2	26	0	52
			NGO Watersheds							
33.	Andhra Pradesh	Anantapur	28	72	32	94	18	60	18	84
34.	Maharashtra	Ralegon Siddhi	60	88	60	100	40	62	20	82
35.	Tamil Nadu	Dharma puri	12	38	8	86	10	30	8	84
36.	Uttar Pradesh	Faizabad	16	82	38	88	12	78	18	82
37.	West Bengal	Purulia	18	52	16	78	14	38	6	66

*S: Seed; F: Fertilizer; P: Plant Protection ; W: Weeding

It is clear from results in the Table 35.1 that the adoption of technologies by farmers varied from location to location and even it can happen among the farmers within the location. Moreover, it is clear that more farmers adopted technologies in watershed area compared to non-watershed villages due to exposure technologies and confidence developed by farmers in watershed area.

Adoption of Improved Seed

The adoption of improved/hybrid seed was more in watershed area farmers compared to non-watershed area farmers. However, the variation of adoption of technologies by farmers varied widely over locations in the country. The percentage of farmers adopted improved seed varied from 6 per cent at Ramanthapuram (Tamil Nadu) to 92 per cent at Solapur in watershed villages as against 2 per cent at Ramanthapuram (Tamil Nadu) to 86 per cent at Indore in non-watershed villages. The adoption rate of improved seed was more in commercial crops followed by coarse grains at all locations in watershed and non-watershed villages. The availability of adoption of improved seeds is very high over locations.

The variations in using latest seed/variety over locations is due to fast changes of availability of different seed varieties with marginal advantage of latest seed. However, hybrid seeds were reported to be costlier and could not sustain under unfavourable (low/high erratic rainfall situations etc.) conditions. Improved seed was reported to be required fertilizer plant protection chemicals, improved agronomic practices. Whereas local seeds could be produced by farmers in their fields serving and providing yields under adverse situations unlike hybrid seeds. Rabi sorghum (M-35-1) has been in the field for the past 50 years due to no other variety replaced as yield advantage variation received fast adoption. Thus the variability was seen based on socio-economic factors and agro-climatic conditions. The adoption rate was slightly low in non-watershed areas.

Adoption of Fertilizers

Among the four components of technology, fertilizers is the costliest. Therefore majority of farmers did not apply recommended doses for crops. The application of fertilizers by farmers varied in between 10 per cent to 60 per cent in case of grain/ pulses crops and 20-75 per cent of recommended doses in case of oilseed crops/ commercial crops. In case of cotton in Vertisols and soybean in Indore area, a very few farmers apply even more than recommended doses. Nevertheless, majority applies fertilizers to all crops but the quantity varied based on economic conditions of farmers and expected income of crops. Even farmers, by and large, applied more quantity of fertilizers to latest varieties. The application of fertilizers over location was found to be more due to non-availability of Farm Yard Manure, which was utilized for irrigated crops which receives not only FYM but also higher fertilizer doses due to assured yields/income. Thus Indian farmer is clever enough to use his resources more effectively for maximizing his income. The adoption of fertilizers in number of farms and quantity is low in non-watershed areas farmers.

Plant Protection Chemicals

The use of plant protection chemicals for dryland crops was reported to be low. However, commercial crops receive more attention. Pigeonpea, chickpea, castor etc. are sprayed with plant protection chemicals. Even dryland cotton receives more attention in this regard and the cost of plant protection is reported to be higher in case of cotton. Seed treatment is also becoming common practices at all locations in most of the crops. Thus all hybrids are taken care of by plant protection chemicals in all crops over all locations. Thus even dryland crops have been receiving the attention of plant protection chemicals use. However, it is low in non-watershed villages.

Improved Weeding

Family labour could be used for weeding. Even marginal and small farmers have exchange of labour practice at all locations. Thus it is considered as low monetary input. The adoption levels were by and large, higher except in crops grown under broad casting. All line saving crops are given more importance in improved weeding.

However, farmers in non-watershed area also practice improved weeding practice to large extent. However, adoption was marginally slightly low in non-watershed area compared to watershed area farmers

Extent of Area under Adoption of Improved Technologies

The percentage of area adopted under four crops production technologies are presented in Table 35.2.

Table 35.2: Percentage of Area Adopted Under Different Crop Production Technologies at Different Locations in India

S.No.	State	Watershed Location	Watershed Development Programme Villages				Non-watershed Villages			
			*S	F	P	W	S	F	P	W
			NWDPRA Watersheds							
1.	Andhra Pradesh	Chittoor	8	40	3	90	6	30	4	82
2.	Assam	Jorhat	38	58	8	82	30	50	10	60
3.	Bihar	Ranchi	6	68	10	70	4	62	12	65
4.	Chattisgarh	Raipur	30	72	18	80	26	58	22	72
5.	Gujarat	Banaskantha	52	66	18	82	32	48	18	82
6.	Karnataka	Chitradurga	12	56	12	96	8	44	6	85
7.	Madya Pradesh	Indore	70	82	45	70	68	72	50	70
8.	Maharashtra	Solapur	72	78	35	76	70	72	32	74
9.	Orissa	Pulbhani	10	30	2	50	6	18	2	60
10.	Punjab	Relmajra	58	68	28	72	62	54	18	90
11.	Tamil Nadu	Madurai	42	48	16	70	22	36	20	64
12.	Uttar Pradesh	Agra	20	60	46	76	18	48	46	88
13.	Uttar Pradesh	Mirzapur	16	30	10	72	10	22	6	68
			MRD Watersheds							
14.	Andhra Pradesh	Medak	16	52	8	86	8	50	4	68
15.	Karnataka	Gollahally	18	60	36	88	22	52	30	88
16.	Maharashtra	Kolhapur	22	58	20	72	18	50	8	66
17.	Orissa	Koraput	18	52	20	72	16	40	15	64
18.	Punjab	Hoshiarpur	32	72	52	78	30	38	45	82
19.	Rajasthan	Bhilwara	25	50	12	68	18	48	7	76
20.	Tamil Nadu	Coimbatore	16	52	6	66	14	38	5	70
			ICAR Watersheds							
21.	Haryana	Bhiwani	25	60	15	88	16	60	15	78
22.	Karnataka	Bellary	10	62	12	80	6	52	5	69
23.	Karnataka	Bijapur	12	58	8	80	7	52	4	71
24.	Karnataka	Kolar	8	49	6	82	6	43	6	80
25.	Maharashtra	A'nagar	33	68	12	88	20	45	8	78
26.	Rajasthan	Kota	25	60	9	85	14	44	5	76
27.	Uttar Pradesh	Jhansi	26	69	22	86	15	56	12	75

Contd...

Table 35.2–Contd...

S.No.	*State*	*Watershed Location*	*Watershed Development Programme Villages*				*Non-watershed Villages*			
			**S*	*F*	*P*	*W*	*S*	*F*	*P*	*W*
		International Agency supported Watersheds								
28.	Haryana	Adilabad	22	60	16	87	9	49	6	75
29.	Karnataka	Maheswaram	23	62	17	82	20	52	12	78
30.	Karnataka	Junagadh	52	72	48	95	41	66	18	76
31.	Karnataka	Tiptur	28	62	16	82	15	50	10	78
32.	Tamil Nadu	Ramanathapuram	4	33	3	48	2	21	0	50
		NGO Watersheds								
33.	Andhra Pradesh	Anantapur	23	70	22	91	13	38	13	81
34.	Maharashtra	Ralegon Siddhi	50	68	56	100	30	45	17	75
35.	Tamil Nadu	Dharma puri	8	30	6	81	8	28	7	75
36.	Uttar Pradesh	Faizabad	8	60	32	85	9	70	12	80
37.	West Bengal	Purulia	10	38	12	75	10	35	5	63

*S: Seed; F: Fertilizer; P: Plant Protection ; W: Weeding

It is clear from the results of Table 35.2 that the percentage of area covered under different components of technology be quite close to the percentage of farmers adopting the technology. Therefore extent of area under adoption of technology varied from location to location and also under two different situations (watershed and non-watershed). The trend as in case of percentage of farmers adopted emerged in case of extent of area under adoption of improved technology also. The trend also was more or less in case of watershed area vis-à-vis non-watershed area. The adoption of improved seed in commercial crops was more compared to pulse/grain crops. However, adoption of area with improved varieties of sorghum at different locations in Vertisols was also more. Improved varieties of soybean have become more popular in Madhya Pradesh due to higher productivity and income. Similarly fertilizers were also applied by majority of farmers. However, the area covered under fertilizers was also less in non-watershed areas compared to watershed areas. Plant protection measures have gained momentum even in non-watershed areas in case of commercial and hybrid varieties. Thus farmers turned towards adoption of larger area for income. However, adoption rate in relation area also was low at all locations compared to watershed area.

The improved weeding was widely accepted by the farmers for all line saving crops in all regions, as it is not a totally cash input. The near saturation levels of adoption was found except in minor millets/coarse grain crops/minor pulse crops because family labour with the addition of labour on exchange/wage basis could complete the weeding operation.

It is clear from the adoption of four technologies by number of farmers and the extent of area under adoption was certainly higher in watershed area compared to non-watershed areas in arid/semi-arid regions in India. However, adoption was low/slow in arid regions compared to semi-arid regions. The adoption levels depend on rainfall and its distribution, marketing ability of crops, type of soil, and socio-economic conditions of farmers. Generally adoption rate is higher in Vertisols compared to Alfisols/Aridisols due to obvious reasons.

Constraints in Adoption

Among the four components of technology, the adoption of improved seed was also low, as there were many seed varieties on par with latest varieties under marginal/normal conditions where the risk of adoption/crop failure is low. Improved seed requires better agro-climatic/favourable conditions for higher productivity. Hence, the majority of farmers are adopting improved but locally available seeds whereas some farmers go for hybrid varieties. Hence the adoption rate varied. However, the major constraint for non-adoption by majority of farmers at different locations was climatic factor including rainfall, which is not under the control of anybody. The second reason reported was not only costly but also non-availability at the right time. The third reason was reported to be the economic condition of farmers who cannot spend on fertilizers/plant protection chemicals due to uncertainty.

In case of fertilizer, the price was reported to be major constraint, as the prices of fertilizers have been increasing steeply and prices of output of dryland crops were more or less stagnant during the past ten years. Hence farmers suggested having rationale in price fixing for fertilizers and agricultural produce, which has not been there for the past 10 years.

The adoption of plant protection chemicals has been gaining momentum but it has caught the attention of farmers for all crops due to lack of knowledge, no action against dealers for supplying spurious chemicals etc. Farmers reported that Government should stipulate strict regulation in fixing prices, sale etc. in addition to popularizing through proper education/extension methods.

A majority of farmers do not adopt the new technology as a package. The investigation to identify the constraints to adoption of different components of technology revealed that these ranged from primary reasons such as lack of awareness of the practice to such rational reasons as the risk involved in switching over to improved technology. However, economic factors seem to dominate the spectrum of constraints.

Conclusion

Thus, it can be concluded that the fear of loss due to failure of rains/monsoon was attributed to be the major constraints in adoption of improved seed and fertilizers in addition to lack of awareness on improved seed and plant protection measures among farmers. The economic factor also plays a major role. These constraints, individually or in combination were found to be responsible for non-adoption of technology by farmers.

Chapter 36

Extension Gap and Impact of FLD on Oilseed and Pulses Crops Conducted by KVK Janjgir-Champa: A Case Study

☆ ***R.N. Sharma, Manish Kumar, N.K. Toorray, M.K. Chandraker and S. Shantaiya[1]***

Introduction

The Janjgir–Champa district is situated 21.6° to22.4° N latitude,82.3° to83.2° E longitude with an altitude of 294.4 M. The total geographical area of the district is 486674 ha. The district is surrounded by Raigarh in east, Bilaspur in west, Korba and Bilaspur in north and Raipur in south.The average rainfall of the district is 1147.4 m.m while normal rainfall is 1477.8 mm. The maximum and minimum temperature of the district is 49° c and 8° c respectively. The area under *kharif* crops is 261323 ha and in *rabi* 90000 ha. The total irrigated area of the district is 134905 ha.

Oilseed and pulse crops observe as rich source of energy, employment and commerce. Being a rich source its required high attention for taking a bumper crop. But unfortunate part is that it is grown under energy-starved conditions of uncertain and low moisture availability, low fertility and lack of knowledge about suitable agro-technique.

---

1 Krishi Vigyan Kendra, Janjgir-Champa , Indira Gandhi Agricultural University , Raipur – 492 006.

In Year 2005- 06 the FLD of oilseed and pulse crops were conducted under the supervision of KVK scientists in different villages of Jangir- Champa district in both *kharif* and *rabi* seasons.

Yield of *Kharif* Oilseed Crops (Ground Nut and Soybean)

The variety of Ground Nut ICGS. 44 was sown in 5 farmers field having an area 2-5 ha. The average yield obtained was 20.5 q/ha as compare to 16.36 ha with local variety (Table 36.1). The increase in yield was 25 per cent more.The variety of soybean J.S.335-95 was sown in four farmers field. However, for Soybean the average yield was 16.22 q./ha as compared to local check 12.16 q/ha (Table 36.1). The increase in yield was 33 per cent more with improved package and practices.

Yield of Kharif Pulse Crop (Urd)

The crop Urd was grown under rainfed condition. The variety P.U 30 was sown in 7 farmer's field having an area 5 hect. The yield obtained was 5.16 q/h as compared to local check variety 3.26 q/ha (Table 36.1). Although the yield was poor but economically beneficial to the farmers because it was grown under light textured soil. Further,earlier rice crop was grown in that field with very low productivity.

Yield of Rabi Oilseed Crop (Sunflower)

The crop Sunflower was sown in the second fortnight of January in nine farmers field in the village Parshahi Nala. The Sunflower hybrid varieties used were Mahabeeg and 933. The yield obtained between 12.40 to 13.60 q/ha with having average yield 12.95 q/ha (Table 36.2).

Table 36.1: Results of FLD of Oilseed and Pulse Crops Conducted During *Kharif* 2005 by KVK, Janjgir-Champa

1.1 Ground Nut

Sl.No.	*Name of Farmer's*	*Village*	*Yield (q/ha.)*
1.	Shri Santu Ram	Barha Gura	21.4
2.	Shri Dhuki Ram	–do–	19.8
3.	Shri Panath Ram	–do–	20.8
4.	Shri Jagat Ram	Vadhi Kona	20.3
5.	Shri Rohit Kumar	–do–	20.2
	Average Yield : 20.5 (q/ha.)		

1.2 Soybean

Sl.No.	*Name of Farmer's*	*Village*	*Yield (q/ha.)*
1.	Shri Hari Singh	Tharowed	15.50
2.	Smt kalpana chandel	–do–	16.20
3.	Shri Gopal Souriyabansi	–do–	17.30
4.	Shri Vishram Gaurh	–do–	15.90
	Average Yield : 16.22 (q/ha.)		

Contd...

Table 36.1–Contd...

1.3 Urd

Sl.No.	Name of Farmer's	Village	Date of Sowing	Date of Harvesting	Yield (q/ha.)
1.	Shri Gupi Lal	Nandali	12.07.05	6.10.05	4.33
2.	Shri Satayabhan	–do–	13.07.05	7.10.05	6.4
3.	Shri Prabhat	–do–	22.07.05	18.10.05	4.4
4.	Shri Buwan	–do–	28.07.05	26.10.05	6.0
5.	Shri Khansi ram	–do–	27.07.05	25.10.05	5.2
6.	Shri Manharan	–do–	25.07.05	23.10.05	4.8
7.	Shri Guray Lal	–do–	24.07.05	22.10.05	5.2
	Average Yield : 05.18 (q/ha.)				

Table 36.2: Results of FLD of Oilseed and Pulse Crops Conducted During Rabi 2005–06 by KVK, Janjgir-Champa

2.1 Sunflower

Sl.No.	Name of Farmer's	Village	Date of Sowing	Date of Harvesting	Yield (q/ha.)
1.	Shri Nilamber	Parshahi Nala	16-1-06	2-5-06	12.80
2.	Shri Paras	–do–	16-1-06	2-5-06	12.40
3.	Shri Ratiram	–do–	17-1-06	3-5-06	12.60
4.	Shri Swetamber	–do–	17-1-06	3-5-06	13.00
5.	Shri Anurag	–do–	18-1-06	1-5-06	13.20
6.	Shri Bhagvat	–do–	18-1-06	1-5-06	13.60
7.	Shri Iandravan	–do–	20-1-06	4-5-06	12.80
8.	Shri Roshan	–do–	20-1-06	4-5-06	13.0
9.	Shri Prem Lal	–do–	20-1-06	4-5-06	13.20
	Average Yield :12.95 (q/ha.)				
1.	Shri Krishna Kumar	Kotami	19-12-05	19-4-06	14.00
2.	Shri Bodhan	Kotami	19-12-05	19-4-06	13.00
3.	Shri Sursalan	Khatola	20-12-05	21-4-06	13.20
4.	Shri Naval Prasad	Khatola	20-12-05	21-4-06	16.40
5.	Shri Narendra	Khatola	20-12-05	21-4-06	15.00
6.	Smt. Ahilaya	Barganwa	25-12-05	27-4-06	15.60
7.	Shri Bhoosan	Barganwa	25-12-05	27-4-06	15.20
8.	Shri Ranavir Sing	Barganwa	26-12-05	28-4-06	14.00
9.	Shri Vishanu Kumar	Kotami	31-12-05	29-4-06	15.80
10.	Shri Kamal	Kotami	31-12-05	29-4-06	14.80
11.	Shri Vijay Kumar	Kotami	31-12-05	29-4-06	14.20
	Average Yield : 14.6 (q/ha.)				

Yield of *Rabi* Pulse Crop (Gram)

The crop Gram was sown in eleven farmers field having an area 5 ha in four villages. The variety Vijay was used for sowing purpose. The crop was sown in second fortnight of December 2005. The yield was obtained between 13 to 16q/ha (Table 36.2).

The above results revealed that the percentage increase in yield of various oilseed and pulse crops is up to 60 per cent and it is a huge gap between recommended practices and farmers practices. This is a clearcut indication still there are lot of scope for enhancement of yield. The number of FLD's should be increased so that more area come under benefit of improved techniques.

Chapter 37

Adoption and Level of Knowledge of Farmers Regarding Biofertilizer in Agra District (U.P.)

☆ ***Rajeev Chauhan[1], J.P. Singh[2] and R.P. Singh[3]***

Introduction

The Indian agriculture from mid 60s has made significant strides in food grain production which has increased from a level of 72.35 million tones to 212 million tones in 2005 thus despite of ever growing population and need for import of food grain for domestic use has virtually been dominated. During 1991-2000 our population growth *i.e.* 1.91 verus the 2.11 annual food grain growth which are satisfactory situation. But till 2030 the population of India will increase on a level of 1.3 billion till then, the India will have to maintain 2.1 to 3.1 per cent continuous annual growth in food grain production.

Along with Green Revolution in the Indian farming system, the departure amount of nutrients from soil have been always high against the provided nutrients and this story seems to continue, so our soil fertility has been degraded to great extent.

---

1 Research Scholar, Department of Agricultural Extension, R.B.S. College, Bichpuri, Agra.

2 Head, Department of Agricultural Extension, R.B.S. College, Agra.

3 Associate Professor, Department of Agricultural Communication and Rural Development, SVBP University of Agriculture and Technology, Modipuram, Meerut.

Under an estimate of Planning Commission and Ministry of Agriculture found that this situation will be continue in the future circumstances with a increase level. And there is hope to increase in production by using more unit of chemical fertilizers and we cannot use this amount of chemical fertilizers because such kind of production economy will be put adverse effect on our production cost. Other scenario such as recent energy crises rapid deplation of non renewable energy resource like naptha, natural gas and sulphur etc. may not permit high production of chemical fertilizers. Moreover, import of fertilizers to meet the growing demands will impose heavy foreign exchange burden on the country. Further in U.P. the majority of the farm holdings are operated by small and medium level of farmers it is difficult to them to purchase and use recommended chemical fertilizer doses at current prices. Therefore, they need to exploit other less expensive nutrients sources to the maximum in this context we have to think for alternative and or supplementary sources of nutrients supply to crops and emphasis be given to the use of fertilizers of biological origin biofertilizers like rhizobium azotobacter blue green algae etc because they are based on renewable energy source and cost effective supplement to chemical and help to economics high revertment needed for fertilizer used as far as N and P are concerned.

They are used to fix atmospheric nitrogen to solubulize and make available nutrient like phosphorus and to synthesize other growth promoting substances.

As per report (Motsara *et al.*, 1995) Rhizobium and Azotobacter energy upto 50-100kg and 20-25kg n/ha under field conditions rhizobium can increase the yield of pulses and cereals upto 10-35 and 7.52 per cent respectively (raj et al.) keeping in view of the importance of bio fetilizer, the study was undertaken with the following specific objectives.

Objectives

1. To study the level of knowledge of farmers and adoption of biofertilizers.
2. To identify the major constraints experienced by the farmers in the adoption of biofertilizers.

Methodology

The study was conducted in C.D., Block, Bichpuri of U.P. state. Four villages namely Chauhatna, Sadarban, Sahara and Anoothi were selected randomly. Out of those selected villages 80 respondents were selected through random sampling technique. The data were collected with the help of well structured interview schedule. Independent variables namely, education, land holding, knowledge and respondent use of biofertilizers were related for the purpose of the study. The collected data were analysed with the help of using percentage and chi-square methods.

Results and Discussion

The important outcomes of the study are presented in the Tables 37.1 to 37.14.

Table 37.1 demonstrate that only 2.5 per cent respondents have low level of education, while 15 per cent respondents have medium level of education and

82.50 per cent respondents have high level of education. Thus study concludes that the majority of the respondents under study have high level of education.

Table 37.1: Distribution of the Respondents According to their Education Level

Sl.No.	*Educational Category*	*Score Range*	*No. of Respondents*	*Percentage*
1.	Low	<2	2	2.50
2.	Medium	3-4	12	15.00
3.	High	>4	60	82.50
	Total		80	100.00

Table 37.2 reveals that 45 per cent respondents are having medium size of holding, while 35 per cent having their large holding and only 20 per cent respondents have small size of land holding.

Table 37.2: Distribution of the Respondents According to their Size of Holding

Sl.No.	*Size of Holding*	*No. of Respondents*	*Percentage*
1.	Small Holding	16	20.00
2.	Medium Holding	36	45.00
3.	Large Holding	28	35.00
	Total	80	100.00

Table 37: 3 Overall Association Between Education Level/Land Holding with Knowledge about Biofetilizer Doses, Time of Application, Method of Application of Different Categories of Respondents for Selected Crops

Sl.No.	*Selected Crops*	*Education Level with Knowledge*			*Land Holding with Knowledge*		
		X_2 Values (Overall)			*X_2 Values (Overall)*		
		Doses	*Time of Application*	*Method of Application*	*Doses*	*Time of Application*	*Method of Application*
1.	Arhar	26.310	22.391	19.496	35.364	30.052	26.525
2.	Moong	38.130	34.250	32.719	49.752	45.507	44.100
3.	Wheat	32.156	26.057	23.339	36.231	36.361	34.287
4.	Barley	29.139	26.620	24.681	31.840	29.035	22.231
5.	Mustard	16.965	13.404	9.673	21.357	11.808	8.647

The results incorporated in Table 37.3 shows that significant association were found with education and knowledge of respondents related with selected practice of biofertilizer application except the method of application in case of mustard which means that knowledge of respondents has increased with increase of level of education of subsequently adoption of biofertlizer practices were also enhanced. Table 37.3 also reveals that association between land holding and knowledge about selected practices of biofertilizers application was found significant except in case of mustard it means that knowledge of respondents has increase with increase in size of holding it might be due to the reason that large size of holdings make farmers innovative, resourceful, risk oriented and provided crave of him to make progressive farmers so they acquire more knowledge for proper adoption.

The Table 37.4 depict that education level and size of land holding was found significantly associated with extent of adoption of biofertilizer by the overall respondents in all respective crops except crop of mustard.

Table 37.4: Overall Association Between Education Level/Land Holding with Adoption of Biofertilizers of Different Categories of Respondents for Selected Crops

Crops	*Educational Level with Adoption X_2 Value Overall*	*Land Holding with Adoption X_2 Value Overall*
Ahar	9.662	14.919
Moong	22.082	31.513
Wheat	16.831	25.838
Barley	16.228	12.066
Mustard	4.258	3.805

It was reflecting that the adoptions of biofertilizers by respondents there was increase with increase of level of education and size of land holding. The reason for this might be that educated farmers had progressive outlook and ability to adopt innovations than little literate and illiterate farmers. For size of land holding it was found that adoption of biofertilizer increase with increase of level of education and size of land holding. The reason for this might be that educated farmers had progressive outlook and ability to adopt innovations than little literate and illiterate farmers. For size of land holding it shows that adoption of biofertilizers by the respondents increase with increase in size of land holding it might be due to that large size of holding increase the risk bearing ability of the farmers and they adopt new technologies.

Conclusion

From the above findings the main conclusion are drawn here as under:

1. A very high majority *i.e.* 82.50 per cent respondents had high level of education belonged to medium and large size of family.
2. There was significant association with education and knowledge of respondents and selected biofertilizer practices in case of all the five selected crops (Arhar, Moong, Wheat, Barley and Mustard).
3. The overall reason concludes that education and size of holdings were significantly associated with the adoption of biofertilizer in respect of related crops of pulses.
4. Majority of the respondents have experienced the problem of regarding adoption of biofertilizer technology.

References

Punia and Punis, 2003. Report on the biofertilizers adoption and demand is limited because of unawarenesss of farmers about the advantages of biofertilizer and inadequate knowledge about the use of biofertilizer.

Zora, 2001. Concluded the rhizobium inoculation in leguminus saved 50 per cent N_2 requirement of the crop.

Section IV:

ENTREPRENEURSHIP DEVELOPMENT AND INNOVATIVE APPROACHES IN EXTENSION EDUCATION

Chapter 38

Strategies for Accelerating Technology Adoption to Improve Rural Livelihood: A Prospect of North Bihar

☆ *G. Trivedi*[1] *and Ashok K. Singh*

Introduction

Bihar's economy is primarily agrarian in nature where 80 per cent of the population is dependent on agriculture and its allied sector. The dominance of agriculture in Bihar's economy has become more apparent as a result of bifurcation of the state in 2000. After the bifurcation of the Bihar state, no industry is left under its possession, so agriculture is now become the mainstay of the people of Bihar. The natural resource endowments of the state are dominated by land and water with a meager presence of forests and absence of mineral resources. The rate of transformation of Bihar economy in terms of growing importance of non-farm and non-agricultural sector is much lower. Lack of opportunities in industry, poor rural-urban growth, lack of linkages and underdeveloped infrastructure has contributed to continued dependence on agriculture. Moreover, the population pressure on natural resources is also exceeding high. These attributes amply have the dominance over low- income agricultural sector and, consequently, low per capita income and high poverty,

---

1 Former Vice-Chancellor, RAU, Pusa.

particularly in rural areas calling for immediate attention by the planners and policymakers who are involved in development process.

Although, several intervention have been made through the various governmental and non-governmental agencies in addition to the many International Organizations to evolve and develop appropriate management strategies to increase the efficiency and productivity of the crops in this region yet a range of promising or 'best bet' technologies are still lying for the benefit of small and resource poor farming groups for these agro-ecological systems in order to enhance their livelihood capabilities. Keeping these issues in mind, the present project aims to demonstrate the "Prototype" sustainable technology by using farmer's participatory approach for the benefit of farmers at the community level. To achieve this end, the 'prototype' technologies were disseminated at the community level for the benefits of the small and resource poor farmers in order to accelerate the adoption of selected star technologies.

Objective of the Project

The goal of the project is to contribute to the reduction in rural poverty, generate employment and improve rural livelihood by enhancing productivity and other income generating activities among the resource poor farmers through the adoption of technologies.

Methodology

Locale

In recent years the production system in Eastern India has gone under rapid changes with growing opportunities in the non-farm sector. The understanding of these dynamic forces is essential to accelerate the technology adoption process in order to enhance the rural livelihood. In view of the objective a village Ladaura of Samastipur district and other village Pipraha of Muzaffarpur district were selected purposely on the basis of various assumptions such as agro-climatic and socio-economic condition, cropping pattern and other demographical characteristics.

Selection of Farmers

The selection of farmers was made as per their land holding and other socio-economic conditions of the village. This information was generated through Participatory Rural Appraisal (PRA) method. Here, the efforts were made to select only those farmers who were found proportionally suitable for each technology demonstration. Thus, there were thirty cooperators and equal numbers of non-cooperator from each village for the purpose of technology dissemination and its acceleration. Thus, all together one hundred twenty households were participated during the project activities.

Problems-Cause Analysis and Identification of Potential Opportunities

Based on Participatory Needs and Opportunities Assessment (PNOA), which is a component of PRA tool, several problems faced by different farm families in both the villages were identified.

Further some potential opportunities were also identified in terms of best matching technologies, which were implemented during the intervention for enhancing rural livelihood through acceleration of adoption of the pertaining technologies. The identified 'best bet' technologies for present project were;

- ☆ Enhancing income through mushroom production by using paddy straw.
- ☆ Enhancing crop production by using QPM + Potato intercropping system; and
- ☆ Application of Zero tillage for timely sown wheat in order to reduce the cost of cultivation and get better wheat yield.

Process of Technology Acceleration

For acceleration of selected technology adoption and its validation various activities were carried out such as; *awareness camps* in the selected villages related with identified technologies, which were being pushed for acceleration among the farm communities, frequent *field visits* were made by groups of scientist along with the concerned KVKs functionaries of the area. Several *field demonstrations* were also carried out in the adopted villages for acceleration of the adoption related with selected technologies. The capacity building approach was also used through several specializes *training programmes*. On several occasions *Kisan Gosthi* were organized for the benefits of the selected farm families of the adopted villages.

Salient Achievements

Mushroom Production Technology

As per an estimate, in Bihar state one fourth of total agricultural waste may provide 250g mushroom/per week/per family. The different varieties of mushroom such as oyster, button, and milky white can be successfully grown in the state. The Oyster mushroom becomes the most popular sub-tropical variety, which is widely grown by the farmers of the state followed by the button mushroom. These varieties are performing well during the September to March on farmer's location under natural setting. The paddy straw and milky mushroom are also being grown by farmers in summer season.

During the first year of the project activity, the mushroom production has got unique importance in North Bihar region with the help of regular trainings in terms of mushroom and its quality seed production. Due to its lower production, it is available @ Rs. 50 to 80/per kg. As such rural women, rural youth, unemployed human resource may generate extra income through the mushroom production, which is low cost technology, made available to them. Even retired and under employed rural mass may also utilize their leisure time and harvest the mushroom crop leading to earn a handsome amount with the minimum investment.

Keeping in view, the acceleration and validation of mushroom production technology, efforts were made to disseminate this low cost technology among the landless, marginal and small farmers involving their youth and women folk to get the mushroom production at a large scale. This income generating enterprise has got a tremendous response among them in order to boost their socio-economic condition. With this objective, several mushroom awareness and training progrommes were organized at both of the locations. During the year 2004-2005 a great consciousness has been emerged among the rural youth as well as in women folk and they have started to produce the mushroom in both of location of the project. Now a day, the mushroom production considered as an income generating enterprise among the landless and marginal group of armers residing in these areas.

During the year (2005-06) of the project activity, many entrepreneurs came forward and they have started to produce mushroom for their livelihood. One mushroom producer Sri Kameshwar Singh has undertaken this work at the large-scale basis and two women entrepreneurs also came forward and they have started to produce mushroom for commercial purposes. At present, a large scale of women are eager to involve themselves in the mushroom production system and efforts are being made to involved them in this enterprise through the formation of Self Help Group (SHG) in order to provide them quality seeds and available technology in a very easy mode. The women/rural youth of nearby villages are also being attracted to have the benefit from this technology.

Performance of QPM as Sole Crop

The farmers of the both selected villages were growing local composite varieties of maize, which have not sufficient amount of balanced amino acid composition in its grain. Recently, released Q.P.M. hybrids *viz.* Shaktiman-1, Shaktiman-2, Shaktiman-3 and Shaktiman-4 have high content of lysine and tryptophane. The essential amino-acid is also available in these varieties in order to prevent from protein calorie and mal–nutrition along with drastic increase in yield.

In order to assess the efficiency of Q.P.M. hybrids as sole crop during the year 2005 of Rabi season, ten farmers of both the sites were given seeds of Q.P.M. hybrids with the help of front line demonstration project of the Rajendra Agricultural University. The efficiency of Shaktiman-2 and Shaktiman-3, (Q.P.M. hybrids) was tested along with popular maize variety of Luxmi, Dewki and other local cultivars at the farmers field of both the village and it was found average good yield (71q/ha) with Q.P.M hybrids against the average yield of Luxmi, Dewki and local cultivars (48 q/ha) in the village of Ladaura while the average yield of 70 q/ha of QPM hybrids against the average yield of local cultivars (46 q/ha) was reported in the village Pipraha. The benefit cost ratio was found 1:2.4 with this practice. The details of result are being displayed here through Table 38.1.

Table 38.1: Performance of Q.P.M Hybrids as Sole Crop (Yield of *Rabi* Season)

Variety	*Ladaura (Yield q/ha)*			*Pipraha (Yield q/ha)*		
	2005	*2006*	*Average*	*2005*	*2006*	*Average*
Local cultivars (Luxmi/Dewki)	48	47	47.5	46	47	46.5
Q.P.M hybrids (Shaktiman-2& 3)	71	72	71.5	70	71	70.5

This Table 38.1 shows the relative strength of the QPM hybrids across the local cultivars of maize, which were generally practiced by the farmers of both the locations. This technology intervention has received a great popularity among the farmers of the project area due to its better taste and protein quality in grain leading to the high market price as compared to the local cultivars. With the attraction of this system, during the *Rabi* season of 2006 the area under Q.P.M hybrids has been extended at large extent in both sites of project in order to enhance their productivity and profitability. Now there is a great demand for the QPM hybrids seeds. Recently, some progressive farmers are engaged in seed production process with the collaboration of the university in order to supply quality hybrid seeds to the farmers as per their requirement

Performance of Q.P.M Hybrids with the Potato Inter Cropping System

Maize + Potato intercropping system is unique introduction in the operational area of the project villages. It is a generally practiced in irrigated up and medium land of the area. The maize is shown 8 to 10 days after the potato cultivation in the space available with the rows. Due to lack of selection of compatible varieties of maize and potato, the farmers were getting low productivity in their maize + potato intercropping system. But with the introduction of Q.P.M. hybrids (Shaktiman-1, Shaktiman-2, Shaktiman-3) along with suitable varieties of potato (Kufri Sinduri, Kufri Jyoti, Rajendra Aaloo), the farmers of this area have started to get higher production and better profit. The drastic increase in yield of both the crops with the help of single input system has given a attractive boost to the farmers of this region. The strength of this technology can be judged with help of growing areas with this system along with their perceived benefits, which are being received by them under the maize + potato intercropping system. During the *Rabi* season of 2005, the performance of some farmer's field trials on both the location strongly supports this assumption. The details of the findings are given here in Table 38.2.

Table 38.2: Performance of Q.P.M+ Potato Intercropping System (Yield of Rabi Season)

Variety	*Ladaura (Yield q/ha)*				*Pipraha (Yield q/ha)*			
	2005		*2006*		*2005*		*2006*	
	Maize	*Potato*	*Maize*	*Potato*	*Maize*	*Potato*	*Maize*	*Potato*
Farmers practice (*Local cultivars of maize and potato*)	37	128	38.6	132.5	32	130	41	131
Introduced practice (*QPM hybrids + suitable improved variety of potato*)	48	140	49.5	150.5	47	145	51	146

Performance of QPM+Potato

In the village Ladaura the farmers were able to receive average higher yield of maize crop as compared to the farmers of Pipraha villages (48 q/h in Ladaura village while 47 q/ha yield of maize was found in Pipraha village). With regards to potato yield the farmers field of Pipraha village was found little bit superior as compared to the farmer's yield of Ladaura village. The last *Rabi* season of 2006, again found more suitable for the Maize+ Potato intercropping system and this year too, the farmers who has gone with this system received better yield of maize + potato as compared to their traditional practice. It shows the potentiality of this maize+ potato-intercropping system. In fact, this area is very suitable for this type of intercropping. At present, several farmers were having their new innovation related with intercropping system along with maize crop especially in *Rabi* season. The intercropping system of Rabi maize along with coriander, sugarcane and peas are also becoming very popular in the project villages. Thus, the capability of Q.P.M hybrids along with potato intercropping is becoming more suitable and beneficial for this region. That is why; the farmers are very much attracted towards the Q.P.M. hybrids + potato intercropping system. The farm community of project sites has also started to make practicing it in larger area for getting double yield with the help of single input system.

Resource Conserving Technology (Zero Tillage)

The wheat sowing is generally delayed under low land situation of the project area where local/tall varieties of rice is grown. The selected villages have about one fourth of their arable area under low land ecosystem. Here unworkable conditions of soils are prevailed due to excess moisture in land after harvest of rice. It makes delay in land and seedbed preparation and ultimately wheat sowing often gets delayed. Earlier wheat sowing was carried out even up to the 20th January in low land areas of the project sites and late sown wheat becomes the victim of low temperature at the time of germination and hot westerly desiccating wind at the time of flowering and milk stage. These problems are being resolved now with the intervention of resource conserving technology *i.e.* Zero Tillage Technology. In order to evaluate the performance of Zero Tillage machine for wheat sowing as compared to the conventional method initially demonstration trials were performed on project sites with the help of C.A.E., R.A.U., Pusa, during the *Rabi* season of 2005. The six farmers from Ladaura and only four farmers of Pipraha village have made the demonstration in their approximately 8 hac. of land. Seeing the advantages and good crop establishment the other farmers of adopted village along with their neighboring villages came forward to adopt this resource conserving technology in their wheat sowing in order to save the cost of cultivation as well as to get the good yield in comparison to their traditional system. During the *Rabi* season of 2006 altogether sixteen farmers were participated in the adoption of this technology, as they were able to demonstrate the potentiality of this technology through the zero tillage wheat sowing under the approximately fifteen hac. of their land. The details of yield data of both the villages are presented here through the Table 38.3.

With the introduction of Zero till drill machine farmers were able to plant their wheat about 10-15 days earlier then the normal conventional practice. The early planting of wheat resulted about 18 per cent additional grain yields over the conventional practice. It demonstrates the potential strength of this technology at both of the adopted villages.

Table 38.3: Performance of Zero Tillage Machine for Timely Sown Wheat

Treatments	*Ladaura (Yield q/ha)*		*Pipraha (Yield q/ha)*	
	2005	*2006*	*2005*	*2006*
Conventional Method	32.40	31.86	31.06	30.80
Zero Tillage Method	38.70	38.80	36.29	37.12

Thus, the wheat is being sown timely under the low land ecosystem of the adopted villages. This intervention is creating good impact over the neighboring areas in view of its lower cost of cultivation and better yield. Now, several service providers are coming fore in this region. The most of the tractor owners started to provide this tillage facility to the farmers on hire basis. The rental charges for zero tillage wheat sowing were reported in the range of Rs. 1400/- to 1600/- per hac. in this area.

Conclusion

During the process of accelerating the adoption of selected stare technologies the responses of farmers were found very positive. Since, these demonstrated technologies were able to give more benefit and higher yield hence, they are gaining popularity and being treated as most suitable combination. The main problem confronted by them are, the mushroom spawn, supply of QPM hybrid seed and easy availability of zero-tillage drill. For the purpose of mushroom spawn and Q.P.M. hybrid seeds efforts are being made to involve progressive farmers in production system of quality seed production programme so that these technologies can attract farmers of other neighboring areas of the State and provide more profit to them. With respect to supply and availability of zero-till drill the State Government has started to provide incentive to farmers. Apart from providing subsidy on purchase of zero tillage machine efforts are also being made by the Government to make available these drills at block levels so that farmers can take maximum benefits out of it.

References

Abedin, M.Z., 2004. Participatory agricultural research and technology transfer: Challenges and issues. *Working Paper of Social Science Division,* IRRI, Manila, Philippines.

Ellis, F., 2000. *Rural Livelihood and Diversity in Developing Countries.* OUP, Oxford.

Hobbs, P.R., Giri, G.S. and Grace, P., 1997. Reduced and zero tillage options for the establishment of wheat after rice in South Asia. *Rice-Wheat Consortium Paper Series 2,* New Delhi, India.

Chapter 39

Constraints Faced by the Programme Coordinators and Suggestions Offered by them for Effective Management of KVKs in India

☆ *N.M. Chauhan*[1]

Introduction

At present, more than 564 KVKs working in India have been designed to have vocational training to the practicing farmers, extension personnel and rural youths as well as for all those who want to seek self-employment. The KVK as a grass root level institution has to meet this need, which is so vital for technical literacy and rapid transfer of technology. The Krishi Vigyan Kendra is an organization, which is concerned with social system having several groups of people like; specialists, administrators, technicians, farmers, extension workers, labourers and other beneficiaries Anonymous (2005). Each group has its own characteristics, attitudes, traditions, values and forms of behaviour. The people for whose benefits, the extension activities are designed may not be readily interested due to multiplicity of factors like; likeness, inhibitions, taboos, suspiciousness, reservations and prejudices Bhople and Palaspagar (1996). Looking to the importance of KVKs in effective TOT System, it

---

1 Programme Coordinator, KVK, Vadodara, Gujarat, India.

was found worthy to study the constraints faced by Programme Coordinators of KVKs in the management of different activities of KVKs and seek their suggestions. To fit in the changing scenario of the global agriculture the acclimatization of the Programme Coordinators is a current need of the time for sustainable agriculture development Van de Ban (2004). The constraints faced by Programme Coordinators in performing extension management and suggestions offered by them to overcome some of the practical problems will also be useful in structuring effective strategy for future. The Programme Coordinators have to perform multifold roles of planner, organizer, director, communicator, controller, supervisor, coordinator, leader and decision-maker. It was, therefore, felt necessary to examine constraints faced by them in effective management of their KVKs and also to collect their valuable suggestions to overcome the same in effective Transfer of Technologies in the field of Agriculture Extension management Chitra Pathak *et al.* (2005). For this, present study was undertaken with following objectives:

Objectives of the Study

1. To study the constraints faced by Programme Coordinators of KVKs in the management of different activities of KVKs.
2. To seek their valuable suggestions based on practical experience to overcome the constraints for efficient Transfer of Technology through KVKs.

Methodology

Keeping in view the objectives of the study, the present study was carried out in all the eight zones of India where Krishi Vigyan Kendras are in operation in transfer of technology process. The Ex-post facto research design was applied for the study. Out of total 320 KVKs in India, with five years of functioning experience with well established, full-fledged activities, about fifty per cent (*i.e.* 160 KVKs) KVKs were selected purposively for the study. A random sample of the total 160 programme coordinators of KVKs from the different zones of India was selected for this study. A random sample of 12 to 20 programme coordinators of those KVKs which have completed at least 5 years of working were selected from all the eight zones. The responses were collected in three point continuum *viz.* "Most important"," Important" and "Not important". In the present study, constrains faced by the programme coordinators in managing KVK were studied. They were requested to express the constraints faced by them. These constraints were ranked according to their importance based on mean score and are presented in Table 39.1. In the present study, after studying some constraints faced by the programme coordinators in managing KVK, they were also requested to express their valuable suggestions based on their practical experiences. These suggestions were felt important to study to provide feed back to the concerned authority for the effective results of existing KVKs as well as to provide some guideline for those KVKs, which are under considerations to establish all over India. The suggestions were collected in three continuums and their mean scores were worked out to understand its importance. The suggestions were also ranked according to their mean score and are presented in Table 39.2.

Table 39.1: Constraints Faced by Programme Coordinators in Managing KVK (n = 160)

Sl.No.	Constraints	Mean Score	Rank
1.	Many posts of the subordinate staff lying vacant.	1.79	II
2.	Majority of the subordinate staff not fulfilling the required educational qualification	1.31	IX
3.	Frequent transfer of the staff from KVK to other department	1.52	V
4.	Fund for the extension activities/FLDs is not released in time by the council	1.58	IV
5.	Most of the staff remains engaged with local organizational work and less for the KVK activities	1.41	VI
6.	No promotion channel for the KVK staff	1.93	I
7.	Inadequate infrastructure facilities such as building, fencing, tubewell, vehicle etc.	1.38	VII
8.	Unnecessary involvement of organizational head in taking decision regarding planning, execution and implementations of the KVK activities	1.35	VIII
9.	Unfair interference of the local organizational head should be removed	1.29	X
10.	Intermittent flow of the funds from organization to KVK head for carrying out mandated activities	1.59	III
11.	Pressure from the local politician to fetch more benefits from KVK schemes to their own jurisdiction	1.13	XI
12.	Local organizational rules do not allow the KVK staff to work freely according to the guidelines of council	1.31	IX

Table 39.2: Suggestions Offered by Programme Coordinators to Overcome the Constraints in Managing KVK (n = 160)

Sl.No.	Suggestions	Mean Score	Rank
1.	Vacant posts especially technical including programme coordinators should be filled up immediately	1.63	I
2.	Disciplines as well as qualification should be considered while appointing technical staff	1.52	II
3.	Career advancement scheme for the KVK staff should be implemented and frequent transfer of the KVK technical staff should be avoided	1.40	III
4.	Demonstration unit/training laboratory as well as infrastructural premises should be well equipped	1.36	IV
5.	KVK grant should be directly released to the KVK account for timely operation. Additional grant should be provided to the KVK for carryout extension activities	1.32	V
6.	Sufficient grant especially for extension activities should be provided and such grant should be released timely	1.30	VI

Contd...

Table 39.2–Contd...

Sl.No.	Suggestions	Mean Score	Rank
7.	Power and authority should be delegated to the programme coordinators adequately, considering their responsibility	1.27	VII
8.	Vehicle facility especially for carrying extension activities in village situation is highly essential	1.24	VIII
9.	The KVKs should be restricted to perform only extension tasks. Administration and extension work should be separated accordingly.	1.21	IX
10.	Modern information and communication technology should be made available to all KVKs	1.19	X
11.	A strong coordination and linkage on permanent basis should be established between research institutions and KVKs	1.17	XI
12.	All the necessary infrastructural facility and educational facility for the staff should be made easily available	1.14	XII
13.	There should be a provision for work motivation and incentives for the staff of KVKs	1.11	XIII
14.	Frequent transfer of the staff should be restricted.	1.09	IXV
15.	Mechanism of strong linkages and coordination with other extension agencies is necessary	1.07	XV
16.	An adequate personnel management policy should be adopted	0.62	XVI
17.	Standard and uniform evaluation system for extension work should be suggested by the ICAR to all the KVKs	0.38	XVII

Findings and Discussion

Constraints Faced by Programme Coordinators in Managing KVK

In the present study, some constrains faced by the programme coordinators in managing KVKs were studied. They were requested to express the constraints faced by them.

These constraints were ranked according to their importance based on mean score and are presented in Table 39.1. A glance of the Table 39.1 shows that no promotion channel has been arranged for the staff of KVK, many post of the subordinate staff lying vacant, intermittent flow of the funds from organization to KVK head for carrying out mandated activities, fund for the extension activities and FLDs is not released in time by the council Jondhale *et al.* (2000)., frequent transfer of the staff from KVK to other department are the major constraints faced by the programme coordinators in managing the KVK. Further they opined that most of the staff is engaged with organizational work or say other than KVK activities, inadequate infrastructure facilities such as building, fencing, tube well, vehicle etc., unnecessary involvement of organizational head in taking decision regarding planning, execution and implementations of the KVK activities, inadequate required educational qualification among the subordinate staff, local organizational rules do not allow the KVK staff to work freely according to the guidelines of ICAR were the other constraints experienced and expressed by the programme coordinators to restrict their extension management ability. The same results were reported by Patel (2001). Promotion encourages employees to view the organization as one offering them career growth. An incentive like promotion increases participation rate of employees and helps to create healthy completion to develop organization. It is needed to adhere healthy behaviors of employee. The purpose of the incentive is to encourage employees to adopt positive behaviors or maintain an existing positive behaviour. Many organizations also provide incentives merely for active and healthy participation of employee in the development of organizations. Promotion helps in making organization more effective and enhances responsiveness of the employees. For this reason, if channel of promotion does exist in KVK setup, authority should take care of this constraint expressed by the programme coordinators.

The another important constraints faced by the programme coordinators, was issue of the vacant posts, it is natural that to perform all the activities of KVK effectively, sufficient number of subordinate staff is needed by the manager or any organizers, understanding this; urgent appointment of subordinate staff on the vacant posts was realized by the programme coordinators. The high performance team of human is one of the most important resources in development of any public dealing organizations. The organization like KVKs are involved in many activities, so to carry out work as per the mandates, programme coordinators need necessary active human resources. One should not expect best result from programme coordinators without enough and skillful human resources. The organizations that pay attention to the key factor of giving professional force of co-workers to the managers can create a positive and productive environment for their organizations that will drive overall performance and long-term success.

The high performance teams with the quality of autonomous team- work attitude and self-directed work nature can perform their various roles effectively. High-performance teams of co-workers usually achieve a high level of motivation. Motivation, in this regard is translated into zeal, drive or enthusiasm. They are impelled, intrinsically to improve their performance. Realizing this, constraints like inadequate number of subordinate staff, lack of qualified staff and unnecessary engagement of staff in other than organizational activities were faced by the programme coordinators of KVKs.

All over India, KVKs are being run by SAUs, NGOs, Vidyapiths and ICAR research stations, it is therefore, programme coordinators have to depend on the concern authority to utilize funds of KVK because funds are not being given directly to them. Within this circumstance, untimely availability and improper utility of needed funds were understood as constraints by the programme coordinators Patel *et al.* (1999). In addition to this, engagement of KVK staff to do less work for KVK and more for local organizations, intermittent flow of the funds from organization to KVK head for carrying out mandated activities, untimely release of fund for the extension activities/FLDs, pressure from the local politician to fetch more benefits from KVK schemes to their own jurisdiction and dependency on the administrative procedure of local organizational to work freely according to the guidelines of council were another constrains faced by the programme coordinators.

Suggestions Offered by Programme Coordinators to Overcome the Constraints in Managing KVKs

In the present study, after studying some constraints faced by the programme coordinators in managing KVKs, they were also requested to express their valuable suggestions based on their practical experiences. These suggestions were felt important to study to provide feed back to the concerned authority for the effective results of existing KVKs as well as to provide some guideline for those KVKs which are under considerations to establish all over India collected in three continuums and their mean scores were worked out to understand its importance. The suggestions were also ranked according to their mean score and are presented in Table 39.2.

Table 39.2 reveals major suggestions given by programme coordinators in performing their work as extension mangers of KVKs. The major suggestions given by programme coordinators in performing their work as extension mangers of KVKs were related to appointment of staff, sufficient and timely major and additional grants, all of the necessary infrastructural facility for the staff should be made easily available which indirectly influences in a positive manner on extension management work of KVKs (Chauhan and Chauhan) facility, career advancement, restriction on frequent transfer which adversely affects the planning and implementation of pre-decided extension programmes. Modern information communication technology such as computer, internet, fax, zerox, mobile phone, etc. should be made available to all KVKs, which are very essential for the development of extension management Sawant, A. G. (2004). It will reduce the communication gap between them, linkage with other organizations and personnel management policy, is necessary for effective management of extension programme. The KVKs should be restricted to perform only extension tasks, administration and extension work should be separated accordingly. Decentralization of power and authority may be helpful to maintain the standard/ quality or work and to carry out the responsibilities effectively. To get best results through KVKs, their practical suggestions should be considered by the concerned authority. In addition to this, while establishing new KVKs all over India, programme coordinators of new KVKs should be provided all the facilities to run the activities of KVKs as per the desired goals and mandates.

Conclusion

The Constraints Faced by the Programme Coordinators

1. Lack of promotion channel for the staff of KVKs, many post of the subordinate staff lying vacant, intermittent flow of the funds from organization to KVKs head for carrying out mandated activities, pressure from the local politicians to fetch more benefits from KVKs schemes to their own jurisdiction, funds for the extension activities and FLDs are not released in time by the council, frequent transfer of the staff from KVKs to other department are the major constraints faced by the programme coordinators in managing the KVKs.

2. Further they opined that most of the staff is engaged with organizational work or say other than KVK activities, lack of adequate infrastructure facilities such as building, fencing, tube well, vehicle, unnecessary involvement of organizational head in taking decision regarding planning, execution and implementations of the KVK activities, inadequate required educational qualification among the subordinate staff, local organizational rules do not allow the KVK staff to work freely according to the guidelines of council are the other constraints experienced and expressed by the programme coordinators to restrict their extension management ability.

The Suggestions Expressed by the Programme Coordinators to Overcome Constraints

The major suggestions given by Programme coordinators in performing their work effectively as extension mangers of KVKs were related to appointment of qualified technical staff, sufficient and timely availability of main and additional grants, infrastructural facility, incentives, vehicle facility, career advancement, restriction on frequent transfer, modern information communication technology, linkage with other organizations and personnel management policy. To get best results through KVKs, their practical suggestions should be considered by the concerned authority. In addition to this, while establishing new KVKs all over India, Programme coordinators of new KVKs should be provided all the facilities to run the activities of KVK as per the desired goals and mandates.

References

Anonymous, 1999. *Guidelines for the KVK Manager*, New Delhi.

Anonymous, 2005. *Annual Report 2004–05*, Department of Agricultural Research and Education, Ministry of Agriculture, Govt. of India, ICAR, New Delhi.

Bhople, R.S. and Palaspagar, V.W., 1996. Appraisal of managerialability of farm supervisors. *PKV. Res. J.*, 20(2): 161–162.

Chauhan, N.M. and Chauhan, N.B., 2007. Extension management ability of programme coordinators of KVKs in India. *International Journal of Extension Education*, 3(December 2007): 52–60.

Jondhale, S.G., Ingle, L.A. and Pathak, U.N., 2000. Impact of krishi vigyan kendra training on adoption of improved practices of summer groundnut. *Mah. J. of Ext. Edu.*, 19: 109–112.

Pathak, Chitra, Kumar, Manish and Kwatra, J.K., 2005. Extension training for agricultural development. *Agri. Ext. Review*, 17: 17–22.

Patel, A.A., Patel, M.R. and Desai, C.P., 1999. Job related and socio-psychological factors affecting research management ability of the heads in Gujarat Agricultural University. *GAU Res. J.*, 23(2): 77–80.

Patel, G.J., 2001. A study on extension management ability of training organizers on KVKs. *Ph.D. Thesis (Unpub.)*, G.A.U., Sardar Krushinagar.

Sawant, A.G., 2004. Role of informatics and bioinformatics in extension education in India. *Asian J. of Ext. Edu.*, 23(1): 22–41.

Van de Ban, A.W., 2004. Changes in society require changes in agricultural extension: Some observations. *Asian J. of Ext. Edu.*, 23(1): 1–17.

Chapter 40

Improving Livelihood Security through Farmer Participatory Pulses-based Interventions: Experiences from Rainfed Bundelkhand Region of India

☆ ***Shantanu Kumar Dubey*[1], *S.K. Singh, Uma Sah and A.K. Singh*[2]**

Introduction

The significance of pulses in livelihood and nutritional security, sustainable crop production and conservation of valuable natural resources is unequivocally accepted. In rainfed production situation, where crop productivity often suffers with low and unstable productivity, farmers' resources are not adequate and thereby bringing nutritional security at stake. Under such circumstances, pulses have greater mitigating capacity. However, the benefits of research outcomes have not been harnessed by farmers from rainfed areas primarily due to their nil to least involvement in technology development besides the other related factors. Recognizing the significance of pulses production in rainfed areas of India and for three-fold objectives

---

1 Indian Institute of Pulses Research, Kanpur.

2 Zonal Coordinator, ZC Unit, Kanpur.

of meeting nutritional security, sustainable agricultural and livelihood improvement of rural sectors vis-a-vis the scope of farmer participatory approach in this endeavour, two participatory projects namely IVLP-TAR (NATP funded) and IFAD (IFAD TAG 532 ICRISAT Project) were implemented by Indian Institute of Pulses Research, Kanpur (UP) during the period of 2000-2005 among 1000 farm families of four villages namely Vidokhar, Helapur, Kalauli Tir and Amirta under Hamirpur district of Bundelkhand region of India. The main objectives of the project were to make the recommended pulses production technologies appropriate to the prevailing production situations of the area with active involvement of farmers at every stages of project.

After a gap of two years of the termination of the project, it was attempted to understand what impact these participatory projects gave on the productivity and profitability of pulse crops and through the usable technologies in the areas of crop production and protection.

Methodology

Impact study was conducted concurrent as well as after the termination of above project. The main parameters involved for impact studies were changes and trends of pulses productivity and profitability on longitudinal basis (2004-05 to 2006-07) for integrated crop management as well as for the technological components. Farmers' perception/criteria were also the important parameter to assess the impact of the above project. Pre-structured interview schedule supported with interactive and informal discussion with the partner farmers were utilized to elicit the information related to research variables. Enhancement in productivity and profitability and their longitudinal trends for the three important crops namely chickpea, lentil and pigeonpea were ascertained for the integrated crop management as well as for the components of technologies like varietal, integrated nutrient management, integrated pest and diseases management. Extent of acceptance and continuance of the appropriate technology as well as the adoption gap in the recommended pulse production technologies were also documented for the both types of the projects and contrasted with the top-down projects implemented by the institute. Extent of percolation of the technology in the neighbouring villages and farmers' criteria for the rapid adoption of validated technologies were also analyzed.

Results and Discussion

Productivity and Profitability Enhancement through ICM

It is evident from the Figure 40.1, that during the period of 2004-05 to 2006-07, there was overall average enhancement of 33.80 per cent, 32.83 per cent and 39.78 per cent in the yield over the pre-project status for the crop like chickpea, lentil and pigeonpea respectively under the participatory projects. During the same period, however, there was declined trends in the productivity improvement for the above crops and there was –13.67 per cent, –11.44 per cent and –10.14 per cent average annual growth rate in the crop productivity for chickpea, lentil and pigeonpea respectively during the period of 2003-04 to 2006-07.

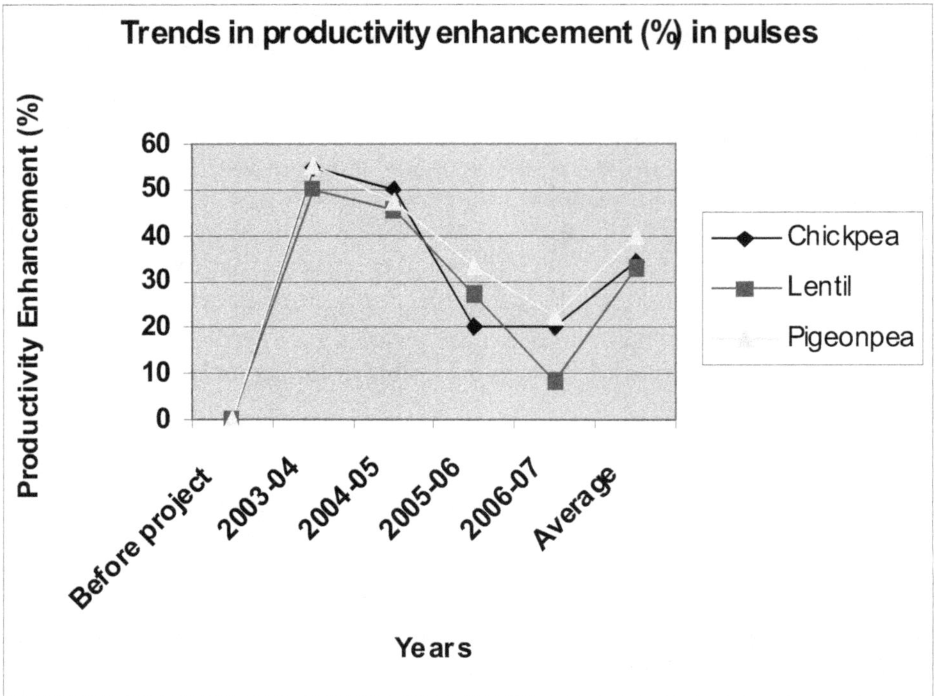

Figure 40.1

The scenario of profitability enhancement also showed similar trends. The average annual increase in profitability (incremental B:C ratio) were computed to the extent of 28.57 per cent, 10.71 per cent and 23.07 per cent respectively for the crops like chickpea, lentil and pigeonpea respectively during the period of 2004-05 to 2006-07 as compared to pre-project status (Figuer 40.2). The growth rate of incremental B:C ratio during the same period for the above crops also showed the decline trends *i.e.* –6.90 per cent, –6.06 per cent and 19.97 per cent respectively. The findings imply that even with the greater decline rate in the productivity of the crop, profitability did not affect much. This may be probably because of the fact the even on account of the gradual reduction in total production level of pulses in the region during the period, there was correspondence increase in the demand leading to increase in the market price thereby less adversely affecting the profitability.

Impact of Location Specific Assessed and Validated Technologies

Varietal Impact

It was revealed from the discussion with the farmers that in the production situation having clay and clay loam soil, rainfed and monocropping system (situation I), JG 315, DPL 62 and JL 1, and NA 1 varieties of chickpea, lentil and pigeonpea were found as the best fit giving about 42 per cent, 30 per cent and 52 per cent yield advantage respectively over the local varieties with their respective yield level of 904,

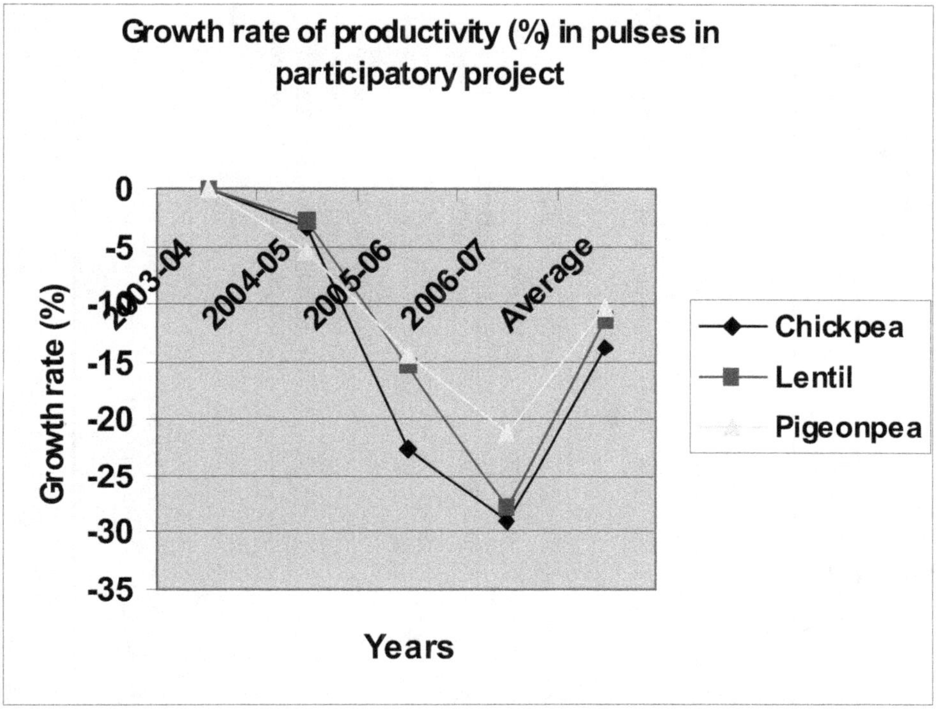

Figure 40.2

1100 and 918 kg/ha (Table 40.1). Similarly, in the situation, where soil is clay loam and clay, mono-cropping and having facility for pre-sowing irrigation (situation II), DCP 92-3 and JG 322 varieties of chickpea said as the best varieties giving, 400 kg/ha more yield than the local one (951 kg/ha). Further analysis of data showed that 40 per cent and 80 per cent of the chickpea growers were still continuing to use JG 315 and DCP 92-3. Similarly, DPL 62 was being used by as many as 90 per cent of the lentil growers in the area.

Impact of Integrated Nutrient Management (INM)

Integrated Nutrient Management package was the validated technology that contained 50 per cent of the recommended NPS and seed treatment with rhizobium culture. It was found that INM had more visible impact in case of pigeonpea for both productivity (42 per cent) and profitability (44 per cent) enhancement followed by chickpea and lentil for which increase in productivity (25 per cent and 12 per cent respectively) and profitability (43 per cent and 20 per cent respectively) were also encouraging (Table 40.1). Farmers also disclosed the fact that the impact of INM more visible in the clay and clay loam soil with rainfed condition and monocropping system. INM was found partially continued by the 90 per cent of the of pulse growers (Table 40.2).

Table 40.1: Impact of Different Components of the Assessed and Validated Technology for Different Pulse Crops at Farmers' Field

Sl.No.	*Components of Assessed and Validated Technologies*	*Chickpea*		*Lentil*		*Pigeonpea*	
		Increase in Yield (%)	*Increase in B:C Ratio*	*Increase in Yield (%)*	*Increase in B:C Ratio*	*Increase in Yield (%)*	*Increase in B:C Ratio*
1.	Varietal						
	(*i*) Situation I	904–1256 kg/ha (39 per cent)	2.0 to 2.4 (20 per cent)	1100–1425 kg/ha (30 per cent)	2.6 to 3.15 (21 per cent)	612–929 kg/ha (52 per cent)	2.0 to 3.08 (54 per cent)
	(*ii*) Situation II	951–1355 kg/ha (42 per cent)	2.1 to 3.0 (42.85 per cent)	–	–	918–1329 kg/ha (45 per cent)	2.2 to 3.5 (56.82 per cent)
2.	Integrated Nutrient Management (50 per cent NPS and RC)						
	(*i*) Situation I	890–1112 Kg/ha (25 per cent)	1.7 to 2.3 (34 per cent)	1175–1318 kg/ha (12.17 per cent)	2.5 to 3.0 (20 per cent)	718–1019 kg/ha (42 per cent)	2.1 to 3.1 (44 per cent)
	(*ii*) Situation II	1025–1286 kg/ha (25 per cent)	2 to 2.8 (43 per cent)	–	–	1022–1200 kga/ha (22 per cent)	2.2–3.3 (44 per cent)
3.	Integrated Pest Management (5 per cent NSKE +Endosulphan)						
	(*i*) Situation I	1023–1220 kg/ha (19 per cent)	1.8 to 2.4 (33 per cent)	–	–	720–980 kg/ha (36 per cent)	2.0–2.8 (39 per cent)
	(*ii*) Situation II	–	–	–	–	1110–1436 kg/ha (29 per cent)	2.3–3.1 (34 per cent)
4.	Integrated Disease Management (Seed treatment with Vitavax and *Trichoderma herzianum*)						
	(*i*) Situation I	–	–	1130–1250 kg/ha (11 per cent)	2.6 to 2.9 (12 per cent)	–	–
	(*ii*) Situation II	–	–	–	–	–	–

Situation I: Clay and clay loam soil, rainfed and monocropping system

Situation II: Clay loam and clay, mono–cropping and having facility for pre-sowing irrigation.

Figures in parentheses show percentage increase over the baseline *i.e.* pre-project situation.

Table 40.2: Extent of Acceptance and Continuance of the Adopted Technologies by the Farmers Under Different Projects

Sl.No.	*Components of Technology Refined and Demonstrated*	*Participatory Project (IVLP-TAR and IFAD-TAG 532 ICRISAT: 2000-2005)*		*Top-Down Project (UNDP Funded Project (1997-2001)*	
		Extent of Acceptance (%)	*Extent of Continuance (%)*	*Extent of Acceptance (%)*	*Extent of Continuance (%)*
1.	Improved varieties				
	(a) Chickpea	JG 315: 70% DCP 92-3; 100%	40.00 80.00	Avrodhi: 25% K-850: 10%	80.00% 40.00%
	(b) Lentil	DPL 62: 80% JL1: 20%	80.00 100.00	DPL 62: 10% DPL 15: 15%	90.00% 25.00%
	(c) Pigeonpea	NA 1: 80% Amar: 05% Bahar: 10%	70.00 35.00 40.00	UPAS 120: 10%	35.00%
2.	Integrated Nutrient Management (50% NPS and RC)	90.00%	70.00 (Only with 50 kg of DAP, no S and no RC)	–	–
3.	Integrated Pest Management (One spray of 5% NSE and one spray of endosulphan 0.07%)	90.00% (Only one spray of endosulphan)	06.00% (NSKE + endosulphan)	–	–

Impact of Integrated Pest Management

Infestation of the pulse crops by pod borer and pod fly was one of the major causative factors for yield reduction. Under the participatory project, therefore, integrated pest management protocol was standardized. Data showed that pulse growers expressed that the one spray of 5 per cent NSKE followed by single spray of endosulphan (0.07 per cent) resulted in considerable decline of pest infestation, thereby leading to yield enhancement from 1023 kg to 1220 kg for chickpea and from 720 kg/ha to 980 kg/ha for pigeonpea. There was partial level of acceptance of this technology and it was found that as high as 90 per cent of the respondents especially from large farmer categories were using only one spray of endosulphan and as low as only 20-25 per cent of the small farmers were found applying the correct regime of IPM (Table 40.2).

Farmers' Perception for the Rapid Adoption of the Validated Technologies

For the validated pulse varieties, farmers felt that because of the criteria like greater relative advantage in terms of more yield, greater profitability and simplicity in use in that order, there was greater percolation of these varieties in the project villages as well as the neighbouring villages. INM was partially adopted by the farmers that too at relatively lesser pace (Table 40.3). The criteria like greater cost involved, relatively lesser immediacy of effect of INM in pulses as compared to cereal crops, trialibility and productivity enhancement were the reasons for partial acceptance of the technologies. For IPM, the time and cost made good trade-off for the successful adoption of application of NSKE. However, application of NSKE was perceived as immediate and better results giver than the application of only endosulphan. Criteria based ranking revealed that farmers' preferred more of the improved variety followed by integrated nutrient management, IPM and integrated disease management in that order. Also, simplicity-complexity, immediacy of effect, productivity enhancement and trialability were the criteria as perceived by the farmers for rating validated technology for rapid adoption.

Table 40.3: Farmers' Rating of Criteria for the Rapid Adoption of the Validated Technologies

Sl.No.	*Aspects of Technologies*	*Simple-Complex to Use*	*Cost*	*Immediacy of Effect*	*Trial-ability*	*Productivity Enhance-ment*	*Total Score*	*Overall Rank*
1.	Improved variety	9.2	5.4	8.3	9.0	8.4	40.3	I
2.	INM	7.0	4.4	7.2	7.7	6.4	32.7	II
3.	IPM	5.7	7.6	6.3	5.3	6.5	31.4	III
4.	IDM	64	5.0	6.0	5.1	6.0	28.5	IV
	Total Score	28.3	22.4	27.8	27.1	27.3		
	Overall Rank	I	V	II	IV	III		

Scores are obtained on the rating scale 1-10.

Score 1: Minimum positive or greater negative rating.

Score 10: Maximum positive or minimum negative rating.

Conclusions and Implication

Above findings helped to make strong conviction that participatory projects are the viable means for validation and dissemination of recommended pulses production technologies as per the agro-ecological situation of Bundelkhand region of India. The emanated location specific usable technologies needs large scale dissemination through available technology delivery mechanism of the state department of agriculture, Krishi Vigyan Kendra in the location and other related institutions. Moreover, Farmer-Farmers spread of these technologies also require constant support.

Chapter 41

Extent of Adoption of Recommended Production Technology of Pigeonpea by Farmers

☆ *A.P. Dwivedi*[1], *B.P. Singh*[2], *R.P. Singh*[3] *and Mamta Singh*[4]

Introduction

India is largest Pigeonpea growing country in the world. The area, production and productivity of the pulses in the country are 24.0 millian hectare, 25.23 millian tonn. and 11 q./ha, respectively. Out of which the area, production and productivity of Pigeonpea in our country are 31.58 lakh ha., 25.5 Lakh tonn. and 7.6 q/ha, respectively. Pigeonpea ranks II after chickpea among all the pulses in the country. The area, production and productivity of pigeonpea in district Ghazipur of U.P. are 6206 ha, 10143 metric tonn. and 16.34 q/ha; respectively.

The pulse production in the country can be sustained through productivity growth. All the research efforts to develop a technology are useless unless farmers

---

1 Subject Matter Specialist (Agronomy), KVK, P.G. College, Ghazipur, U.P.

2 Reader, Department of Agronomy, P.G. College, Ghazipur, U.P.

3 Subject Matter Specialist (Plant Protection), KVK, P.G. College, Ghazipur, U.P.

4 Subject Matter Specialist, Plant Breeding and Genetics, RARS, KVK, Sagar, M.P.

adopt the technology. The productivity can be increased with the increase of the level of adoption of recommended technology. The technology evolved for Pigeonpea crop is intended to get spread among Pigeonpea cultivators to accelerate production process. An important task in the development programme for agriculture is to create awareness among the farmers about the improved agricultural practices. Building up a sound agricultural information and extension education programme does this.

In order to ascertain the extent of recommended technology in Pigeonpea cultivation, the present study was done in Ghazipur district of U.P.

Methodology

The present investigation was based on an intensive study of farmers interactive group of Pigeonpea growers of selected cluster village of Ghazipur district of Uttar Pradesh. Cluster villages of district Ghazipur of U.P. has been purposively selected for this study, looking to its typical and actual representation of the district with respect to Pigeonpea production. Three villages namely, Sabua, Laxamanpur and Chochakpur of Karanda block of district Ghazipur were selected randomly. The final selection of the farmers was done from each selected village based on proportion to its size (No. of Pigeonpea growers). The no. of farmers selected were hundred with medium farmers (Medium land holding) for present study.

The field data along with other required information pertaining to the selected holdings were collected through structured schedules by personal interview/method, where farmers were asked to give the account of package of practices they followed in Pigeonpea cultivation in the study area.

Results and Discussion

The collected information were analysed for ascertaining the level of adoption of recommended technology in Pigeonpea cultivation. The results are presented were as given below:

Adoption Level of Respondents about Recommended Technology in Pigeonpea Cultivation

The data presented in Table 41.1 revealed that overall 52 per cent of the total respondent were found to be in the medium adoption group, whereas 33 per cent respondent were reported from the low adoption group and only 15 per cent respondent could be placed in the high adoption group.

Table 41.1: The Level of Adoption of Respondent about Recommended Technology in Pigeonpea Cultivation

Sl.No.	*Adoption Level*	*No. of Farmers (100)*	
		Farmers	*Per cent*
1.	Low	33	33
2.	Medium	52	52
3.	High	15	15

Table 41.2: Extent of Adoption of Recommended Technology in Pigeonpea Cultivation

Sl.No.	*Extent of Adoption*	*Farmers (100)*	
		No. of Farmers	*Per cent*
1.	Preparatory tillage		
	(*i*) As recommended	20	20
	(*ii*) Above than the recommendation	80	80
2.	Line sowing/bed planting		
	(*i*) As recommended	10	10
	(*ii*) Nil (Broad Casting)	90	90
3.	Varieties		
	(*i*) As recommended	75	75
	(*ii*) Non-descriptive local varieties	25	25
4.	Chemical fertilizers		
	(*i*) As recommended	20	20
	(*ii*) Below than the recommendation	80	80
5.	Time of sowing		
	(*i*) Before recommended time	20	20
	(*ii*) In recommended time	70	70
	(*iii*) After recommended time	10	10
6.	Seed rate		
	(*i*) Recommended	20	20
	(*ii*) Above than the recommendation	80	80
7.	Seed treatment		
	(*i*) Treated	20	20
	(*ii*) Untreated	80	80
8.	Weed management		
	(*i*) Manual	15	15
	(*ii*) As chemical recommendation	40	40
	(*iii*) No weed management	45	45
9.	Plant protection measures		
	(*i*) As recommended	20	20
	(*ii*) Below than the recommendation	40	40
	(*iii*) Nil	40	40

Extent of Adoption of Recommended Technology in Pigeonpea Cultivation

Adoption of Field Preparation

Four or five ploughing for pigeonpea crop is recommended. The information presented in Table 41.2 reveals that only 20 per cent farmers ploughed their field as recommended while rest of majority(80 per cent) of the farmers over ploughed their land. The reason for this over ploughing is the common practice by almost all the farmers, to grow Pigeonpea after wheat or keeping fallow land in *Rabi* or prominent the Pigeonpea-Pigeonpea rotation.

Adoption of Recommended Varieties

The recommended improved varieties of Pigeonpea are Malviya Arhar-13 (Malviya Chamatkar), Narendra Arhar-1, Narendra Arhar-2, Malviya Arhar-6 etc. A perusal of the data in Table 41.2 on this particular aspect reveals that almost 75 per cent of the farmers adopted recommended varieties Malviya Arhar-13. Lower adoption of other recommended varieties *i.e.* NA-1, NA-2, Malviya Arhar-6 was due to non availability of seeds in time and lack of knowledge about other improved varieties. It can be said that there was good adoption of improved varieties as almost all the farmers were found using the recommended varieties. About 25 per cent farmers were adopted or using non-identified/non descriptive varieties due to lack of knowledge about importance of varieties.

Adoption of Method of Sowing

Line sowing by 60-75 cm x 30 cm for Pigeonpea cultivation behind the plough or by raised bed techniques is recommended. But majority (90 per cent) of the formers are not using the recommended sowing method due to lack of knowledge and lack of raised-bed planter. Only 10 per cent farmers are using line sowing techniques.

Adoption of Fertilizer Use

As per recommendation, 18 kg nitrogen and 46 kg phosphorus per hectare should be applied as a basal dressing at the time of sowing. Only 20 per cent farmers followed the recommendation while 80 per cent farmers were not using fertilizers. The reason of not using recommended dose of chemical fertilizers were mostly attributed by the farmers to the lack of knowledge behind the using of fertilizers in pulse crop and high risk involved in Pigeonpea crop, pulses crop treated as a secondary crop by the farmers.

Adoption of Time of Sowing

The recommended time of sowing of Pigeonpea crop is 25 June- 26 July. Table 41.2 expresses that 20 per cent of farmers were sown their crop before time, 70 percent in recommended time, while rest 10 per cent have been sown after recommended time. The decision of sowing before time was observed to be based on the continuous rain fall in the month of July and there was no chance for sowing for some time.

Adoption of Seed Rate

The recommended seed rate for Pigeonpea is 12-15 kg/ha. The data in Table 41.2 indicated that the majority of farmers (80 per cent) used seed rate more than the recommended. It may be due to sowing by broad casting method resulted poor germination and plant population.

Adoption of Seed Treatment

Seed is to be treated by bio-fuyngicide *i.e.* Trichoderma and Rhizobium culture for wilt management and better nodulation (nitrogen fixation); respectively. The analysis reveals that majority of the farmers (80 per cent) might not have done seed treatment as per the recommendation. It is because of lack of knowledge about advantage of seed treatment and lack of bio-fungicides and Rhizobium culture.

Adoption of Weed Management

The crop plant grow very slowly during their early growth period of 40-45 days and the crop suffers from a severe weed infestation which causes in drastic reduction in grain yield. Therefore, it is advisable to keep the field free from weeds and a weed free condition may be achieved by giving two hand weedings once about 25-30 days and another about 45-50 days after sowing of the crop. Where hand weeding is not possible use of Lasso(Alacholor) at the rate of 3.0 kg/ha as pre-emergence or basaline at the rate of 1.0 litre/ha as pre-sown incorporation in soil may be done. By using herbicides the field will be weed free upto 50 days of sowing and least damage to the crop will take place. Other herbicides are Metalachlor, Oxadiazon and Pendimethalin @ 1.0 kg 500 ltr./ha as a pre emergence.

The data from Table 41.2 reveals that 40 per cent farmers are adopting recommended weedicides as a chemical weed control, 15 per cent farmers are applying manual weed management by khurpi (a hand weeding small equipment) and majority of the farmers (45 per cent) not applying any method for weed management in Pigeonpea crop; it is because of lack of knowledge about losses in productivity due to weed problem in Pigeonpea crop.

Adoption of Plant Protection Measures

Wilting, pod fly, pod borer are major pest problem in Pigeonpea crop. To control them, it is recommended that use of Trichoderma @ 5gm/kg of seed as a seed treatment. For pod fly and pod borer, spraying of Endosulfan @ 1-2 ml/litre has been recommended for control of the pod fly and pod borer. An ideal IPM module for Pigeonpea crop has been also developed.

As regards plant protection measures the data revealed that 40 per cent farmers followed plant protection measures below than the recommended, 20 per cent formers followed plant protection measures as recommended and 40 per cent farmers did not follow any recommendation for plant protection measures in Pigeonpea crop. The high cost, non availability of effective biofungicides, insecticides, sprayer might have important reason. It is a problem to operation of sprayer in long standing broadcasted Pigeonpea crop. The lack of knowledge about pest control and IPM module in Pigeonpea crop was the important reason behind this.

Conclusion

In terms of the adoption of the various component technologies of the package of practices of the farmers were studied and it can be concluded that the adoption of the improved variety has been 75 per cent amongst all the component technology followed by time of sowing. Most of the farmers were not using seed treatment due to lack of knowledge about advantage of treatment, plant protection measures due to high cost, non-availability of biofungicides, sprayer, duster and problem in using sprayer in long standing crop to control pod borer and pod fly. Therefore, for enhancing the production and productivity of pigeonpea crop, strategy should be made for getting the more and more recommended technologies adopted by the farmers.

References

Ali, A., 1998. Research, development and management for production of pulses. In: *IPM System in Agriculture, Vol 4: Pulses,* (Eds.) Rajeev K. Upadhyay, K.G. Mukherjee and R.L. Razak. Aditya Book Pvt. Ltd., New Delhi, India, pp. 1–40.

Singh, S.S., 2002. *Crop Management Under Irrigated and Rainfed Condition*. Kalyani Publishers, New Delhi, pp. 181–191.

Singh, U.N., 2006. *Crop Production and Field Experimentation*. Khitij Prakashan, Varanasi, India, pp. 186–194.

Chapter 42

Entrepreneureship Development: Concept and New Approaches

☆ *B.K. Mandal[1], Sarvesh Kumar and P. Saroj*

Introduction

The term entrepreneurship originated from the French word "entreprendre" and the German word "unternehmen" both mean "to undertake". The earliest attempt to define entrepreneurship was by the French writer Bernard F. de Belidor, who defines entrepreneurship "as buying labor and materials at uncertain prices and selling the resultant output at contracted prices." However,the concept gathered prominence in economics literature mainly through the writings of the Richard Cantillon (1680-1734),who gave concept some analytical treatment and assigned the entrepreneur an economic role by emphasizing on "risk" as a prominent entrepreneurial function (Gopakumar,1995). According to encyclopedia Britannica entrepreneur means individual response for the operation of a business including the choice of a product, the mobilization of necessary capital, decisions on product prices and quantities,the employment of labor and expanding or reducing the productive facilities.

Entrepreneurship development is an approach of developing human resources.It is concerned with the growth and development of people towards high level of competency, creativity, and fulfillment.This approach helps people to grow in self

---

1 Research Scholar, Department of Extension Education, IAS, BHU, Varanasi, U.P.

control, responsibilities and other abilities and then try to create a climate in which all clients may contribute to the limits of their improved abilities.

It is assumed that expanded capabilities and opportunities for the people will lead directly to improvement in operating effectiveness.Thus entrepreneurship development as an approach does not confine itself in setting up enterprises but surpasses their limit in creating conducive climate for optimum utilization of limited and scattered resources and making people functional in all walks of life.

Entrepreneurship is the creative response to an environment that combines innovativeness,readiness to take risk,sensing opportunities, heightened initiative, standard of excellence, persistence in achieving the goal, positive orientation to problem solving and constant striving for growth and excellence.When all attributes are developed in a person,the person can be found in any field of activity such as industry, business,education,public or professional bodies etc.But wherever they go they create a land mark, turn the direction of the side attain heights.During the last three decades academicians and practitioners contributed significantly theorizing and applying entrepreneurship concept which resulted into few important generalizations.

- There is positive linkage between entrepreneurship and economic prosperity.
- Entrepreneurship is instrumental in achieving progress in all walks of life.
- Entrepreneurship can be developed through planned training intervention/efforts.
- Entrepreneurship is a process consisting of stimulatory, support and sustaining activities.
- It requires to be viewed in terms of integration of activities and coordination among different departments.
- Entrepreneurial culture need to be inbuilt mechanism of development process.

Recent Approaches for Entrepreneurship Training Programme

With the increased emphasis of developing entrepreneurs, witnesses a phenomenon change in its approach to create entrepreneurial awareness and generate entrepreneurship in society. Recent approaches to entrepreneurship training may be seen as:

- Target oriented
- Location specific
- Product and process oriented

Target Oriented

Such entrepreneurship development programmes are directed and planned for a specific group which has distinguishing features as compared to other groups.It is because of the fact that target audience vary in terms of their background, experiences, training and exposure to the business world.Moreover, the size of the proposed enterprise may also vary from one target group to others.These variations demand for the matching training inputs with varying degree of intensity to bring these entrepreneurs to the threshold from where they can launch their enterprise and be able to manage them successfully.

Location Specific

In order to reduce regional imbalances and the imbalances that may exist between progressive and non-progressive areas, it is of considerable importance to plan EDPs depending on the characteristics of the area itself.Location in terms of urbanization, concentration of specific target group in a particular area it may be considered under such specific EDPs.

Product and Process Oriented EDPs

The late eighties is experiencing a shift from target oriented EDPs. Such EDPs are organized for a group of perspective entrepreneurs who opt for enterprises having set product lines or process such as plastic or electronics or construction materials of food technology etc.It seems to have a great future but some of the efforts made recently were found suffering from the lack of appreciation of the role played by inputs, other than product and process orientation. In the absence of such comprehensive programme, the acquired knowledge and skills about product and process will not automatically result in development of entrepreneurial quality, competence for enterprise launching and ability to manage which are crucial to start and ensure success.

All these approaches need to be viewed in terms of certain activities which also in a way, indicates as how such programmes could be conducted. These activities are as follows:

- ☆ Documentation of entrepreneurial opportunities in the area.
- ☆ Information about the possible supply of entrepreneurs.
- ☆ Motivational campaign for attracting and identifying potential entrepreneurs.
- ☆ Selection of potential entrepreneurs.
- ☆ Identifying training need to the target group.
- ☆ Training.

Emerging Trends

The scope of wage employment is not only limited but decreasing very fast. Even the nature of wage employment is undergoing drastic transformation from a security oriented job to performance based contractual assignment. Such a situation has virtually compelled people in general to think more for entrepreneurship/self employment.In this process, an analysis of the situation indicates clearly that today's environment is pointing concern towards the aspects like:

- ☆ Highly competitive market

- ☆ Competition between public and private sector
- ☆ Quality consciousness and standardization of products
- ☆ Security/guarantee.
- ☆ Fast changing technologies
- ☆ Technological collaboration
- ☆ Liberalization in policy and action
- ☆ Health/hygiene
- ☆ NRI's readiness to invest in India.
- ☆ Privatization of public sector
- ☆ Increased export opportunities and mental readiness of other countries to buy products from India.
- ☆ Emphasis on self-sustenance of supporting/promotional organization leading to better services from them.

There fore, the environment in totality appears to be quite challenging.In such a direction the question arises as how to help people sense business opportunities which fulfill the global expectations in general and local needs in particular. For this an attempt is made here list out possible models in operation in one hand and emerging demands on the other hand.

Models

In order to operate effectively in context of the above approaches a few models in operation are as follows.

Market Driven Model

Demand estimation and demand forecasting are the best ways through which demands of various products are reflected in the market.while the demand is estimated on the basis of the gap between the demand and supply, the forecasting of the demand is made on the basis of future projection of possible needs of the concerned products.Therefore economic activity is generated related to this model to fulfill the requirement of the market.

Need Based Specific Model

Generally, we find two types of needs *i.e.,* felt and unfelt needs.It is possible to understand the felt need on the basic requirements and the gap but it seems difficult to visualize the unfelt need *i.e.* the need bellow the surface.This is one area on the basis of which various products find place in the market

Related to this is another concept which is known as need creation in the society. There are number of examples where in deferent products have been introduced through this process in a particular social system like emergence of fast food in traditionally strong society specially in African and south Asian countries.

Uniqueness Model

A variety of products have been introduced in the market by projecting unique features of particular product for specified segment of the society. This may be seen quite clearly in automobile industry *i.e.* cars, two wheelers etc.

Conclusions and Suggestions

In this context, it seems important to take advantages of the earlier experiences and research findings which may be summarized as follows:

- ✰ Promotion and development of entrepreneurship is more developmental in nature and is not administrative or target oriented.
- ✰ Entrepreneurship development is not a programme but a process whereby people in general are stimulated, supported and their enterprising activities are sustained.
- ✰ Accumulated experience has brought maturity in approaches and strategy of developing entrepreneurship.
- ✰ Experience suggests that entrepreneur could be developed through a well planned integrated approach.
- ✰ A set of competencies is already identified to help people become entrepreneur.These competencies could be developed through the training intervention.
- ✰ Competencies could be broadly grouped as,
  - ✰ Behavioral
  - ✰ Enterprise launching and resourcing
  - ✰ Enterprise management
- ✰ New economic opportunities are identified and available which may be integrated in entrepreneurship development programmes.
- ✰ Special attention may be made to promote marketing entrepreneurs and R&D entrepreneur.
- ✰ Emergence of women entrepreneur through women empowerment seems to be an established fact.
- ✰ Micro enterprise creation and development through micro finance has shown a very significant result.
- ✰ These findings could be used to promote entrepreneurship professionally in all the sectors including agriculture.This deserves a mention here that farm entrepreneurship need to be promoted keeping in view in shaping the farmers as farm entrepreneurs capable of looking farming not an hereditary vocation but as business venture.

Chapter 43

A Comparative Analysis of Entrepreneurial Behavior of Farmers in Irrigated and Dry Land Areas of Raichur District of Karnataka: An Analytical Study

☆ *S.K. Meti*[1]

Introduction

Rural development is more than even before linked to entrepreneurship; Entrepreneurship can be defined as a creative and innovative response to the environment. Entrepreneurship as a composite variable means that the, person possessing entrepreneurship behavior, has some specialized characteristics. Entrepreneurship has been therefore, accepted as a function of several factors. All round development of agriculture is possible with the effective exploitation of human as well as material resources. In our country, where human resources are found to be plenty, we can identify individuals in all segments of population who have the requisite entrepreneurial skills. At this juncture, it is logical to analyze to what extent

---

1 Department of Agricultural Extension Education, College of Agriculture, Raichur, Karnataka.

our farmers are progressive, forward looking and willing to diversify their occupations. Keeping the above facts in view, the study was designed to critically analyze the entrepreneurial behavior of farmers with the specific objectives, To study the entrepreneurial behavior of farmers in Raichur district of Karnataka and to compare the entrepreneurial behavior of farmers in irrigated and dry land areas.

Material and Methods

The study was conducted in Raichur district of Karnataka during the year 2004–05. For this study Ex-post facto research design was followed. Two taluks *viz.*, Sindhanur and Deodurg were selected based on highest irrigatable and dry land cultivable areas, respectively. From each taluk four villages were selected based on highest cultivable area. From each village 15 farmers–5 each of small, medium and big categories were selected by using random sampling procedure. Thus, the total sample size constituted 120 respondents.

Results and Discussion

Entrepreneurs is the central figure of economic activity and positive manner of development. They are persons who initiate, organize, manage and control the affairs of in enterprise that combine the factors of production to supply goods and service in any sector, As such, the development of under development of entrepreneurship in the country. Entrepreneurial skill therefore is to be regarded as the most needed component for the development (Porchaehian *et al.*, 1998).

Development of economy of any nation depends primarily on the important role played by entrepreneurs. The role played by such entrepreneurs is of vital important in developing country like India.

Table 43.1: Distribution of respondents according to their entrepreneurial behavior

Sl.No.	*Category*	*Small Farmers*		*Medium Farmers*		*Big Farmers*		*Total*	
		f	*Per cent*	*f*	*Per cent*	*f*	*Per cent*	*f*	*Per cent*
1.	Low	17	42.50	11	27.50	10	25.00	39	32.50
2.	Medium	15	37.50	15	37.50	13	32.50	47	39.17
3.	High	8	20.00	14	35.00	17	42.50	34	28.33
	Total	40	100.00	40	100.00	40	100.00	120	100.00

X = 88.130 SD = 25.663

The results from Table 43.1 revealed that majority (39.17 per cent) of the farmers belonged to medium entrepreneurial behavior category followed by 32.50 per cent and 28.33 per cent of the respondents belonging to low and high entrepreneurial behavior categories, respectively. Among categories, it is observed that majority of small farmers (42.50 per cent had low entrepreneurial behavior compared to medium farmers (37.50 per cent) and big farmers (42.50 per cent), who belonged to medium and high entrepreneurial behavior, respectively. The possible reasons for high entrepreneurial behavior of big farmers might be due to their sound financial condition, large size of land holding to take risk and adopt new technologies besides high innovativeness, leadership ability, higher knowledge level and information seeking behavior. The results are in conformity with the finding of Pandya (1997), Patil *et al.* (1999) and Rao and Dipak (2003).

The results from the Table 43.2 revealed that there was a significant difference between, the entrepreneurial behavior of farmers in irrigated (Sindhanur) and dry land (Deodurg) areas. As the mean entrepreneurial behavior score was found to be comparatively higher (96.56) in Sindhanur taluk than that of Deodurg taluk (79.70). One of the possible reason for significant difference in the entrepreneurial behavior of farmers in Sindhanur taluk contributed for higher annual income and facilities, which seems to have influenced the entrepreneurial behavior. Other reason might be the comparatively higher assistance of management services to farmers in Sindhanur taluk due to the active involvement of private organizations.

Table 43.2: Comparison of Entrepreneurial Behavior of Farmers in Irrigated and Dry Land Areas

Category	*Mean*	*SD*	*Sample Size*	*"Z" Test*
Irrigated (Sindhanur taluk)	96.56	217.87	60	4.26**
Dry land (Deodurg taluk)	79.70	242.82	60	

**–Significant at 1 per cent level of probability.

Completely rain fed agriculture in Deodurg taluk might have influenced their entrepreneurial behavior in terms of decision making and risk abilities. Further, continuous droughts in Deodurg taluk might also be one of the reason which contributed for difference in the entrepreneurial behavior of farmers in two taluks, the results are in conformity with the findings of Anitha (2004).

Table 43.3: Comparison of Entrepreneurial Behavior of Different Categories of Farmers in Sindhanur Taluk

Category	*Mean*	*SD*	*Sample Size*	*"t" Test*
Small farmers	85.55	112.99	20	1.87 NS
Medium farmers	93.85	190.55	20	
Medium farmers	93.85	190.55	20	1.61 NS
Big farmers	101.3	242.22	20	
Small farmers	85.55	112.99	20	3.73**
Big farmers	101.3	242.22	20	

NS: Non-significant; **: Significant at 1 per cent level of probability.

It is observed from Table 43.3 that in Sindhanur (irrigated) taluk, no significant difference was observed between small and medium farmers (t value 1.87) and also between medium big farmers (t value 1.61). One of the possible reasons for this might be that Sindhanur taluk being highly irrigated; the farmers were aware of the improved agricultural technologies and have good adoption level, because of resource availability and assistance of management service. However, A significant difference was observed between the entrepreneurial behavior of small and big farmers (t value 3.73). This might be due to low innovativeness and risk taking ability coupled with low leadership ability of small farmers unlike big farmers.

From Table 43.4, it is found that in Deodurg (dry land) taluk, significant difference was observed between the entrepreneurial behavior of small and medium farmers (t value 3.74) and also between small and big farmers (t value 5.17). But there was no significant difference between entrepreneurial behavior of medium and big farmers. The possible reason for this might be that small farmers in contrast to medium and big farmers were having smaller land holding and low annual income. Also majority of small farmers had low social participation, mass media participation, extension participation and level of adoption. Other possible reason might be that majority of small farmers were illiterates coupled with low level of aspiration compared to medium and big farmers.

Table 43.4: Comparison of entrepreneurial behavior of different categories of farmers in Deodurg taluk

Category	*Mean*	*SD*	*Sample Size*	*"t" Test*
Small farmers	65.85	253.92	20	3.74**
Medium farmers	82.6	146.25	20	
Medium farmers	82.6	146.25	20	1.92^{NS}
Big farmers	90.65	204.66	20	
Small farmers	65.82	253.92	20	5.17**
Big farmers	90.65	204.66	20	

NS: Non-significant.

Conclusion

Further, it may be concluded that majority of farmers having medium entrepreneurial behavior is a clear indication of progressiveness of the farmers. But, the comparatively lower level of entrepreneurial behavior of farmers in dry land areas particularly the small farmers reveals the fact that much attention is required to improve their entrepreneurial behavior. Effort at involving farmers in the mainstream of development must include their training as entrepreneurs. Such programmes, apart from concentrating on entrepreneurial awareness, training and credit reach for farmers must also concentrate upon bringing the new technology within the reach of farmers and acquaint them with the new avenues of entrepreneurship.

References

Anitha, B., 2004. A study on entrepreneurship behavior and market participation of farm women in Bangalore rural district of Karnataka. *M.Sc.(Agri.) Thesis*, University of Agricultural Sciences, Bangalore.

Pandya, R.D., 1997. Entrepreneurial behavior of inland fisherman. *Journal of Extension Education*, 8(2): 1717.

Patil, V.G., Mahadik, R.P. and Patil, A.S., 1999. Entrepreneurial behavior of litter gourd growers. *Maharashtra Journal of Extension Education,* 18: 240–243.

Porchezhain, M.R., Vijayaraghavan, R., Babykumari, P. and Shusheelakumari, T., 1998. Entrepreneurial behavior of farmers. *Journal of Extension Education*, 9(1): 1963–1964.

Chapter 44

Productivity Enhancement in Soybean-Wheat Cropping Pattern through on Farm Trials in Vertisols of Malwa Plateau

☆ *U.S Gautam*[1] *and D.K. Paliwal*[2]

Introduction

Poverty and unemployment are major problems of the increasing population of India. More than 65 per cent Indian population is presently living in villages and about 35-40 per cent families classified as poor. Unemployment is one of the major causes of the rural poverty. The other factors are small land holding, poor quality natural resources, harsh agro-climatic condition, poor infrastructure and limited opportunities for rural industrial development, poor health care, illiteracy, social suppression, addiction to alcohol and exploitation by vested interested. Agriculture is the major source of livelihood but most of the illiterate farmers have not been successful in using their available resources. Most of them treated agriculture as a family tradition following age old practices and they adopted new changes only after observing the success of their neighbors. Over 12-15 per cent of the rural families are landless and among the land holders, 69per cent are marginal farmers with less than 1 ha holding (17 per cent of the total land) and about 21 per cent are small farmers

1 Zonal Coordinator, Zonal Coordinating Unit–VII, ICAR, Jabalpur – 482 004, M.P.

2 SMS (Agronomy), KVK, Kasturbagram, Indore – 452 020, M.P.

with 1-2 ha holding (34 per cent of the total land). Thus about 90 per cent families own less than 51 per cent lands; which are mostly owned by the poor families. As the chances of crop failure on this land are very high, the farmer generally does not invest in external inputs like improved seeds, fertilizers and plant protection measures and end up with poor yields.

Malwa plateau, which is a prominent agro-climatic zone of the central India and was known for its prosperity and resources, had good acreage of cotton, jowar, arhar (in *Kharif*) and sugarcane, wheat (in *Rabi*) crops. Now the *Kharif* crops have been replaced by soybean due to its wider adaptability and profitability. Economically feasible cropping systems are a requirement for a farm to survive as a business. Soybean is well suited for multi-crop systems. Soybean has the ability to perform well under a wide range of soil types and planting dates (Farno *et al.,* 2002). Soybean–wheat cropping system has emerged as a predominant system in the region. This system has emerged as an important cropping system only after 1980 with the introduction of soybean as a *Kharif* crop in wheat growing areas of the country particularly under irrigated ecosystem. This cropping system is a major cropping system in the Vertisols and associated soils of Central India under regime. Soybean being a rainfed crop in this system, preferred by the farmers due to its low input requirement and wider adaptability to climate and soil conditions as well. Majority of soybean area (about 70 per cent) in India is in Madhya Pradesh and in Malwa region 95-98 per cent area is under the crop by and large the crop is grown under rainfed situations. At present, it covers above 7.44 million hectares and is likely to produce above 7.0 mt. Wheat, which is another crop of this system in Rabi, grown in a very large area. These two crops constituted the most predominating system of central India.

The productivity of the cropping system is very low as compared to the yields achieved in various research trials. There could be a number of factors, which may be associated with the low productivity of the crops, imbalance nutrients in soils and imbalance use of fertilizers, heavy weed infestation, incidence of several diseases and pests and no use of proper control measures, lack of quality seed causing sub-optimal plant population and poor growth, poor drainage etc. Excess and deficient moisture in soil environment in the same season is also one more reason for it. This low productivity created many problems for the farming community. Due to the low productivity and low income farmers have to migrate towards cities for search of temporary employment and getting food and clothes. This migration checked their growth and down their livelihood to bottom. This situation can be controlled by increasing productivity of the crops and increasing employment opportunities and income of the farmers.

Methodology

Krishi Vigyan Kendras (KVKs) can play a successful role in this direction. Krishi Vigyan Kendras (KVKs) in the country are the primary links for the farmers to know about the agricultural technologies being generated. They act as the training centres for the transfer of the technology with an aim to reduce the time lag between technology generation and their transfer (Prime Minister in National Conference of

KVKs–2005). The activities of the KVK include on-farm testing to establish the location specificity of agricultural technologies under various farming situations, frontline demonstrations to establish its production potentials on the farmers' fields, training of farmers to update their knowledge and skills in improved agricultural technology, and training of extension personnel to orient them in the frontier areas technology development. The Malwa plateau is covered by 9 KVKs, which are Indore, Dhar, Ujjain, Mandsaur, Neemuch, Ratlam, Shajapur, Raigarh and Dewas. The KVKs under Malwa plateau conducted several in farm trials for overcoming the situations explained above. The productivity of the system has been increased significantly due to these on farm Trials.

Keeping in view the reasons associated with the low productivity of the system, these KVK conducted 19 on farm trials in this cropping system for enhancing productivity of the crops and return from the system in year 2007-08. Total 10 On Farm Trials on soybean were conducted. The major themes were integrated nutrient management; weed control, integrated pest management and integrated disease management. These On Farm Trials were taken on the basis of priority and intensity of the problem in the area. Similarly a number of 9 On Farm Trials were conducted on wheat crop. The major Thrust areas covered were integrated nutrient management; varietal evaluation, integrated crop management water management, water management and integrated disease management. It was observed that the varietal replacement has outstanding impact on productivity. These On Farm Trials (OFTs) were conducted involving farmers perspective, farmers' participation, farmers' management status and suitability of sites according to Singh (1999). Under each OFTs Total 5 number of replication were taken. For conducting these OFTs innovative and receptive farmers were selected to ensure their active participation. Each OFT consisted of two treatments *viz.* farmers' practices *i.e.*, their traditional practices and scientific practices recommended for the location.

Findings and Discussion

Being a kharif season crop, the weed and pest infestation is a major problem in soybean. Under On Farm Testing the enhancement in yield was recorded highest under weed control which was 29.81 per cent higher than the yield obtained under farmers practices, followed by integrate pest management (27.08 per cent). The management of diseases enhanced the production by 24.67 per cent while the yield enhancement was 22.74 per cent under integrated pest management. Technology wise share in productivity enhancement is showing in Graph 44.1. Overall 24.18 per cent higher production was obtained as compared to the farmers practices. The details of OFTs on soybean in the Malwa plateau are given in Table 44.1.

The problem of low productivity of soybean exists in an area of more than 1500 thousand ha in Malwa plateau. The results of the on farm trials showed a tremendous increase in the production. If the technologies adopted by the farmers there should be a great increase in production. The technologies increased the net return also. Due to adoption of these technologies, farmers earned an additional net return of approximate Rs. 6079 per ha, this shows the profitability of the technologies. Besides that the

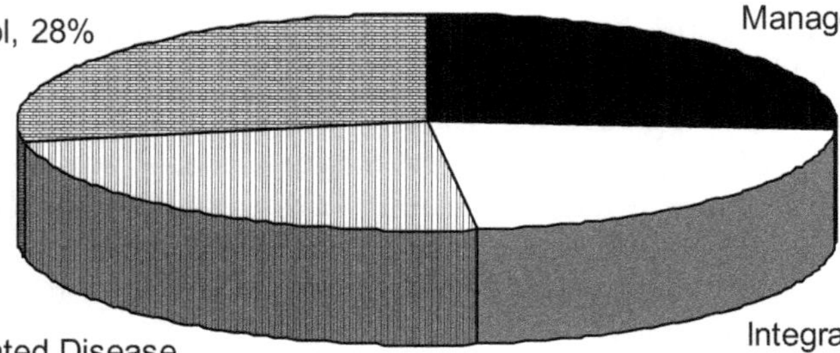

Graph: 44.1: Share of Different Thrust Areas in Productivity Enhancement of Soybean in Malwa Plateau

technology also enhanced employment by 15.57 man days per ha. Hence, the technologies will be sustainable and the analysis of yield obtained under the two micro farming situations reveals that the yield obtained under recommended practices was substantially higher over farmers' average yield, so the technology seems to be true representative of the problem.

Table 44.1: Implementation of On Farm Trials on Soybean in Malwa Plateau (2007-08)

Thrust Area	*No. of OFTs*	*No. of Replication*	*Yield (q/ha)*		*Increase in Yield Over Farmers Practice*	*Additional Income (Rs/ha)*
			Recommended Practice	*Farmers Practice*		
Integrated Disease Management	2	10	18.04	14.47	24.67	5712
Integrated Nutrient Management	6	30	17.42	14.20	22.74	5152
Integrated Pest Management	1	5	16.52	13.00	27.07	5632
Weed Control	1	5	16.50	12.71	29.81	6064

Similarly, the results obtained by the OFTs on wheat showed a considerable increase in the production. The recommended practices here increased the production by 14.89 per cent as compared to that under farmers' practices. The implementation details of OFTs on wheat in Malwa plateau is given in Table 44.2.

The highest increase in production was recorded under thrust area varietal substitution. Farmers of this area are using their own stored grains as seed, varieties are very old. This reduced the production of the crop. The on farm trials in wheat on thrust area varietal evaluation or replacement showed a significant increase in the production. It increased the production by 23.38 per cent, while the increase in production was only 1.14 per cent in weed management. Integrated crop management practices increased the crop yield by 17.09 per cent and the nutrient management practices increased the production by 16.76 per cent. The problem of low productivity of wheat exists in an area of about 550 thousand ha in the Malwa plateau. There should be a tremendous increase in the production, if the technologies adopted by the farmers of the agro-climatic zone. The on farm trials conducted in the zone not only resulted into a considerable additional income but more employment generation also. The additional income due to adoption of new technologies was recorded as Rs. 3552 with additional employment generation of about 6.86 man days per ha. The substantial improvement in productivity and profitability under assessed technology at selected farmers' field showed that the technology is true representative of the problem.

Table 44.2: Implementation of On Farm Trials on Wheat in Malwa Plateau (2007-08)

Thrust Area	*No. of OFTs*	*No. of Replication*	*Yield Under Recom-mended Practice*	*Yield Under Farmers Practice*	*Increase in Yield Over Farmers Practice*
Integrated crop management	1	5	36.3	31	17.10
Integrated disease management	1	5	43.8	38.4	14.06
Integrated nutrient management	3	15	43.74	37.46	16.76
Weed management	1	5	35.2	34.8	1.14
Water management	1	5	35.1	33.6	4.46
Varietal assessment	2	10	42.29	34.275	23.38

The share of different thrust areas in productivity enhancement of wheat crop is shown in Graph 44.2.

The increase in productivity of the soybean–wheat cropping system was 17.72 per cent in the cropping pattern and this gave an additional net income of Rs 9631 per ha. As the problem of low productivity of the cropping system exists in an area of about 450 thousand ha in Malwa Plateau, these technologies spread in an area of 150 ha during the year which is expected to increase the income by about Rs 14.44 lakhs. Further, spread of technology will help to solve the problem and may increase the applicability on an area of 450 thousand ha in ensuing years.

This has resulted in increasing yield of cropping system by 17.72 per cent and gave additional net income of Rs 9631 per ha, this showed that the technologies are profitable. These profitable technologies can be adopted by the farmers. The enhanced yield achieved through adoption of improved production technologies increased the income of the farmers. This has cumulatively been able to raise living standard of the farmers.

As the sustainability is concern, the technologies also enhanced the employment by 22.37 man days per ha. If the technologies adopted in all problematic area a huge

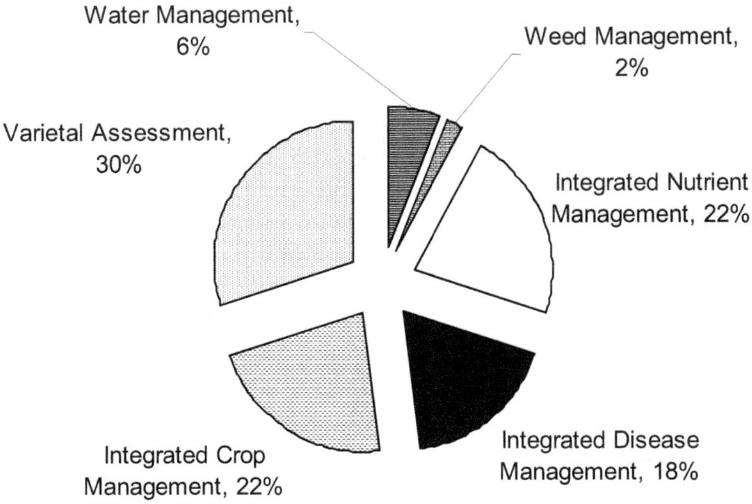

Graph: 44.2: Share of Different Thrust Areas in Productivity Enhancement of Wheat in Malwa Plateau

amount of huge employment can be generated, which checks the migration of rural labour towards cities. This showed the sustainability of the technologies.

The on farm trials on soybean-wheat cropping system had fabulous impact not only on the farmer adopting but also on nearby farmers. There is a need of adoption in large scale through networking of developmental agencies for enhancement in production per unit area.

Conclusion and Suggestions

Soybean has shown a spectacular growth in area and production in Malwa plateau. Soybean–wheat cropping system has emerged as an important cropping system in the agro-climatic zone. A major concern of planners and scientist has been its low productivity under real farm conditions. Some major constraints associated with its low productivity are inadequate transfer of appropriate production technology, In order to overcome these lacunae, the programme of on farm trials were formulated and conducted for assessing the scientific and recent technologies on farmers' field to develop faith of growers in the technologies. It can be concluded from the above study, that the on farm trials conducted on improved production technologies of soybean wheat cropping system farmers field resulted in increasing productivity of the system significantly. About 24.18 per cent increase in soybean production, 14.89 per cent increase in wheat production and about 17.72 per cent increase in the productivity of the system has been recorded under the on farm trials, which shows that still there is a tremendous scope for increasing productivity of cropping pattern in Malwa plateau. The enhanced yield achieved through adoption of improved production technologies increased the income of the farmers. This has cumulatively been able to raise living standard of the farmers.

There is a need of adoption in large scale through networking of developmental agencies for enhancement in production per unit area. Till now, technologies generated at the research stations were thrust upon the farmers in the technology transfer programme without understanding their existing situation. It therefore, required a complete reversal of on farm testing programme, where farmers are fully involved at all the stages.Under the programme there should be a in built mechanism of capacity building for technical and managerial empowerment of farmers like field days, training programme etc. farmers participation in various activities provide them opportunities for decision making and work participation.

References

Singh, S. P., 1999. Methodology of on farm trials. In: *Krishi Vigyan Kendra: A Reality*, (Eds.) P. Das and B. Hansra. Division of Agriculture Extension, ICAR, pp. 127–134.

Farno, L.A., Edwards, L.H., Keim, K. and Epplin, F.M., 2002. Economic analysis of soybean-wheat cropping systems. Online. Crop Management doi:10.1094/CM-2002-0816- 01-RS (http://www.plantmanagementnetwork.org/pub/cm/research/soywheat).

Chapter 45

Constraints with Regard to Improved Vegetable Production Technologies

☆ ***R.S. Suman***[1]

Introduction

The vegetables are the most important to the human diet for better health, because they possess high nutritive value and are rich source of carbohydrates, proteins, vitamins and minerals. The selection of research area in Kullu Valley of Himachal Pradesh was due to the very good conditions for temperate vegetable production and seed production of the same. Hence, the area was suitable for both that the vegetable production and their breeding for most of the temperate types of vegetables. The summer being mild was suitable for many sub-tropical important vegetables. Before the establishment of Katrain station, there was absolutely no vegetable growing in the valley, but there after it has picked up. Still in this area the vegetable production is low, because Improved Vegetables and Seed Production Technologies (IVSPT) are not fully adopted by the farmers at their own field (self reviewed). Even if they produce, the marketing problems are faced by them. Therefore, this research programme is aimed to find out the current adoption level of the IVSPT at the farmers field, technological gap between the farmers and Institute, constraints faced by the farmers in adoption of IVSPT and to suggest the solution regarding the constraints.

---

1 Scientist (Sr. Scale) IARI Regional Station, Katrain – 175 129, Kullu Valley, H.P.

Materials and Methods

The necessary information for the project (number of Blocks in the Kullu Valley for the research and number of villages in each Block) were collected from the Block headquarters and as well district headquarters. After the collection of the information as above, selection of the Blocks, Villages and the Farmers (Large, Medium and Small) based on stratified random sampling would we done. The schedule for the data collection from the selected farmers with regard to the objectives were prepared and also data collected, compiled, tabulated, analyzed and interpreted with statistical methods mean and S.D. were applied.

Results and Discussion

Improved Vegetable Production Technologies and Seed Production Technologies are not fully adopted by the farmers at their own field (self reviewed). Even if they produce, the some problems were faced by them. It was found that the current adoption level of the farmers with respect to Improved Vegetables Production Technologies and Seed Production Technologies were very low and it was need for improvement in the vegetable and seed production by the farmers.

Table 45.1: Distribution of the Farmers with Regard to Age

Sl.No.	*Age Groups*	*No. of Respondents*	*Percentage*
1.	< 32	43	14.33
2.	32–59	202	67.33
3.	> 59	55	18.33
	Total	300	100.00

It is revealed from the Table 45.1 that most of the farmers in the valley represented the age group of 32–59 years (67.33 per cent) followed by 18.33 per cent belong to above 59 years group.

It was found that the percentage of cultivable land was the highest (61.16 per cent) followed by under buildings, road etc. (30.66 per cent). Only 2.94 percent land is uncultivable (Table 45.2) whereas 5.24 per cent of the land has been left fallow.

Table 45.2: Distribution of the Farmers with Regard to Land Position of the Area

Sl.No.	*Categories*	*Percentage*
1.	Cultivable Land	61.16
2.	Fallow Land	5.24
3.	Un-cultivable Land	2.94
4.	Building, Road etc.	30.66
	Total	100.00

Table 45.3: Analysis of the Constraints Faced by the Farmers in Vegetable Production

Sl.No.	Items	Mean	S.D.	Rank
1.	Availability of recommended technologies	125	8.5557	I
2.	Availability of package of practices	110	8.4617	II
3.	Proper price of the crops	240	7.9246	III
4.	Proper transportation available	186	7.7201	IV
5.	Labour shortage as per needs	219	7.5895	V
6.	Availability of insecticides/pesticides	40	7.3756	VI
7.	Availability of good quality of seeds	236	7.1833	VII
8.	Proper knowledge about disease or insect-pests	60	6.4498	VIII
9.	Proper methods applied for harvesting	300	6.4498	VIII
10.	Fully satisfied by crop production	68	6.3246	X
11.	Proper market for sold the crops	119	5.6921	XI
12.	Crop damaged by diseases or insect-pests	284	5.0596	XII
13.	Proper channel for communication	121	4.6043	XIII
14.	Availability of proper irrigation facilities	55	4.5607	XIV
15.	Seeds did not available at sowing time	29	3.8471	XV
16.	Shortage of money for vegetable production	54	3.5777	XVI
17.	Interference of the middleman in the marketing	11	2.6077	XVII
18.	Availability of the seeds	15	2.6067	XVIII
19.	Proper knowledge about insecticides/pesticides	0	0.0000	XIX
20.	Proper knowledge about herbicides	0	0.0000	XIX
21.	Availability of herbicides	0	0.0000	XIX
22.	Give the preference to vegetable production than other crops	0	0.0000	XIX

It is indicated from the above table that an availability of recommended technologies to the farmers ranked first with 8.5557 S.D. followed by availability of the package of practices, proper price of the crops, proper transportation facilities, labour shortage at the time of requirement and availability of insecticides/pesticides with 8.4617, 7.9246, 7.7201 and 7.58.95 S.D. by ranked second, third, fourth and fifth, respectively. The some constraints were not effective to the vegetable production, these were give the preference to vegetable production than other crop followed by proper methods applied for harvesting the crop, availability of herbicides and proper knowledge about herbicides etc by 0.0000 S.D. with ranked last.

Conclusion

In the study 41 per cent farmers applied agronomical practices properly and rest of 59 per cent used local practices. It is indicated from the above table that an availability of recommended technologies to the farmers ranked first followed by availability of the package of practices, proper price of the crops, proper transportation facilities, labour shortage at the time of requirement and availability of insecticides/ pesticides second, third, fourth and fifth, respectively. The some constraints were not effective to the vegetable production, these were give the preference to vegetable production than other crop followed by proper methods applied for harvesting the crop, availability of herbicides and proper knowledge about herbicides etc with ranked last.

References

Salam, A., 1985. Socio-economic and institutional factors influencing fertilizers use in Punjab. *Pakistan Dev. Rev.*, 14(1).

Salasya, B., Mwangi, W., Hugo, V., Odendo, M. and Dona, J., 1998. An assessment of the adoption of seed and fertilizers packages and the role of credit in smallholder maize production in Kakamega and Vihiga District. CYMMYT, Nairobi.

Williams, M.S., 1958. Farmers decision in the use of fertilizers. *J. Farm. Econ.*, Vol. 40.

Chapter 46

Entrepreneurial Behavior of Mango Orchard Growers in Junagadh District of Gujarat State

☆ *N.B. Jadav[1], M.N. Popat and D.V. Patel*

Introduction

Management is the process essential to accomplish enterprise goal and objectives. In general we say management as the process of working with and through individual an groups and other resources such as equipment capital and technology. Every enterprise basically is interested in increasing productivity. The agriculture being an enterprise is not an exception to this. The farmers as the manager of the agricultural enterprise are expected to bring about maximum profit with available resources. The progress, prosperity and success of agricultural enterprise mainly depend on the managerial role played by the farmers.

Managerial ability in present study has been defined as the degree or ability to which an individual acquires and adopts effective factors in an enterprise to reach higher level of performance (Belshaw, 1974) rightly stated that today farming enterprise is becoming more complex and complicated and therefore, management is a key to face this problem. Mango is one of the ancient fruit of India. It undoubtedly

---

1 Krishi Vigyan Kendra, Millet Research Station, Junagadh Agricultural University, Jamnagar.

deserves to be national fruit. In area, production, nutritive value and popularity of apple, no other fruit can compete with it. Among all the fruit crops, mango is being cultivated commercially in number of countries of the world, but no where does it achieve the same premier position as in the subcontinent of India where it is actually the king of all fruit. Orchard growing enterprise also requires the skill and knowledge pertaining to the latest technological know how in the context of management aspect. There are several factors affecting to the productivity of orchard crop. So, it is worthwhile to study the related factors responsible for the production of the mango orchard cultivation. The specific objectives of the study are as under:

1. To measure the managerial ability of the mango grower about scientific cultivation of mango orchard.
2. To explore the relational analysis of selected variables of mango orchard growers.

Research Methodology

The present study was conducted in Talala and Vanthali block of Junagadh district in Gujarat state, where proportionately more number of mango orchard had cultivated and the study was confined to 'ex post–facto' research design. The multistage purposive sampling technique was used for the selection of districts, talukas and villages and it was based on the highest area under mango cultivation. The total numbers of 200 respondents were selected from the list of 20 selected villages by proportionate random sampling methods. For the selection of independent variables, which were found most relevant were finally selected for the study. The tools and techniques used in the present study was interview schedule along with the suitable scales and indices for measurement of dependent and independent variable. For the data collection pretested structured interview schedule was prepared after discussing with a group of expert and necessary modification were made. In relational analysis the coefficient of correlation was find out to know the relationship between each of the independent variable and dependent variable. To know the combined effect of the independent variables in explaining the variation in dependent variable the multiple regression analysis was done and stepwise multiple regression analysis was carried out to know the important variables with their predictive ability in explaining the variation in the dependent variables.

Results and Discussion

Measurement of Managerial Ability of Mango Orchard Growers

It can be seen from the Table 46.1 that majority of respondents (60.00 per cent) were observed in the medium managerial ability category, while 21.50 per cent respondents fall under the category of low managerial ability. The remaining 18.50 per cent respondents possessed high managerial ability. Thus, the managerial ability of the respondents was predominantly medium. It can be concluded from the above finding that the managerial ability of mango growers about scientific cultivation of mango orchard was medium.

Table 46.1: Distribution of Respondents by Managerial Ability (n = 200)

Sl.No.	Category	Frequency	Percentage
1.	Low MA(below 46.94)	42	21.50
2.	Medium MA (46.94 to 70.12)	120	60.00
3.	High MA (above 70.12)	37	18.50
	Mean = 58.53	S.D. = 11.58	C.V.= 19.79

This may be due to the medium education level, less training received on management aspects and having medium level of farm inputs with majority of the mango growers. Further, more than two-thirds respondents had medium level of attitude towards modern agriculture resulting into medium level of managerial ability.

The finding was in the conformity with the finding of that Patel and Patel (2000).

Relational Analysis between Selected Variables

Correlation Analysis

To ascertain the association between managerial ability of mango orchard growers (dependent variables) and their selected characteristics (independent variables) correlation analysis was carried out. On the basis of operational measure developed for the variable null hypothesis were stated for testing the relationship and their significance on zero order correlation. The zero order correlation are given in Table 46.2.

Age and education are the personal characteristics of mango orchard growers in which age was negative and significantly associated with managerial ability of mango orchard grower. This might have happened because young farmers having higher level of managerial ability as compare to old age farmers'. Similar finding was observed by Patel and Patel (2000). The education had positive and significant relationship with managerial ability. It means farmers having higher level of education had higher level of managerial ability. This finding was in conformity with the findings of Rao (1995), Sumathi (1987) and Nagarajan (1989).

In case of operational management variables there was significant correlation existed between the level (degree) of managerial ability of mango orchard growers and their level of adoption, annual income, farm mechanization index and mango yield index. It implied that increase in mango orchard growers level of adoption annual income, farm mechanization index and mango yield index resulted in increase of positive managerial ability about scientific cultivation of mango orchard. While, remaining variables likes, size of land holding, mango crop intensity, irrigation potentiality, borrowing of the total credit and level of farm wage were not significantly related with managerial ability of mango orchard grower.

Strategic management variables, experience as a mango growers, attitude towards modern agriculture and risk orientation were significantly related with managerial ability of mango orchard growers and social participation, personality, achievement motivation and level of aspiration were non significant with managerial ability of mango orchard growers. This finding of Vyas (1995) was supported this finding While findings of Bora (1986) and Nagarajan (1989) differed with this investigation.

Table 46.2: Zero Order Correlation Coefficient of Independent Variables with Managerial Ability of Mango Growers

Sl.No	*Independent Variables*		*'r' value*
I	**PERSONAL**		
1	X_1	Age	-0.25699**
2	X_2	Education	0.52564**
II	**OPERATIONAL MANAGEMENT**		
3	X_3	Adoption Index	0.58392**
4	X_4	Size of land holding	0.11279 ns
5	X_5	Annual income	0.15768*
6	X_6	Farm mechanization index	0.27693**
7	X_7	Mango crop intensity	0.06648 ns
8	X_8	Irrigation potentiality	0.04587 ns
9	X_9	Mango yield index	0.19406**
10	X_{10}	Borrowing of total credit	0.07263 ns
11	X_{11}	Level of farm wage	0.10268 ns
III	**STRATEGIC MANAGEMENT**		
12	X_{12}	Experience as a mango growers	0.34711**
13	X_{13}	Social participation	0.12288 ns
14	X_{14}	Personality	0.06587 ns
15	X_{15}	Achievement motivation	-0.06154 ns
16	X_{16}	Orientation towards the competition	0.09234 ns
17	X_{17}	Attitude towards modern agriculture	0.25488**
18	X_{18}	Level of aspiration	0.11565 ns
19	X_{19}	Risk orientation	0.30592**
IV	**EXTENSION–COMMUNICATION**		
20	X_{20}	Extension participation	0.04616 ns
21	X_{21}	Training received	0.22302**
22	X_{22}	Mass media exposure	0.24046**
23	X_{23}	Personal guidance on better farming	0.1191 ns

*: Significance at 0.05 level (0.138); **: Significance at 0.01 level (0.181).

ns: Non significance.

In extension communication variables the half of variables like, training received and mass media exposure were positively and significantly associated with managerial ability of mango orchard growers (Nagrajan, 1989) and half of the characteristics like extension participation and personal guidance on better farming were not significantly related with managerial ability of mango orchard growers.

Based on the coefficient of correlations, ten independent variables viz, adoption index, education, annual income, experience as a mango growers, farm mechanization index, mango yield index, training received, attitude towards modern agriculture, mass media exposure and risk orientation were found having significant and positive relationships with managerial ability at 0.01 level of significance. While age had significant but negative relationship with managerial ability at 0.01 level of significant.

Extent of Variation

Multiple Regression Analysis

All the independent variables mentioned in Table 46.3 explained as much as 63.70 per cent of total variation in the managerial ability. The unexplained variation was of 36.30 per cent may be due to the factors outside the scope of the study.

Table 46.3: Multiple Regression Analysis of Managerial Ability (n = 200)

Sl.No	*Variables*	*Regression Coefficient (bi)*	*S.E. of bi*	*'t' value*
1.	Age (X_1)	–0.2483	0.0665	–3.7338**
2.	Education (X_2)	0.9760	0.1833	5.3246**
3.	Adoption Index (X_3)	0.1702	0.0339	5.0206**
4.	Size of land holding (X_4)	–0.3358	0.2951	–1.1379
5.	Annual income (X_5)	0.00061	0.0019	0.6027
6.	Farm mechanization index (X_6)	0.0841	0.0461	1.8243
7.	Mango crop intensity (X_7)	0.0041	0.0318	0.1289
8.	Irrigation potentiality (X_8)	0.0429	0.0339	1.2655
9.	Mango yield index (X_9)	0.0033	0.0397	0.0831
10.	Borrowing of total credit (X_{10})	2.8939	6.3808	0.4535
11.	Level of farm wage (X_{11})	0.00769	0.1100	0.0700
12.	Experience as a mango growers (X_{12})	0.8233	0.1996	4.1247**
13.	Social participation (X_{13})	0.2918	0.2126	1.3725
14.	Personality (X_{14})	0.2418	0.2867	0.8434
15.	Achievement motivation (X_{15})	–0.4718	0.2580	–1.8287
16.	Orientation towards the competition (X_{16})	–0.0807	0.1418	–0.5691
17.	Attitude towards modern agriculture (X_{17})	0.0933	0.0716	1.3031
18.	Level of aspiration (X_{18})	0.1790	0.0714	2.5070*
19.	Risk orientation (X_{19})	0.6976	0.2435	2.8649**
20.	Extension participation (X_{20})	–0.0084	0.0613	–0.1370
21.	Training received (X_{21})	–0.3176	0.7676	–0.4138
22.	Mass media exposure (X_{22})	0.4864	0.1843	2.6392**
23.	Personal guidance on better farming (X_{23})	–0.0696	0.0935	–0.7444

*: Significance at 0.05 level (1.960); **: Significance at 0.01 level (2.576).

R^2: 0.6370.

It can also be revealed that the 't' values of six variables viz; adoption index, age, education, experience as a mango growers, mass media exposure and risk orientation were significant at 0.01 level of significance and the 't' value of one variable *i.e.* level of aspiration was significant at 0.05 level of significance. These seven variables significantly contributed in explaining the variation in managerial ability.

Remaining variables have failed to contribute significantly in managerial ability of mango growers about scientific cultivation of mango orchard.

It can be concluded that 63.70 per cent total variation in managerial ability was explained by set of 23 independent variables together. It can also be concluded that out of 23 variables, seven variables viz; adoption index, age, education, experience as a mango growers, mass media exposure, level of aspiration and risk orientation had significant contribution in managerial ability. This result provides evidence about the overwhelmingly important role of seven significant variables in achieving managerial ability.

Stepwise Multiple Regression Analysis

It is clear from Table 46.4 that the seven variables *viz.* adoption index, age, education, experience as a mango growers, mass media exposure, level of aspiration and risk orientation put together explained as much as 60.67 per cent of total variation in the managerial ability. The unexplained variation was 39.33 per cent may be due to factors other than above mentioned seven factors.

It can also be seen from the Table 46.4 that the 't' values and 'f' values for all the above seven variables were found significant at 0.01 level of significance indicating significant contribution of these seven variables on managerial ability. The partial regression coefficients indicated that one unit change in adoption index, age, education, experience as a mango growers, mass media exposure, level of aspiration and risk orientation would change 0.1813 unit,–0.2042 units, 1.0789 units, 0.9220 units, 0.5142 units, 0.2060 units and 0.6393 units in the managerial ability, respectively.

Table 46.4: Stepwise Multiple Regression Analysis of Managerial Ability (n = 200)

Sl.No	*Independent Variables*	*Partial Regression Coefficient (bi)*	*S.E.*	*'t' Value*	*'f' Value*	*Standard Partial Regression Coefficient (bi')*	*Rank of bi*
1.	Age (X_1)	–0.2042	0.0590	3.4610**	3.4610**	0.20973	IV
2.	Education (X_2)	1.0789	0.1734	6.2220**	6.2220**	0.35525	I
3.	Adoption Index (X_3)	0.1813	0.0320	5.6656**	5.6656**	0.29735	II
4.	Experience as a mango growers (X_{12})	0.9220	0.1664	5.5409**	5.5409**	0.26813	III
5.	Level of aspiration (X_{18})	0.2060	0.0668	3.0838**	3.0838**	0.15892	V
6.	Risk orientation (X_{19})	0.6393	0.2085	3.0662**	3.0662**	0.15480	VI
7.	Mass media exposure (X_{22})	0.5142	0.1710	3.0070**	3.0070**	0.13990	VII
	CONSTANT =18.04	Multiple R = 0.7789				R^2 = 0.6067	

*: Significant at 0.01 level significance.

It is clear from Table 46.5 that the variable, adoption index alone account 34.10 per cent variation in managerial ability followed by adoption index + education (46.22 per cent), adoption index + education + experience as a mango growers (53.64 per cent), above three + mass media exposure (55.66 per cent), above four + risk orientation (57.41 per cent), above five + age (58.72 per cent), above six + level of aspiration (60.67 per cent).

Table 46.5: Stepwise Variation Accounted by Different Independent Variables in Managerial Ability (n = 200)

Step No.	*Variables included*	*Multiple 'R'*	*Total variation Accounted (R^2)*
I	Adoption Index (X_3)	0.5839	0.3410 (34.10 per cent)
II	Education (X_2) + (X_3)	0.6799	0.4622 (46.22 per cent)
III	Experience as a mango growers (X_{12}) +(X_2) + (X_3)	0.7324	0.5364 (53.64 per cent)
IV	Mass media exposure (X_{22}) +(X_{12}) +(X_2) + (X_3)	0.7460	0.5566 (55.66 per cent
V	Risk orientation (X_{19}) + (X_{22}) +(X_{12}) +(X_2) + (X_3)	0.7577	0.5741 (57.41 per cent)
VI	Age (X_1) + (X_{19}) + (X_{22}) +(X_{12}) +(X_2) + (X_3)	0.7663	0.5872 (58.72 per cent)
VII	Level of aspiration (X_{18}) + (X_1) + (X_{19}) + (X_{22}) + (X_{12}) +(X_2) + (X_3)	0.7789	0.6067 (60.67 per cent)

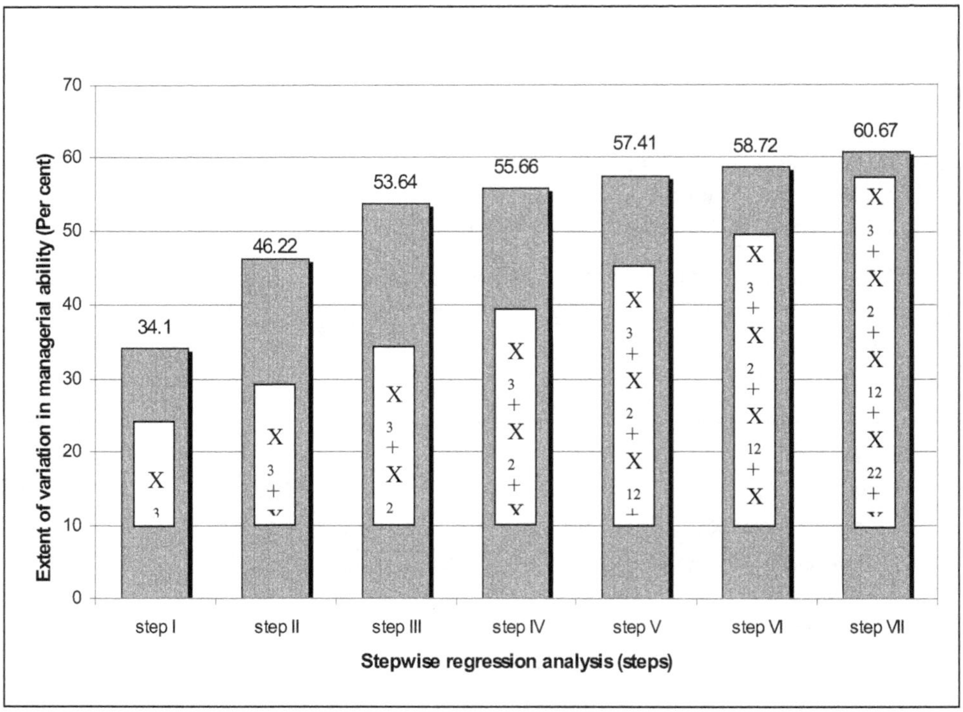

Figure 46.1: Extent of Variation Accounted in Stepwise Multiple Regression Analysis

X_3: Adoption Index; X_2: Education; X_5: Experience as a mango growers; X_{19}: Mass media exposure; X_{22}: Risk orientation; X_1: Age; X_{21}: Level of aspiration

It can be concluded from the above results of stepwise regression analysis that 60.67 per cent variation was accounted by a set of seven independent variables viz, adoption index, education, experience as a mango growers, mass media exposure, risk orientation, age and level of aspiration put together in managerial ability. Adoption index by mango orchard alone was accounted 34.10 per cent variation. The standard partial values also indicated that the education (35.52) and adoption index (29.73) were in highest order of magnitude, which reflected its importance.

Conclusion

It can be concluded from above discussion that two-thirds (60.00 per cent) of the mango orchard growers were from the medium managerial ability category

Based on the above relational analysis, the result leads to the conclusion that the important characteristics of mango orchard growers for selection criteria in relation to managerial ability were education, adoption index, annual income, farm mechanization index, mango yield index, experience as a mango growers, attitude towards modern agriculture, risk orientation, training received and mass media exposure were positively and significant correlation ship with managerial ability of mango orchard growers. While significant contribution in total variation and highest contribution in total variation variables were age, education, adoption index, experience as a mango growers, level of aspiration and risk orientation. Therefore, due weightage should be given to the above characteristics of mango orchard growers to achieve higher managerial ability resulting in to effective management of scientific cultivation of mango orchard.

References

Belshaw, D., 1974. Improving management procedures for agricultural development. In: *Paper Presented at International Seminar on Change in Agriculture*, Reading U.K.

Nagajara, N., 1989. A study on management efficiency and economic performance of sericulturists in Karanataka. *Ph.D. Thesis* (Unpublished), UAS, Bangalore.

Patel, A.A. and Patel, R.K., 2000. Growers managerial ability for plant protection measures in chilli crop. *Gujarat Journal of Extension Education,* 10 and 11: 1–4.

Rao, G.K.V., 1985. A prediction analysis of farming performance of farmers through their entrepreneurial behaviour factors. *Ph.D. Thesis* (Unpublished), UAS, Bangalore.

Sumathi, U.B., 1987. A study on achievement motivation and management orientation of small coffee growers in Chickmagalore district, Karnataka state. *M.Sc. (Agri). Thesis* (Unpublished), UAS, Bangalore.

Chapter 47

Entrepreneurship in Vegetable Growers

☆ ***Rakesh Jain[1] and M.M. Patel[2]***

Introduction

Vegetables enjoy a significant place in our daily diet as they provide essential nutrition including vitamins and minerals. According to ICMR, the per capita requirement of vegetables has been put at 300 g per day. Hence, there is a very high demand for vegetables.

India being primarily a country having agriculture based economy enjoys second position in the world in total vegetables production (with a share of about 14 per cent) next only to China. At present vegetables cultivation occupies 6.09 million hectares area with an annual production of 84.8million tones. The country being blessed with the unique gift of nature of diverse climates and distinct seasons, it makes it possible to grow an array of vegetables whose number exceeds more than a hundred types.

However, potato being the staple food and easy to mix in several preparation ranks first (27.3 per cent) in total production of vegetables followed by the other solanaceous vegetables like brinjal (9.4 per cent) and tomato (9.0 per cent). Cauliflower and cabbage are the most preferred vegetables and their total share in the country's vegetables production is 5.2 and 6.4 per cent, respectively.

In Madhya Pradesh, Malwa Plateau as a whole has been declared Agri-Export Zone, recently by the government for exporting vegetable crops. There is a great

---

1 Ex student.

2 Senior Scientist, Department of Extension Education, College of Agriculture, Indore, M.P.

potential to increase the production of vegetables, if the growers are oriented towards entrepreneurship and adoption of modern production technology.

At present, the farmers are facing the problem of lack of knowledge about modern production technology, insufficient input supply and inadequate regular market for disposal of their production. Hence, there is a need to utilize the available resources wisely, adopt modern production technology for getting maximum profit and search the adequate marketing opportunities. Therefore, an effort was made to know the entrepreneurship of vegetable growers in terms of socio–personal, economic, and psychological attributes.

Methodology

The study was conducted in Indore district of Madhya Pradesh. There are four blocks in Indore district. Out of them, two blocks *viz.* Depalpur and Indore were selected randomly. From each selected block, a list of villages was prepared and five villages from each block were selected randomly. From each selected village, nine vegetable growers were selected randomly. Thus, the total sample size of 90 respondents was covered from the study area. The data were collected with the help of pre-tested interview schedule by using personal interview method.

To measure the extent of entrepreneurship an index developed by Kumar (2005) over Akhouri entrepreneurial economic success index (EESI) was used. The entrepreneurial success index used for the study included six indicators *viz.*, gross return per unit investment (net profit), enterprise diversification, share of profit reinvested, degree of satisfaction, employment of family labour and social prestige gained. The entrepreneurship categories were formulated in low, medium and high on the basis of mean ± 1S.D. The correlation analysis was carried out to find out the relationship between socio–personal, economic, and psychological variables with entrepreneurship.

Findings and Discussion

Entrepreneurial Success in Vegetable Growers

Entrepreneurship of vegetable growers is presented in Table 47.1. A perusal of data reported in Table 47.1 revealed that majority of the respondents (61.11 per cent) had medium level of entrepreneurship while 24.44 per cent had high and only 14.44 per cent had low level of entrepreneurship. Similar findings were also reported by Patil *et al.* (1999).

The Table further revealed that the overall mean entrepreneurship score was 27.93. The mean entrepreneurship score of small, medium and large holding vegetable growers were 25.57, 28.43 and 29.80 respectively. The mean values of entrepreneurship of medium and large farmers were higher than overall mean, while the mean value of entrepreneurship of small farmers was lower than overall mean.

The " t " test revealed that the mean value of small verses medium and small verses large holding farmers varied significantly, where as medium verses large holding farmers did not vary significantly from each other. Similar findings were also reported by Choudhary (2006) in case of potato growers.

Table 47.1: Distribution of Vegetable Growers According to their Entrepreneurship (n = 90)

Sl.No.	Entrepreneurship	Small	Medium	Large	Total
1	Low (<24 score)	10(33.33)	2(6.66)	1(3.33)	13 (14.44)
2	Medium (24-32 score)	17(56.66)	22(73.33)	16(53.33)	55 (61.12)
3	High (>32 score)	3(10.00)	6(20.00)	13(43.33)	22(24.44)
	Total	30	30	30	90
	Mean	25.57	28.43	29.80	27.93
	S.D.	3.52	3.47	2.26	3.82
	"t" value	Small vs Medium = 3.17**			
		Small vs Large = 4.83**			
		Medium vs Large = 1.57			

*: Significant at 5 per cent level; **: Significant at 1 per cent level.

Relationship Between Socio-Personal, Economic and Socio-Psychological Attributes of Vegetable Growers and their Entrepreneurial Success

With the assumption that entrepreneurship of vegetable growers influenced by their socio- personal, economic and psychological attributes, the relationships of various socio-personal-economic and psychological variables with entrepreneurship, coefficient of correlation between entrepreneurship and fourteen selected variables was computed and compared. The results are presented in Table 47.2.

It is observed from the Table 47.2 that, out of nine socio-personal and economic variables, three variables namely material possession, family type and economic status were positively related with entrepreneurship. Similar findings were also reported by Choudhary (2006) in case of family type and material possession, and Patel and Sanoria (1997) reported similar findings in case of economic status.

Table also revealed that of five socio-psychological variables the correlation coefficient of two variables namely economic motivation and risk taking willingness were positively related with entrepreneurship. Similar findings were also reported by Rays and Bora (1991), Choudhary (2006), in case of risk taking willingness,

Constraints Faced by the Farmers in Marketing

It is evident from Table 47.3 that in respect of vegetable production, high price of seeds (91 per cent), non- availability of funds from institutional sources (86.67 per cent), high price of fertilizers, pesticides and fungicides(84 per cent), non- availability of good seeds (83 per cent) and lack of knowledge regarding plant protection (66.67 per cent) were the major constraints perceived by the vegetable growers.

Table 47.2: Relationship Between Socio-Personal, Economic and Socio-Psychological Attributes of Vegetable Growers and their Entrepreneurship

Sl.No.	*Factor*	*Coefficient Correlation (r)*
(A)	**Socio-personal and economic**	
1.	Age	0.177
2.	Education	0.135
3.	Family type	0.249**
4.	Family income	0.088
5.	Land size	0.117
6.	Value of house	0.183
7.	Material possession	0.360**
8.	Economic status	0.337**
9.	Credit worthiness	0.043
(B)	**Socio-psychological**	
1.	Economic motivation	0.266*
2.	Risk taking willingness	0.342**
3.	Market orientation	0.089
4.	Owning Responsibility	0.155
5.	Level of knowledge	0.047

*: Significant at 5 per cent level of probability; **: Significant at 1 per cent level of probability.

Table 47.3: Constraints in Vegetable Production and Marketing Perceived by the Vegetable Growers (N = 90)

Sl.No.	*Constraints Relating*	*No. of Respondents*	*Percentage*
	Production		
1.	Non-availability of good quality seed	75	83.33
2.	High price of seed	82	91.11
3.	High price of fertilizers, pesticides and fungicides	76	84.44
4.	Lack of knowledge regarding plant protection	60	66.67
5.	Non-availability of fund from institutional sources	78	86.67
	Marketing		
6.	Lack of transportation	60	66.67
7.	High charges of transportation	72	80.00
8.	Lack of link road	37	41.11
9.	Lack of market yard	49	54.44
10.	Traders collusion	45	50.00
11.	Malpractices by traders	73	81.11
12.	Higher market charges	70	77.78

Contd...

Table 47.3–Contd...

Sl.No.	Constraints Relating	No. of Respondents	Percentage
13.	No correct weighing	38	42.22
14.	Late payment	46	51.11
15.	Lack of credit facilities	37	41.11
16.	Lack of packing material	41	45.56
17.	Price fluctuation	78	86.67
18.	Lack of information regarding market	72	80.00
19.	Lack of storage in market	75	83.33
20.	Non-availability of adequate storage facilities	82	91.11

These constraints can be overcome by arranging appropriate inputs at reasonable price and conducting training programmes, demonstrations, etc. at village level by extension personnel. Further adequate credit facilities can be provided to the vegetable growers for purchasing of required inputs which will accelerate the adoption of recommended technologies.

In respect of marketing of vegetable, non-availability of adequate storage facilities (91 per cent), price fluctuation (86.67 per cent) lack of storage facilities in market (83 per cent) misconduct by traders (81 per cent), high charges of transportation (80 per cent), lack of information regarding market (80 per cent), high market charges (77.78 per cent), were the key constraints pointed out by vegetable growers.

These problems can be trounced by providing appropriate facilities for storage and transportation by Mandi Board to the vegetable growers.

Conclusion

The study revealed that the majority of the vegetable growers had medium level of entrepreneurship. The study further revealed that the attributes namely material possession, family type, economic motivation, risk taking willingness and economic status were positively related with entrepreneurship. Hence, it can be concluded that the farmers having better material possession, joint family, higher economic status and economic motivation and high risk taking willingness possessed higher entrepreneurship.

The study also revealed that in respect of vegetable production, high price of seeds, non- availability of funds from institutional sources, high price of fertilizers, pesticides and fungicides, non- availability of good seeds and lack of knowledge regarding plant protection were the major constraints perceived by the vegetable growers. In respect of marketing of vegetables, non-availability of adequate storage facilities, price fluctuation, lack of storage facilities in market, misconduct by traders, high charges of transportation, lack of information regarding market, high market charges, were the key constraints pointed out by vegetable growers.

References

Choudhary, R.K.S., 2006. A study on entrepreneurial behaviour of potato growers in Malwa region of Madhya Pradesh, Unpublished *M.Sc.Thesis*, Department of Extension Education, J.N.K.V.V. College of Agriculture, Indore, M.P.

Patel, M.M. and Sanoria, Y.C., 1997. Correlates of Entrepreneurial behavior of sugarcane growers. *Maharashtra J. Extn. Edu.*, 16: 344–346.

Patil, V.G., Mahadik, R.P. and Patil, A.S., 1999. Entrepreneurial behavior of little guard growers. *Maharashtra J. Extn. Edu.*, 18: 240–245.

Chapter 48

Livelihood Security through Adoption of Frontier Technology in Soybean

☆ ***Seema Naberia[1], U.S. Gautam[2] and S.S. Chauhan[3]***

Introduction

Agriculture is the mainstay of the Indian economy, as it constitutes the backbone of the rural livelihood security system. A major farming community of India comes under small and marginal farming community, where the size of the land holding is very small to achieve the standards of livelihood. The daily income of these farmers is not sufficient to get their daily needs. A majority of theses farmers are suffering from poverty and unemployment, which results a failure to achieve necessary households makes a living overtime. About 700 million of India's population lives in villages and their main source of livelihood is agriculture, comprising crop and animal husbandry, forestry, fisheries, agro processing and agribusiness. Therefore, accelerated progress in enhancing the productivity, profitability, stability and sustainability of the major farming system is the best safely net against hunger and poverty.

---

1 Assistant Professor(Agricultural Extension), College of Agriculture, Ganjbasoda, Vidisha, M.P.

2 Zonal Coordinator, Zone-VII, ICAR, Jabalpur, M.P.

3 SMS, KVK, Narsinghpur, M.P.

Increasing the productivity of crops ensure the livelihood security of the farmers because it depends upon the productivity of their fields. To overcome this situation oilseed especially, Soybean plays a vital role besides the cereal crops. Soybean is the rich source of protein quality nutrition and valuable cash also.

Soybean is a major oilseed crop of Madhya Pradesh during *kharif* season and has spearheaded its economic growth. It is grown on 51.434 lakh hectares against 48.792 lakh hectares during last year. The area and production of soybean crop in Narsinghpur district of M.P. is very low as compare to state and National acreage and production. Some major constraints associated with its low productivity are inadequate transfer of appropriate transfer of technology, inherent problems of seed longevity and poor seed quality etc. In order to overcome these problems, the Ministry of Agriculture, Govt. of India taken the innovative methodology to increase the production of oilseed crop *viz.* soybean. The aim of Frontline Demonstration on soybean crop are to demonstrate the superior productivity potentials of various location/region specific technologies to practicing farmers and test their implementability and viability and obtain feedback from the end users and bring about necessary corrections to improve their acceptability and suitability in real farm situations vis-à-vis prevailing additional farmer practices.

Keeping the importance of Front Line Demonstration, the Krishi Vigyan Kendra, Narsinghpur conducted demonstrations on improved agricultural technologies of soybean in scientific manner at farmers' fields during the year 2005-06, 2006-07 and 2007-08 with the following objectives:

1. To demonstrate the performance of recommended fertilizer dose.
2. To compare the yield levels of local check and demonstration field.
3. To collect feedback information for further improvement.

Methodology

Initially the Front Line Demonstration (FLD) programme on soybean was implemented in the villages of Narsinghpur district of Madhya Pradesh during 2005-2006. Farmers of operational area of KVK, Narsinghpur were selected, as per allotment of FLDs to KVK by Zonal Coordinator, Zone-VII. Accordingly, the FLD under *kharif* soybean was laid down in the villages, namely, Chourakheda, Salaiya, Mehmedpur, Khamariya, Bilkhedi and Karakbel. KVK scientists were visited the demonstration field and made to guide them. These visits were also utilized to collect feedback information for further improvement in research and extension programme. Farmers were encouraged to adopt these scientific technologies through organizing field days, group meeting, training programme and farmers' conventions etc. at appropriate time at the demonstration sites. The Krishi Vigyan Kendra supplied the critical inputs accordingly to the farmers. Data were collected from the demonstration fields and analyzed with the suitable statistical tools to compare the yields of local check (Farmers' field) and FLDs fields.

Results and Discussion

Frontline Demonstration on *Kharif* soybean

Performance of Recommended Dose of Fertilizer in Soybean

The progress of frontline demonstration on soybean during *kharif* season, 2005-06,2006-07 and 2007-08 to show the performance of recommended dose of fertilizer (DAP-130.43 kg/ha, $ZnSO_4$–25 kg/ha)in soybean crop is presented in Table 48.1.

Table 48.1: Performance of Recommended Dose of Fertilizer in Soybean

Crop Season	Village	Variety Demon-strated	No. of Demon-stration	Area (ha)	Average Yield (q/ha)			Increase in Yield (%)
					Highest	Demon-stration	Local Check	
Kharif 05-06	Chourakeda,	JS-93-05	13	5.2	20.0	15.00	12.25	22.5
Kharif 06-07	Salaiya,		13	5.0	18.3	13.6	10.2	33.34
Kharif 07-08	Mehmedpur, Khamariya		13	5.2	14.80	11.6	8.55	35.97

It is clear form the data that during *kharif* 2005-06 season, 13 demonstrations were laid down by covering area 5.2 ha in adopted village with variety JS-93-05. The demonstration plots comprised of using DAP @ 130.43 kg/ha and $ZnSO_4$@25 kg/ha where as the farmers practice (local check) was no use of DAP in soybean. Besides that optimum row spacing of 30-45 cm was also given emphasis as against local practice of 22.5 cm. The maximum seed yield under demonstration plots was 20q/ha. The mean seed yield of 15q/ha under demonstration was 22.5 per cent more than the local practice. In *kharif* 2006-07, 13 demonstrations were conducted on soybean covering 5.2 ha in adopted villages with variety JS-93-05.using DAP @ 130.43 Kg/ha and $ZnSO_4$@25 kg/ha and with out DAP and $ZnSO_4$ (Local check). An average yield of 13.6 q/ha of demonstration was obtained, as compared to 10.2 q/ha of local check, where percent increase was accounted at 33.34. During kharif 2007-08, a total number of 13 demonstrations covering 5.2 ha in 3 villages land were laid down. The application of DAP @ 130.43 Kg/ha and $ZnSO_4$@25 kg/ha, as demonstrated factor resulted in 11.6q/ha yield where as the yield obtained by local check was 8.55 q/ha. This accounted for 35.97 per cent average increase in the yield.

Conclusion

Frontline demonstration on soybean showed a significant increase in yield of demonstration over farmers practice, thus a higher income also. In this way the livelihood security of the small and marginal farmers can be improved by increasing the productivity of soybean crop. Obtaining production can solve their problem of malnutrition of food insecurity and money. The enhanced yield achieved through adoption of improved production technologies and increased the income of the farmers. This has cumulatively been able to raise living standard of farmers and overcome the problem of poverty, malnutrition and unemployment.

Chapter 49

Participatory Irrigation Management: An Impact Study of Mehsana District of Gujarat State

☆ *R.F. Thakor*[1] *and P.J. Joshi*[2]

Introduction

The national water policy stresses participatory approach in water resources management.It has been recognised that participation of beneficiaries will help greatly for the optimum upkeep of the irrigation system and effective utilisation of irrigation water. The participation of the farmers in the management of irrigation would include transfer of responsibility for operation and maintenance and also collection of water charges to the water users association in their respective jurisdiction.

Farmers appreciate assured irrigation and water facilities and are prepared to pay appropriate high water rates. It is essential for the irrigation scheme to be efficient and competent in all respects such as beneficiary farmers could be served with sufficient and timely irrigation facilities on the basis of equity with full assurance. Recommendations and deliberation for the national level seminar held at Hyderabad during February, 1992 stipulates and suggests to adopt and develop partnership between government and farmers for the purpose of irrigation management. After

---

1 Programme Coordinator, Gujarat Vidyapith Krishi Vigyan Kendra,Valsad, Gujarat.

2 Programme Assistant, Gujarat Vidyapith Krishi Vigyan Kendra, Valsad, Gujarat.

careful consideration, Government of Gujarat resolves to introduce Participatory Irrigation Management in the state. The term PIM refers to participation of farmers at all levels and aspects of irrigation management. It includes–initial project formulation to all further developmental activities like construction,supervision, financing, resource mobilization, decisions, co-operation, maintenance, conflict resolution, joint management with officials,monitoring and evaluation of irrigation system. One such project was started in 1995 in Mehsana district of Gujarat. Hence, with a view to assess the impact of the project the present study was undertaken.

Objectives

The present study intends to analyze the benefit of participatory irrigation management through 117 irrigation co-operatives functioning in Dharoi irrigation project on Sabarmati river belongs to Mehsana district of Gujarat.

The specific objectives were to study the equitable distribution of water among head richer and tail ender farmers and to study the extent of efficiency and effectiveness in the use of limited water resources. The study was mainly focused on efficiency of canal irrigation, agricultural productivity, conflict resolution and financial management by the farmers group.

Methodology

The study was conducted in Dharoi irrigation project of Mehsana district situated in northern Gujarat. It comprised of 117 irrigation co-operatives with 23,388 members. The programme covers approximately 23290ha. of command area and was jointly implemented by Irrigation department and Development Support Centre-a leading NGO working in the area since 1994 to promote and support participatory management of natural resources and actively engaged in framing policies regarding irrigation management with government. The project of PIM was started in 1995. The study was restricted to the members of 117 ICs only of Dharoi command are of Sabarmati river. The primary information was collected from the members of ICs and secondary data was collected from the records available with the different department.

Results and Discussion

The data regarding impact of Participatory Irrigation Management project on different aspects such as increase in area under irrigation, increase in crop production, participation of farmers in conflict resolution, financial management etc are depicted in Table 49.1.

Increase in Area Under Irrigation

It is observed that the net area under irrigation after PIM has been increased from 6126 ha. to 10,629 ha. shows 74 per cent increase during *rabi* season. About 4503 ha of land has been covered under winter season crop. Effective management of canal, rotational water distribution system, timely availability of irrigation water, reduction in loss of water, proper and regular maintenance of canal structure was some of the reasons expressed by the members of irrigation co-operatives.

Table 49.1: Impact of Participatory Irrigation Management

Sr.	*Parameter*	*Before PIM*	*After PIM*	*Increase in Percentage*
A	**Efficiency**			
1.	Area comes under irrigation	6126	10629	74
B	**Equity**			
1.	Length of canal functioning	145	261	90
2.	No. of farmers received water satisfactorily	7542	15071	100
C	**Finance**			
1a.	Government water rates for wheat	132	482	
1b.	Average water rates fixed by ICS (Rs/ha.)	–	728	51
2a.	Government water rates for castor	240	682	
2b.	Average water rates fixed by ICS (Rs/ha.)		916	34
3a.	Collection of water charges		78.52 lacs	
3b.	Revenue retained by ICs after paying govt. dues		45.95 lacs	
4.	Additional water charges collected by ICs.		24.24 lacs	
D	**Institutional Management**			
1.	Conflict resolution using social pressure (no.)	493	99	
2.	Filed police case (no.)	14	8	
3.	Nos. of operators appointed by ICs	–	202	
4.	Expenditures incurred on canal operators		11.32 lacs	
5.	Total expenditures incurred by ICs for O&M		35.93 lacs	
6.	Production of wheat crop (Kg/ha)	2858	3112	9
7.	Milk production per day/animal in march subject to 52 ICs only	4.07 liters	5.14 liters	

Equal Distribution of Irrigation Water

It was observed that there were two situations: The head reacher farmers were making excess use of irrigation water as it was available frequently and in full quantity in a single rotation due to situational advantage of canal. Continuous use of excess water created problems of water logging and ultimately resulted into poor fertility status of soil. On the contrary the adequate water quantity does not reach to tail-enders. They lost their crops because of want of water even at critical stages of crop growth. It was also noticed that the minor of canal was utilized by 50 percent only of its full capacity.

After PIM the situation was found totally different. It was observed that the canal was utilized at its fullest capacity *i.e.* 261 km length. Earlier only 7542 farmers were received water satisfactorily whereas after PIM it has increased to more than 15,000 farmers. Means about 7500 farmers started getting irrigation water who had not received water for more than 10 years. Even head richer farmers had also fully co-operated in the process because the problem of degradation of soil has reduced drastically as a result of optimum utilization of water by them. Another reason for cooperation by the head reacher was that they bound to pay penalty if not participate in the programme. There was a mechanism developed that such farmers may highlighted on the common notice board of the village. Secondly, the tail-enders were also in position to produce at his best because of availability of water. The fore most objective of this scheme *i.e.* equality in term of water quantity and charges for its utilization was well maintained. Thus remarkable changes were observed in both efficient use of water as well as increase in production of crop during rabi season through maximum use of canal minor.

Financial Management

According to government policies the rate of irrigation charges for wheat crop was Rs. 132 during the year 1997-98. It has been increased to Rs. 482/ha. during the year 2005-06. It is surprising to note that the members of Irrigation co-operatives were paid 51 percent higher charges for irrigation water for wheat crop. Similarly in the case of castor, member farmers had paid 34 per cent higher charges towards irrigation water. This proves that under their own management (Co-operative), farmers were found more satisfactory as compared to the services rendered by government departments. Further, it was observed that all irrigation co-operatives had timely recovered irrigation charges to the extent of cent percent which was not possible under single management system (Government). It is surprising to note that ICs has collected about 48 per cent more revenue in the form of charges. This may be because of sense of belongingness and trustworthy ness amongst members of the ICs as well as executive body.

Production

Production data of wheat crop shows that the yield percent was increase up to 9.00 per after implementation of PIM scheme. This may be due to fact that the tail enders has got the irrigation water throughout the period of crop season. Thus sizable area was brought under cultivation of *rabi* crops (wheat). Therefore, increase in crop area and prompt availability of irrigation water were the important factors in increasing overall production. It was also reported by the members of 52 ICs that cultivation of fodder crops on small piece of fellow land was also made possible.This has also increased milk production.

Institutional Management

Due to effective institutional management under PIM scheme, the number of conflict issues has been reduced drastically from 493 to 99 numbers. This is only because of mechanism developed such as social pressure, cooperative management, sense of belongings amongst members. Higher penalty charges was one of the important factor to address the conflicts if any amongst member farmers. It was also observed that ICs have engaged more operator from their resources to ensure better services to the members.

Women's Involvement

It was also noticed that about one third of the total members of executive body composed of rural women plays equally important role to execute the project.

Constraints Analysis

Analysis of constraints encountered at different level by the ICs showed that lack of technical backstopping in the initial stage of formation of cooperatives, unnecessary delay in the registration of the society, lack of support from the department of cooperatives, in adequate discharge of water from the main outlay, wastage of water due to unlined field chanels, delay in release of the funds from the irrigation authority and development of the illegal outlets etc. are some of the important constraints.

Conclusion

It can be concluded from the study that irrigation cooperatives are very successful village level local institution for management of canal irrigation water.Participatory irrigation management has very good impact on increase in area under irrigation, equal distribution of the available water,increase in crop production, participation of farmers in conflict resolution and financial management as well.

References

Chandregowda, M.J. and Jayaramaiah, K.M., 1990. Impact of watershed development programme on socio-economic status, land productivity and income of farmers. *Indian J. Extn. Edu.*, 26(3): 44–47.

Iyengar, S. and Hirway, I., 2001. In the hands of the people. Selected Papers of A.C. Shah.

Raphael, Jos C., 2004. Irrigation management transfer: A micro level study in Kerala. *Journal of Rural Development*, 23(4): 457–473.

Chapter 50

Innovative Extension Strategies for Farmer's Adoption Towards Salinity Management Practices in Coastal Area of Western Gujarat

☆ *N.B. Jadav*[1], *M.B. Viradiya*[2], *K.A. Khunt*[3] *and R.L. Shiyani*[4]

Introduction

Soil is one of the crucial natural resources essential for the existence of humanity. Contents of the soil profile determine the quality of life as people nourished on it. Changes in the quality of soil led to long run changes in socio-economic and cultural traits of its habitats. The soil degradation affects first on agriculture and then it spreads over the rest of the facets of the society.

---

1 Subject Matter Specialist (Extension Education), KVK, JAU, Jamnagar.

2 SMS (Soil Science), KVK, JAU, Tharghadiya, Rajkot.

3 Research Scientist, Department of Agricultural Economics, JAU, Junagadh.

4 Professor and Head, Department of Agricultural Economics, JAU, Junagadh.

Salinity Scenario in India

The total coastal area in India estimated by Velayutham *et al.* (1998) is 10.78 million hectares. In India, salinity is spread over around 26 m ha of cultivable land. According to the estimates of Global Assessment of Soil Degradation Organisation, 10, 23 and 9 per cent of total irrigated land in India, Pakistan and Sri lanka, respectively converted into "Kharland" and became unfit for cultivation (Anon. 2003). India has 8129 km long coastline.

Gujarat Scenario and Status

The state has a coastline of 1600 km which comprises the Agro-Ecological Sub Regions (AESR) 2.2, AESR 2.4 and AESR 5.3. In the state, recently, the soil degradation through salinity has caught hold over 1.2 m ha of land of which, about 0.3 m ha occurs in the coastal area and rest 0.9 m ha comprises the inland saline area (Rao *et al.*, 2001). About 1125 km long coastal line is in Saurashtra and Kutch which comprises about 0.1079 m. ha of saline land which need to be treated for salinity reduction (Anon., 2000).

Production and Productivity

About 6.9 m ha of coastal Kathiwar (Saurashtra) peninsula comprises 70 per cent net sowing area. Among the major crops, groundnut occupied around 70 per cent of the net sowing area, followed by bajra with high yield of 1520 kg/ha and 1341 kg/ha respectively. The coastal belt of Saurashtra extending from Bhavnagar to Okha is known for its vegetables, fruits, sugarcane and other high value cash crops. But the problem of salinity has created an unfavourable situation for agriculture (Anon., 1978). Even low level of salinity can curtail the yield to the extent of 15 per cent while the medium level salinity can be has brought the productivity to 65 per cent down. The productivity of agriculture has been the slashed down the productivity of major crops by 18 to 50 per cent between Una to Madhavpur during 1971 to 1981 (Golakiya *et al.*, 1999) and by two-third in coastal belt from Bhavnagar to Malia (Gujarat Ecological Commission, 2001).

Much has been done to study the problem of salinity in terms of individual research efforts and Government efforts by way of appointing High Level Commission, or Khar Land Development Board. Moreover, the efforts were also made at different time either to study the specific salinity condition or assessment of some segments of the coastal region. This has yielded a segmentary picture and does not provide holistic approach of the problems. Moreover, the socio-economic impact and adoption of management practices of salinity on coastal agriculture was almost missing, which is highly essential for making comprehensive development projects. The socio-economic assessment and adoption of management practices of the salinity problem may verify the results of previous studies and provide proper perspective useful to formulate an efficient management plan to bear with the menace of salinity in coastal agriculture. With this view, the present study, entitled "farmer's adoption rationale for salinity management practices in coastal area of western Gujarat" was undertaken with following specific objectives:

1. To study knowledge of farmers about salinity management practices

2. To study the extent of adoption of salinity management practices
3. To explore the relationship of salinity management practices with their selected characteristics

Materials and Methods

The whole study area falls under three Agro-Climatic Zones namely; North-West Gujarat Zone V, North Saurashtra Zone VI and South Saurashtra Zone VII. Out of these zones, the present study is confined to the situations 3 and 6 of zone V, situation 10 of zone VI and situations 3, 5, 7 and 8 of zone VII. The area of present investigation is comprised of 20 km wide soil strip along the sea coast of Saurashtra and Kutch. The selection of width of strip and its further division into small strata *viz.*, 0-5 km, 6-10 km, 11-15 km and 16-20 km was made with consultation of soil scientists and past studies (Kanzaria *et al.*, 1985) and also considering the extent of spread of salinity at present to make more homogenous strata for sampling. Two stage stratified random sampling technique was used with the village as first strata and the farmers as ultimate sampling unit.

Along the coastal belt of 20 km width, a total of 9 locations were selected as starting point of village selection. In vertical line from seashore to 20 km interior, one village was selected from each strata of 0-5, 6-10, 11-15 and 15-20 km width. Thus, 36 villages were selected by selecting 9 from each strata, considering the ingress of salinity along the coastal line. Ten farmers were selected randomly from each village and a sample of 360 farmers was acquired for the present study.

The average chemical values of soil samples for both the periods, (September and May) were analysed, and its distribution among the various salinity classes mentioned below, the soil scientists suggested the following salinity classes, which could be more useful particularly for this study area to avoid complexity of interpretation.

ECe dSm^{-1}	*Salinity Class*
<4	Non-saline or negligible saline (A)
4-8	Moderate saline (B)
>8	Strong to very strong saline(C)

The results of soil analysis showed that salinity was not uniform alongwith the coastal belt. No logical inferences could be drawn from the analysis according to distance strata due to agro-climatic condition, soil and cropping pattern as they vary with the locations. Hence, the data were regrouped and reanalyzed considering the soil salinity class and agro-climatic zones (situations). Accordingly, the selected villages and farmers were classified into the different classes of salinity among the agro-climatic zones and situations within the zone to make groups that are more homogeneous and to make the results more comparable.

For measuring the knowledge of respondents about recommended and other salinity management practices, the teacher made knowledge test was developed and used. The knowledge of respondents regarding recommended salinity management practices was measured by asking the questions regarding salinity management practices. A set of objective type questions was prepared by referring related review of literature and in consultation with subject matter specialists.

For measuring the adoption of salinity management practices, the adoption index was developed. From the list of salinity management practices recommended by various Departments of Agriculture, GAU, an index was prepared. The details of the management practices were given in below in Table 50.1. The list was sent to the 25 different scientists of Gujarat Agricultural University. In all, 25 responses were received. The average score for each practice was computed on the basis of sum of the score given by 25 individuals. Thus, an adoption index was developed by distributing 100 marks among various items of salinity management practices, which is given below:

The weightage score was given to the correct answer. The total weightage score by individual respondent for all the statements was calculated. With the help of mean and standard deviation, the respondents were categorized.

Table 50.11: Adoption Index of Recommended and Other Salinity Management Practices

Sl.No.	*Management Practices*	*Weightage (Out of 100)*
1.	Rain water recharge through check dam/well/water harvesting	13
2.	Tank/Canal/River water mixing with poor quality water	6
3.	Khet talavadi	7
4.	Mixing of amendment in water through well	3
5.	Use of drip irrigation system particularly in coconut or wide spaced crop.	7
6.	In furrow, raised bed irrigation (Cotton, Wheat, Gram)	4
7.	In shallow soil, use of poor quality water as protective irrigation to groundnut + FYM every two years	7
8.	Gypsum + poor quality ground water in groundnut–wheat sequence	6
9.	Hedge around the field	3
10.	Primary tillage	5
11.	Mulching : Mulching with wheat straw @ 5 t/ha in pre monsoon groundnut	7
12.	Keeping land fallow	3
13.	Selection of crop/variety	5
14.	Sowing method	4
15.	Inter culturing	4
16.	Application of FYM/cake/organic manure	6
17.	In bajra–wheat sequence, use of N in three split : 1/3 each as basal, at tillering stage and earhead formation stage	6
18.	Utilization of saline waste land: sowing of forest tree spp. *i.e.* Shirish, Indian babul, Casurina, etc.	7
	Total	100

To study the association between dependent (adoption and knowledge) and independent (socio-economic characteristics) variables, the Pearson's product moment method of computing correlation co-efficient, which provides generally accepted means for measuring the association, was used (Chandel, 1975). The correlation co-efficient of the magnitude of the association, gives two kinds of information.

1. An indication of the magnitude of the association.
2. Information about the direction of association.

Pearson's product moment method of correlation.

$$r = \frac{sp\ (xy\ )}{\sqrt{ss\,(x\,) \cdot ss\,(y\,)}}$$

where,

r: Co-efficient of correlation
x and y: Represent the value of two variables under the study
sp (xy): Sum of product of their deviation of x and y from their means
ss (x): Sum of square due to "x" variables
ss (y): Sum of square due to "y" variables

Results and Discussion

Knowledge of Salinity Management Practices

It is evident from the results presented in Table 50.1 and Figure 50.1 that majority of the respondents in class A (non-saline), B (moderate saline) and C (highly saline) have medium level of the knowledge about the salinity management practices. The mean level of the knowledge for the A, B and C class was 22.81, 27.21 and 45.56, respectively. The calculated mean of class C (45.56) is higher than class A (22.81) and B (27.21). This is so, because the respondents in class C are severely affected by the soil and water salinity problem.

Table 50.1: Extent of Knowledge about Salinity Management Practices

Extent of Knowledge	*Salinity class*					
	A (N=200)		*B (N=110)*		*C (N=50)*	
	No. of Respon.	*Per cent*	*No. of Respon.*	*Per cent*	*No. of Respon.*	*Per cent*
Low level of Knowledge	27	13.50	19	17.30	6	12.00
Medium level of Knowledge	145	72.50	71	64.50	36	72.00
High level of Knowledge	28	14.00	20	18.20	8	16.00
Mean	22.81		27.21		45.56	

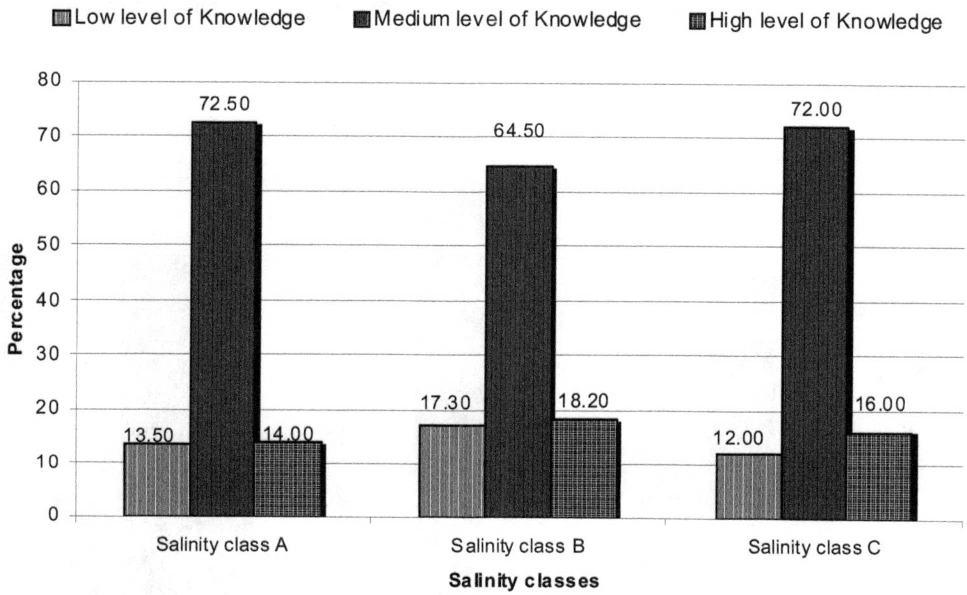

Figure 50.1: Extent of Knowledge about Salinity Management Practices

Adoption of Salinity Management Practices

To ascertain the extent of adoption of the salinity management practices, the data were analysed and presented in Table 50.2 and Figure 50.2. It is clear from the table that 55.00, 78.18 and 58.00 per cent were the medium adopters in A, B and C classes, respectively. This might be due to the fact that majority of the respondents, as discussed earlier, have the medium level of knowledge towards the salinity management practices. The calculated mean of class C was found higher as compared to class A and B. The reason behind this is that class C is highly saline tract as compared to other tracks and they are also more aware of salinity management practices.

Table 50.2: Extent of Adoption of the Salinity Management Practices

Extent of Knowledge	*Salinity class*					
	A (N=200)		*B (N=110)*		*C (N=50)*	
	No. of Respon.	*Per cent*	*No. of Respon.*	*Per cent*	*No. of Respon.*	*Per cent*
Low Adoption	51	25.50	9	8.18	12	24.00
Medium Adoption	110	55.00	86	78.18	29	58.00
High Adoption	39	19.50	15	13.63	9	18.00
Mean	18.48		21.16		36.10	

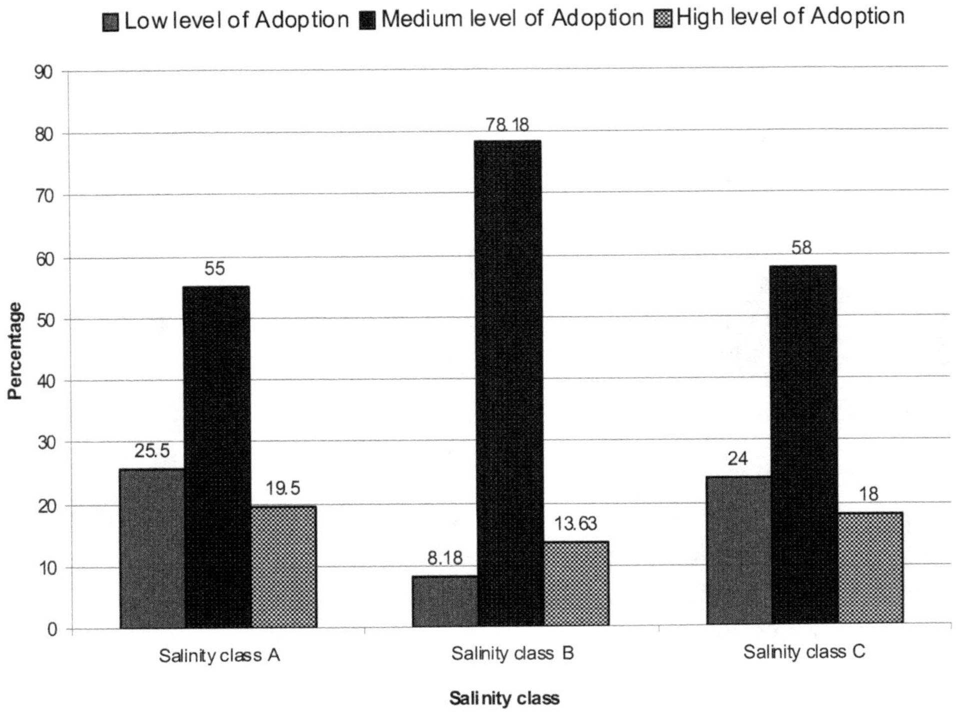

Figure 50.2: Extent of Adoption of Salinity Managment Practices

Practice Wise Extent of Adoption of Salinity Management Practices

On the basis of practice wise scores obtained by the respondents in adopting a particular practice, the mean scores were worked out for all the individual practice. These mean scores were again converted into percentage and ranked all the recommended practices.

The data presented in Table 50.3 clearly indicated that, in class A, medium extent of adoption was found for the practices *viz.*, rain water recharge through checkdam/well/water harvesting, application of FYM/organic manure, interculturing, primary tillage, selection of crop/variety, sowing method, tank/canal/river water mixing with poor quality water, utilization of saline wasteland in their order.

For medium level of adoption of above practices by the respondents, the probable reason is that the first to sixth ranked practices were commonly followed practices by most of farmers in any crops grown in the study area. As rain water recharge through check dam is a group campaign practice carried out by the mass in study, it has ranked first in class A. If it is adopted in village, most of the farmers will automatically be benefited in the village. Lower adoption was found in class A for remaining practices because the farmers in class A were less affected by the salinity problems and these practices were for salinity management.

Table 50.3: Practice-wise Extent of Adoption of Recommended Salinity Management Technology Across the Salinity Classes

Sl.No	Management Practices	Salinity Classes								
		Class A			Class B			Class C		
		No. of Respondents	Weightage Mean Score	Ranks	No. of Respondents	Weightage Mean Score	Ranks	No. of Respondents	Weightage Mean Score	Ranks
1.	Tank/Canal/River water mixing with poor quality water	49	3.19	1	24	2.84	2	21	5.46	1
2.	Khet talavadi	34	1.02	7	22	1.20	8	20	2.4	7
3.	Mixing of amendment in water through well	19	0.38	11	20	0.73	14	19	1.52	10
4.	Use of drip irrigation system particularly in coconut or wide spaced crop	0	0.00	15	0	0.00	16	0	0	17
5.	In furrow, raised bed irrigation (Cotton, Wheat, Gram)	7	0.25	13	13	0.83	12	16	2.24	8
6.	In shallow soil, use of poor quality water as protective irrigation to groundnut + FYM every two years	28	0.56	9	27	0.98	10	18	1.44	11
7.	Gypsum + poor quality ground water in groundnut–wheat sequence	0	0.00	16	20	1.27	6	21	2.94	5
8.	Hedge around the field	13	0.39	10	21	1.15	9	7	0.84	14
9.	Primary tillage	12	0.18	14	18	0.49	15	23	1.38	13
10.	Mulching : Mulching with wheat straw @ 5 t/ha in pre-monsoon groundnut	95	2.38	4	80	3.64	1	34	3.4	4
11.	Keeping land fallow	0	0.00	16	1	0.06	16	2	0.28	15

Contd...

Table 50.3–Contd...

Sl.No	*Management Practices*	*Salinity Classes*								
		Class A			*Class B*			*Class C*		
		No. of Respon-dents	*Weightage Mean Score*	*Ranks*	*No. of Respon-dents*	*Weightage Mean Score*	*Ranks*	*No. of Respon-dents*	*Weightage Mean Score*	*Ranks*
12.	Selection of crop/variety	10	0.15	15	2	0.05	17	2	0.12	16
13.	Sowing method	65	1.63	5	40	1.82	4	36	3.6	2
14.	Interculturing	60	1.20	6	26	0.95	11	18	1.44	11
15.	Application of FYM/Cake/Organic manure	135	2.70	3	35	1.27	5	36	2.88	6
16.	In bajra–wheat sequence, use of N in three split : 1/3 each as basal, at tillering stage and earhead formation stage	101	3.03	2	34	1.85	3	29	3.48	3
17.	Utilization of saline waste land: sowing of forest tree spp. *i.e.* Shirish, Indian babul, Casurina, etc.	10	0.30	12	14	0.76	13	16	1.92	9
18.	Tank/Canal/River water mixing with poor quality water	17	0.60	8	19	1.21	7	10	1.4	12

Similarly in class B (medium saline class), they were medium adoption for salinity management practices *viz.*, primary tillage, rain water recharge through check dam/well/water harvesting, application of FYM/cake/organic manure, selection of crop/variety, intercultruing, in shallow soil use of poor quality water as protective irrigation, utilization of wasteland and water mixing in poor quality water in their order were observed.

The probable reason for the medium adoption of the first to fifth ranked practices is that they were commonly adopted practices. Some practices, *viz.*, mulching and keeping land fallow, mixing of amendment in water through well is not adopted by any farmers. These practices were not feasible as well as it is costly too.

In class C (highly saline class), the practices adopted by farmers *viz.*, rain water recharge through check dam, selection of crop variety, application of FYM and primary tillage were ranked first, second, third and fourth, respectively. The above soil practices are especially recommended for the salinity management so they were ranked higher in class C. Other practices were adopted by farmers *viz.*, in shallow soil, use of poor quality water as protective irrigation with poor quality water (fifth rank), interculturing (sixth rank), tank/canal/river water mixing with poor quality water (seventh rank), use of drip in coconut and wide spaced crops (eighth rank).Since the medium level of adoption of salinity management practices was observed, the third hypothesis is rejected.

Association Between Knowledge and Adoption with Socio-economic Characteristics

Association of respondents' extent of knowledge of salinity management practices with their selected characteristics

The correlation co-efficient was computed to ascertain the association between respondents' extent of knowledge and their selected characteristics, on the basis of the operational measures developed for each variable. The results are given in Table 50.4.

Table 50.4: Association Between Respondents' Extent of Knowledge and Recommended Salinity Management Practices with their Selected Characteristics

Sl.No.	*Name of Variable*	*Salinity Classes*		
		A N=200 'r' Value	*B N=110 r' Value*	*C N=50 r' Value*
1.	Age	–0.0183	–0.2706	–0.2154
2.	Education	0.2507**	0.5266**	0.5126**
3.	Caste	0.0494	0.0223	–0.0163
4.	Social participation	0.2603**	0.3155**	0.3393**
5.	Size of family	0.0689	–0.0884	0.0904
6.	Livestock holding	0.1018	–0.1935	0.1247
7.	Size of land holding	0.0929	–0.1345	–0.1236

***: Significant at 1 per cent level.*

It can be inferred from the results that the education and social participation were positively and significantly associated with the knowledge in class A, B and C. The direction of association was positive and significant which indicates that the respondents' knowledge about management practice can be increased significantly through increasing of education level and social participation. For rest of the characteristics, the correlation coefficients were found non–significant in all the classes indicating no any significant impact on knowledge of farmers.

Association of Respondents' Extent of Adoption of Salinity Management Practices with their Selected Characteristics

The correlation co-efficient was computed to ascertain the association between respondents' extent of adoption and their selected characteristics, on the basis of the operational measures developed for each variable. The results are given in Table 50.5.

Table 50.5: Association Between Respondents' Extent of Adoption and Recommended Salinity Management Practices with their Selected Characteristics

Sl.No.	*Name of Variable*	*Salinity Classes*		
		A *N=200* *'r' Value*	*B* *N=110* *r' Value*	*C* *N=50* *r' Value*
1.	Age	0.1785**	0.5674**	0.4062**
2.	Education	0.0094	–0.2933	–0.3671
3.	Caste	0.0119	0.0355	0.1783
4.	Social participation	0.1834**	0.3731**	0.3313**
5.	Size of family	–0.0159	–0.0888	0.1219
6.	Livestock holding	0.0655	–0.2102	0.1597
7.	Size of land holding	0.0604	–0.1467	–0.1564

**: Significant at 1 per cent level.

It is evident from the data that age and social participation of farmers were positively and significantly associated with the adoption behaviour of farmers in class A, B and C. The direction of association was positive and significant indicating that adoption level of farmers can be raised through increasing social participation. For rest of the characteristics, the correlation coefficients were found non-significant in all the classes indicating that adoption behaviour of farmers is indifferent towards these characteristics.

Conclusion

Knowledge and level of adoption of salinity management were examined for both the saline and non-saline tracts as some common practices are followed in both the tracts. It is evident that majority of the respondents have medium level of the knowledge about the salinity management practices. But, the high mean value was found for salinity tract indicating the higher level of knowledge about salinity management practices. For majority of respondents *i.e.* 55.00 per cent farmers in non-saline tract and 58.00 to 78.18 per cent farmers in saline tract, the medium level of adoption was observed. In non–saline tract, medium level of adoption was found for the practices *viz.*, rain water recharge, use of FYM/organic manure, interculturing and primary tillage in their order. These were the commonly followed practices by most of the farmers in the study area. Similarly in saline tract, medium level of adoption was observed for the practices *viz.*, rainwater recharge, selection of crop variety, application of FYM and primary tillage with first, second, third and fourth rank, respectively. High ranks of these practices were because of their high impact of salinity.

The positive and significant association of knowledge with education and social participation; and of adoption with age and social participation was observed. This has indicated that knowledge and adoption level of farmers can be raised with increasing education and social participation.

References

Anonymous, 1978. Report of the High Level Committee–I to examine the problems of salinity ingress among the coastal area of Saurashtra (Bhannagar-Una Reach). Irrigation Department, Gandhinagar.

Anonymous, 2000. *Gujarat Agrovision 2010: A Working Document*. Agriculture and Cooperation Department, Government of Gujarat, India, pp. 1–375.

Anonymous, 2003. '*Gujarat Samachar*'–April.

Chandel, R.S., 1975. *A Handbook of Agricultural Statistics*. Achal Prakashan Mandir, Kanpur.

Golakiya, B.A. and Gundalia, J.D., 1999. *Ksharmay Jaminma Kheti*. International Potash Institute, Switzerland, Gujarat Agricultural University and Potash Research Institute, Gurgaon, Junagadh, pp. 1–118.

Kanzaria, M.V., Patel, M.S., Gundalia, J.D. and Patel, C.L., 1985. Physico-chemical charactristics of salt affected soils of coastal belt of Saurashtra, Kutch and North Gujarat. In: *Paper Presented in a Seminar on the Management of Saline Alkaline Soils* held in the month of February, 1981, GAU, Junagadh, Gujarat, pp. 1–5.

Rao Gururaja, G., Nayak, A.K., Chinchmalatpure, A.R., Singh, Ravender and Tyagi, N.K., 2001. Resource characterization and management options for salt affected black soils of agro-ecological regions of Gujarat state, CSSRI, Regional Research Station, Anand, Gujarat.

Richard, L.A., 1954. *Diagnosis and Improvement of Saline and Alkali Soils*. United States Salinity Laboratory, Agric. Handbook, 60.

Velayutham, M., Sarkar, D., Reddy, R.S. and Nathrajan, A., 1998. Soil resources and their potentials in coastal area of India. In: *Paper Presented in the National Seminar on Frontiers of Research and its Application in Coastal Agriculture*: 5th National Seminar on Indian Society of Coastal Agricultural Research, 16-20 September, Konkan Krishi Vidyapith, Dapoli, Maharashtra.

Chapter 51

To Study the Extent of Knowledge Level and Profile of the Beneficiaries Towards Sampoorna Grameen Rozgar Yojana (SGRY) Programme

☆ *S.M. Kenchanagoudra, S.K.Meti and S.K. Dolli*[1]

Introduction

The creation of employment opportunities with food security has been an important objective of developmental planning in India. The relatively higher growth of population and labour force has led to an increase in the volume of unemployment and under employment from one plan period to another. The Government of India aims to bring employment through wage and self employment into a larger focus with the goal of reducing unemployment and under employment to a negligible level and of providing food security against hunger. Such an approach is necessary because it is realized that larger and efficient use of available human and other resources is the most effective way of alleviating poverty, reducing inequalities, improving nutritional levels and sustaining a reasonably high pace of economic growth.

---

1 College of Agriculture, UAS, Dharwad.

Since independence, the Government made several attempts to eradicate poverty and unemployment in the country. The Government has initiated several employment programme initiated by the govt, among them Sampoorna Grameen Rozgar Yojana is one of the important recently introduce programme.Keeping the importance and benefit of the Sampoorna Grameen Rozgar Yojana Programme the study has been designed with the following objectives.

Objectives

1. To study the profile of the beneficiaries of Sampoorna Grameen Rozgar Yojana programme.
2. To study the knowledge level of the beneficiaries towards Sampoorna Grameen Rozgar Yojana programme.

Methodology

The study was conducted during 2007 in the Gadag district to know the knowledge level and profile of the beneficiaries towards Sampoorna Grameen Rozgar Yojana programmes and their benefits for the self employment purpose. In the above selected district, 10 villages were selected which are coming under cluster of 5 gram panchayats. From each village, 15 beneficiaries were selected by randomized sampling procedure and total number of sample size was 150 beneficiaries from 10 villages. The information was gathered in pretested interview schedule by asking openend questions. The gathered information was analyzed by using appropriate statistical tools like mean and standard deviation and results were presented below.

Results and Discussion

It is revealed from the Table 51.1 that majority (64.67 per cent) of the respondents belonged to middle age group followed by 32.67 and 2.66 per cent who belonged to young age and old age categories respectively. With respect to caste, majority (42.67 per cent) were belonged to scheduled caste and scheduled tribes categories and only 7.33 per cent of them belonged to other backward (OBC) class category, Regarding education forty per cent of the Sampoorna Grameen Rozgar Yojana beneficiaries were illiterates, while 24.67 per cent and 19.33 per cent were educated up to primary school level and middle school level respectively, among them ten per cent of beneficiaries were educated up to high school level and 5.33 per cent were with PUC education. It was interesting to note that one of the respondents was a graduate, further majority of (74.67 per cent) of the beneficiaries were labourers (main occupation) and remaining 25.33 per cent of the respondents were having subsidiary occupation apart from seeking employment under Sampoorna Grameen Rozgar Yojana programme. With respect to family size, majority (90.00 per cent) of the beneficiaries belonged to nuclear family and only 10.00 per cent of them belonged to joint family, among them majority 63.33 per cent of the beneficiaries belonged to small family, followed by 32.00 per cent belonged to medium size of family and remaining 4.67 per cent were found with big family size.Based on the land holding, majority (47.33 per cent) of the beneficiaries were marginal farmers, followed by 28.67 per cent who were land less and 20.00 per cent belonged to small land holding category. Only 4.00 per cent of them had semi medium land holding. Regarding annual income of the beneficiaries, majority (79.33 per cent) of the beneficiaries had low level of annual income (up to Rs.17, 000/-), followed by 18.00 and 2.67 per cent who had semi-medium level of annual income (Rs.17, 001 to Rs.34, 000/-) and medium level of income (Rs.34, 001 to Rs.51, 000/-) respectively.

Table 51.1: Profile of the SGRY Beneficiaries (n=150)

Sl.No.	Variables	Frequency	Percentage
1.	**Age**		
	Young (18-30)	49	32.67
	Middle (31-50)	97	64.67
	Old (>50)	04	02.66
2.	**Caste**		
	General	75	50.00
	Backward classes (OBC)	11	07.33
	SC/ST	64	42.67
3.	**Education**		
	Illiterate	60	40.00
	Primary school (1-4)	37	24.67
	Middle school (5-7)	29	19.33
	High school (8-10)	15	10.00
	PUC	08	5.33
	Graduate	01	00.67
4.	**Family occupation**		
	Labourer	112	74.67
	Labourer + subsidiary	38	25.33
5.	**Type of family**		
	Nuclear family	135	90.00
	Joint family	15	10.00
6.	**Family size**		
	Small (upto 4 members)	95	63.33
	Medium (5-7 members)	48	32.00
	Big (>7 members)	07	04.67
7.	**Land holding**		
	Landless	43	28.67
	Marginal farmers (upto 2.50 acres)	71	47.33
	Small farmers (2.51 to 5.00 acres)	30	20.00
	Semi-medium farmers (5.01 to 10.00 acres)	06	04.00
8.	**Annual family income**		
	Low (up to Rs.17,000/-)	119	79.33
	Semi-medium (Rs.17,001 to 34.000/-)	27	18.00
	Medium (Rs. 34,001 to 51,000/-)	04	02.67

It is observed from the Table 51.2 that, 44.00 per cent of the beneficiaries belonged to medium knowledge level, followed by 32.67 and 23.33 per cent of the beneficiaries belonged to low and high knowledge of the SGRY programme, respectively.

Table 51.2: Overall Knowledge of Respondents about SGRY Programme (n=150)

Sl.No.	*Categories*	*Frequency*	*Percentage*
1	Low (upto 11.33)	49	32.67
2	Medium (11.33 to 13.42)	66	44.00
3	High (above 13.42)	35	23.33
		150	100

Mean: 12.38; SD: 2.

The reasons for possession of low to medium level of knowledge were low literacy rate, poor social participation and lack of interest to know the details of the programme. The study has revealed elsewhere that as high as 40.00 per cent of the beneficiaries were illiterate and others were having low level of formal education.

For any programme to be successful the beneficiaries should have thorough knowledge of the entire programme. The details like the purpose of initiating the programme, it's objectives, mode of operation, to whom it is going to benefit and all other aspects should be made known to the people. Otherwise, the selection of the beneficiaries and creation of assets/infrastructure becomes very difficult. Though the programme, it is possible to create assets/infrastructure, non-possession of required knowledge by the beneficiaries is a cause for concern. The responsibility of creation of required awareness lies with the programmes implementers, which should be given due consideration.

The above finding of the study is in line with the findings of Usha Rani (1999), Arul Prakash (2004) and Bheemappa (2006).

Table 51.3: Association Between Independent Variables and Knowledge Level of Respondents about SGRY Programme

Sl.No.	*Variables*	*Correlation Coefficient (r-value)*
1	Age	-0.022
2	Caste	-0.023
3	Education	0.082
4	Family occupation	0.006
5	Type of family	-0.118
6	Family size	-0.113
7	Land holding	-0.070
8	Annual family income	0.174*
9	Mass media exposure	0.155*
10	Social participation	-0.046

*: Significant at 5 per cent level.

It is clear from Table 51.3 that, an attempt was made in the present investigation to ascertain the relationship between selected independent variables and their employment generation under Sampoorna Grameen Rozagar Yojana. The findings in this regard are presented in Table 51.3. Socio-economic character namely annual family income and mass media exposure were found to be positive and significantly associated with knowledge of SGRY programme at 5 per cent of probability where as, other characteristics namely age, caste, education, family occupation, type of family, family size, land holding and social participation were found to be non-significantly associated with knowledge of SGRY programme.

Conclusion

The knowledge of the beneficiaries of SGRY programme ranged from low to medium level. This brings home the fact that the beneficiaries possessed inadequate information about the programme.For any programme to be successful and effective, the beneficiaries of the programme should have thorough knowledge of the programme. The present finding calls the attention of the programme implementers. *i.e.;* Panchayat institutions to orient the beneficiaries in the beginning itself regarding complete details of the programme.

References

Arul Prakash, S., 2004. Analysis of Swarnajayanti Gram Swarozgar Yojana in Salem and Thiruvallur district of Tamil Nadu. *M.Sc. (Agri.) Thesis*, Univ. Agric. Sci., Dharwad.

Bharatamma, G.U., 2005. Empowerment of rural women through income generating activity in Gadag district of North Karnataka. *M.Sc. (Agri.) Thesis*, Univ. Agric. Sci., Dharwad.

Bheemappa, 2006. A study on knowledge level of Gram Panchayat members about Sampoorna Grameena Rozgar Yojana in Raichur district of Karnataka. *M.Sc. (Agri.) Thesis*, Univ. Agric. Sci., Dharwad.

Channamma, R.N., 1994. A study on knowledge and attitude of poultry farmers towards poultry insurance. *J. Res.*, Andhra Pradesh Agric. Univ., 22(3&4): 176.

Gayatridevi, S.P., 2000. Impact of selected rural development programmes on self employment of rural women. *M.H.Sc. Thesis*, Univ. Agric. Sci., Dharwad.

Usha Rani, R.C., 1999. A study on opinion of women beneficiaries towards DWCRA and benefits derived in Vizianagaram district. *M.Sc.(Agri.) Thesis*, Univ. Agric. Sci., Dharwad.

Chapter 52

Knowledge and Awareness About Mushroom Cultivation Under Macro-Mode-Management Project

☆ *Dayaram*[1] *and Ashok K. Singh*

Introduction

The demand for mushroom is increasing at the rate of seven per cent per annum resulting further gap between demand and supply in spite of immense potential for seasonal mushroom production. The commercial mushroom production started in our country during the period of 60's when Button mushroom production was introduced in Himachal Pradesh. The seasonal mushroom production began in Northern states of India in mid 90's. At present it is estimated that mushroom production is around 50,000 tonnes/annum. However, there is an increasing demand for quality products at competitive rate both in domestic and export market in spite of having tremendous potential for mushroom production. The seasonal mushroom production is possible in a big way from J&K, H.P., Punjab, Haryana, Uttarakhand, UP, W.B., Bihar, NER, M.P. and in other areas when temperature remains below 20^{0}C. In the state of Bihar recently protocal for different mushroom production have been formulated by Rajendra Agricultural University by which mushroom of different varieties are grown throughout the year in different region of the state (Choudhary,

---

1 Rajendra Agricultural University, Pusa – 848 125, Samastipur, Bihar.

2003; Dayaram and Jha, 2004; Kumar, 2005 and Anonymous, 2006). Still, the production was noticed around 40 tonnes during 2007-08 because of poor adoption of the protocal. Several efforts are being made to popularise the mushroom cultivation through the awareness camps, training and demonstration among the rural communities of the state in order to enhance mushroom production. In order to tackle the problem, regular training programmes are being sponsored by different functionaries of the state and central Government. The Macro-Mode Management on Development of Mushroom Cultivation is also an central sector scheme by which it was planned to train 400 unemployed youth/farmers related with different aspect of mushroom cultivation. They were selected from different district of state during the April 2004–Match 2006. Socio-personal character analysis of the participants is presented here.

Methodology

A total 400 farmers, unemployed youth and rural women were invited from 38 districts of Bihar in training on mushroom cultivation for enhancing their economic and nutritional security under Macro Mode Management Project. The selection of trainees was based on the specific recommendation of District Horticulture Officer, District Agriculture Officer and Director Horticulture, Bihar with respect to their entrepreneurial aptitude. The duration of training was for six days and during the session they were imparted theoretical as well as practical training related with scientific mushroom cultivation. They were exposed with the technique of Oyster mushroom cultivation, button mushroom cultivation (composting, spawning, casing and crop management), paddy straw and milky white mushroom cultivation.

Results and Discussion

Out of 400 participants from 38 districts of Bihar, a total of 246 unemployed youth and women from different social group from 32 districts were participated (Map 52.1). The details of analysis pertaining to their socio-personal characteristics, knowledge and other responses were carried out and results are being presented here through several tables. The Table 52.1 relates with details regarding the age group of selected trainees.

Table 52.1: Age Group of Selected Trainees

Age Group	*Percentage*
Below 20 years	17.36
21–30 years	51.05
31–40 years	22.10
41–50 years	06.84
51–60 years	02.65

It is evident from the Table 52.1 that maximum participation of trainees *i.e.* 51.05 per cent were from the age group of 21-30 years followed by trainees from 31-40 years of age (22.10 per cent). A total 17.36 per cent trainee were having their age less than 20

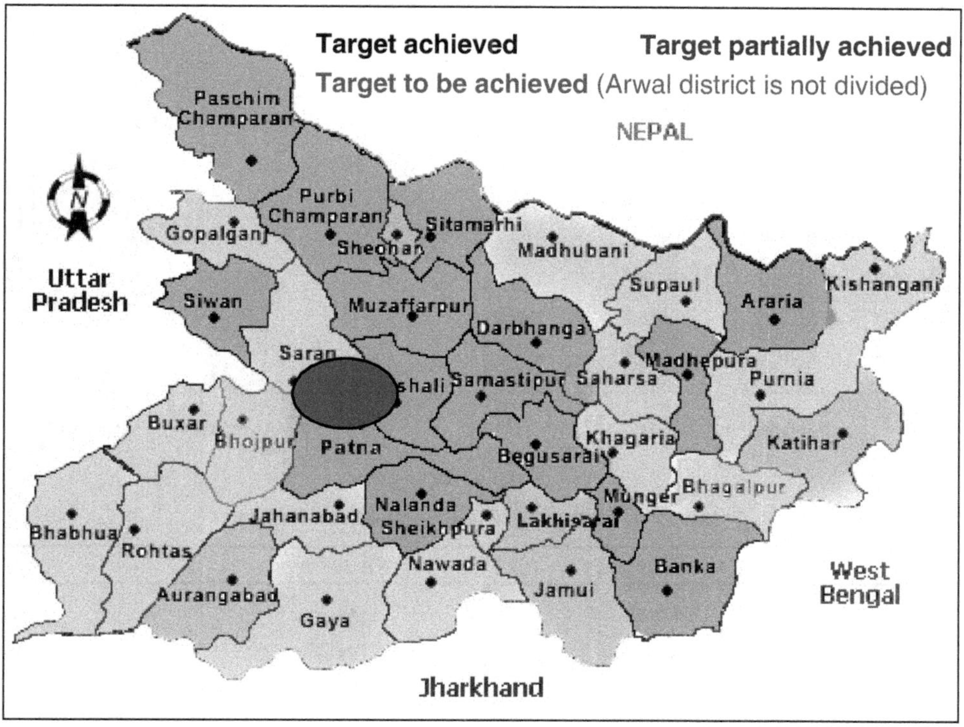

Map 52.1: Participants from Different Districts of Bihar

years while only 6.84 per cent of trainees were possessing their age in between 41 to 50 years. The trend of findings reveal that although the age group of 21 to 30 years were shown their highest involvement with respect to the participation in mushroom training yet the participants of other age group were also shown their interest in mushroom production process.

With respect to the social fabrics, it was found that out of the total trainees 44.85 per cent were belong to general caste followed by 22.33 per cent of trainees from the backward caste community. The 10.75 per cent of the trainees were from the schedule caste group while 16 per cent of the trainees were the women from different caste group. The trainees of Muslim community were found in tune of 5 per cent whenever the scheduled tribe trainees were only 1.07 per cent of the total participants. The Table 52.2 deals the caste wise analysis of the trainees.

With regards to educational qualification of the trainees, it was observed that 31.18 per cent of the total trainees were found Graduate followed by 29.50 of Intermediate and 24.20 trainees were having their Matric level of education. The total 8.5 per cent were found as non-matric whenever out 6.60 per cent was found to have their post graduation degree. The Table 52.3 depicts the educational level of the selected trainees participated in the mushroom cultivation.

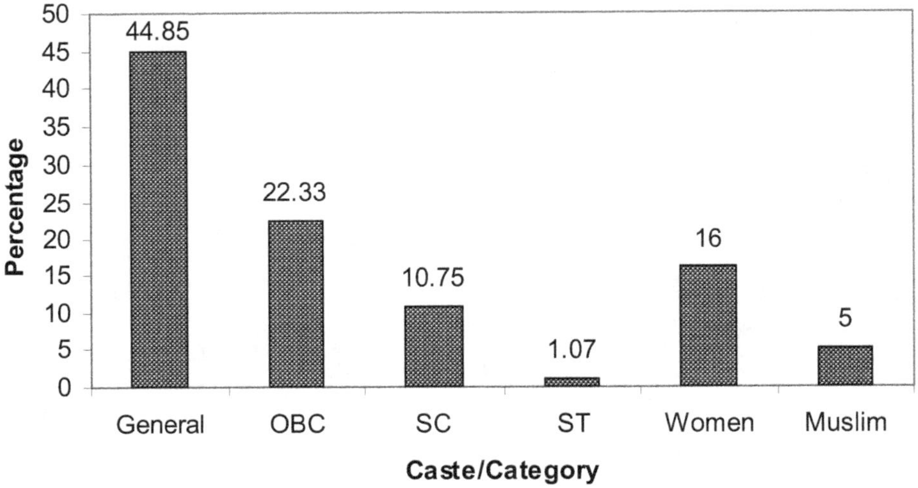

Table 52.2: Category Wise Analysis of Trainees

Caste/Category	*Percentage*
General	44.85
OBC	22.33
SC	10.75
ST	01.07
Women	16.00
Muslim	05.00

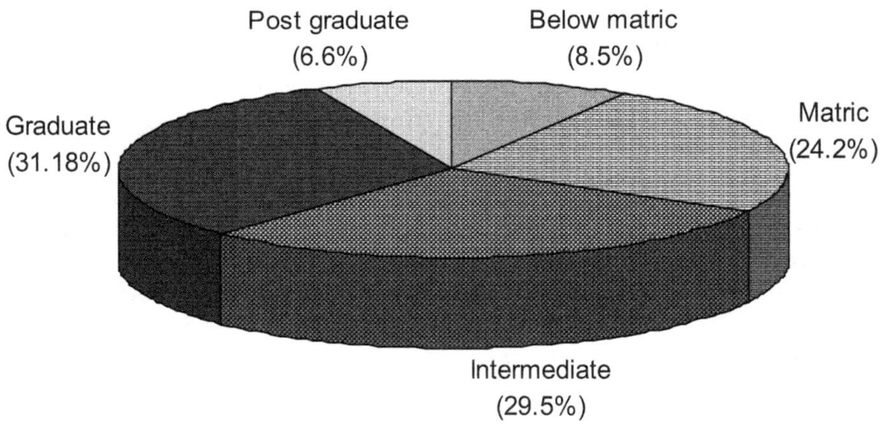

Table 52.3: Educational Status of Selected Trainees

Qualification	*Percentage*
Below matric	08.50
Matric	24.20
Intermediate	29.50
Graduate	31.18
Post graduate	06.60

During the course of training programmes, feedback informations were collected from the participating trainees. It was found that only 8.60 per cent of the selected trainees were well aware about the use of mushroom and they were reported that they frequently use it in their diet as per availability. The rest 91.40 per cent of the total trainees were found to be totally unaware about the mushroom. They were neither seen nor found to be used the mushroom in their diet. The results related with this aspect are being presented here through Table 52.4.

Table 52.4: Awareness/Knowledge About Mushroom

Knowledge/Awareness	*Percentage*
No awareness/knowledge	91.40
Having awareness/Knowledge	08.60

Since, India has the tremendous potential for mushroom production and all commercial edible and medicinal mushroom can be cultivated in our country for the use of domestic and export market hence, it is urgently required to make aware to the masses specially rural communities about the food and medicinal values of mushroom. The plenty of raw materials are easily available in our rural environment. There is need of hour to exploit this potentiality by imparting the training, credit, knowledge and awareness among the rural populations. Keeping these facts in mind, the effort are being made through various projects and programmes to create the awareness and involve the underemployed and unemployed human resource of nation in the area of mushroom cultivation and production. The Macro-mode Management project was also a initiative from the Govt. of India in this line. The major emphasise during the mushroom training programme was given on their self learning through the practical approach of their involvement in mushroom production process. The selected trainings learnt the techniques of Oyster, White button, summer button and paddy straw mushroom cultivation. They were also exposed to the different stages of all mushroom cultivation alongwith the expected problems and solutions which may occur due to their negligence in the process of mushroom production. To achieve the objective in true sense, there is need to formulate continue awareness programmes and the fallow-up of the different programmes related to these trainings.

References

Dayaram and Jha, M. N. (2004). Cultivation of button mushroom on chicken manure based compost in Bihar. *J. Mycol. Pl. Pathol.*, 34: 185–187.

Chaudhary, S. K. (2003). Studies on production and management of oyster mushroom (*Pleurotus* spp.), *M.Sc. (Ag.) Thesis* submitted to the Deptt. of Plant Path., RAU, Bihar (Samastipur), Pusa, Bihar, India.

Kumar, V. (2005). Studies on paddy straw mushroom, *M.Sc. (Ag.) Thesis* submitted to the Deptt. of Plant Path., RAU, Bihar (Samastipur), Pusa, Bihar, India.

Annonymus (2006). Final report of Macro-Mode Management on Development of Mushroom Cultivation (Govt. of India), Deptt. of Microbiology, Compost and Spawn Production Centre, FBS&H, RAU, Pusa.

Section V

GENDER MAINSTREAMING IN FARMING SYSTEM

Chapter 53

Women Empowerment and Gender Mainstreaming

☆ *P.P. Pal, A.K. Singh and S.K. Roy*[1]

Introduction

Of late there has been an increasing consciousness regarding the status of women, which is amply reflected in global debates over the issue of women empowerment. Empowerment of women concerns women gaining power and control over their own lives. It involves awareness raising, building self-confidence, expansion of choices, increased access to and over resources and actions to transform the structures and institutions, which reinforce and perpetuate gender discrimination and inequality. The process of empowerment is as important as the goal. Empowerment comes from within; women empower themselves. Inputs to promote the empowerment of women should facilitate women's articulation of their needs and priorities and a more active role in promoting these interests and needs. Empowerment of women cannot be achieved in a vaccum, men must be brought along in the process of change. This essentially calls for true application of gender mainstreaming concept. Gender mainstreaming is paying the attention to gender perspective as an integral part of all activities across all programmes. This involves making gender perspective-what men and women do and the resources and decision-making processes they have access to-more central to all policy development, research, advocacy, development, implementation and monitoring of norms, standards, and planning, implementation and monitoring of projects. It is contrary to the developing separate women's projects

---

1 Zonal Coordination Unit, Zone-II, Salt Lake, Kolkata.

within work programmes, or even women's components within existing activities in the work programmes.

Gender Issues and Needs in the Agriculture Sector

Women are known to produce two-thirds of the world's food. In general, women provide 50-60 per cent of the labour input, increasing to 80 per cent in paddy and rice production. Due to a rise in the number of female-headed households as substantial numbers of males migrate to cities, more women are becoming functional heads of both households and farming activities. Further, although productions of major food crops have improved in recent times, rapidly growing populations and expanding industries have caused a continuous increase in the demand for food and commercial crops. Thus, the empowerment of women through the provision of training in managerial skills, appropriate technology and resources must receive priority in agricultural development programmes, if such programmes are to contribute significantly to economic development and poverty reduction. This is in addition to the traditional training for the women farmers in crop cultivation, animal rearing and fish harvesting. Agriculture and rural development sector needs to attach significant importance in the following gender issues to achieve the set goal of agricultural development with gender equality. These issues are interlinked and all require social change which needs to have substantial political support, if the limits to growth are to overcome.

- Equal access to land and water resources, and to credit and other support services
- Gender differences in roles and activities
- Gender and agricultural extension and research
- Gender, agricultural biodiversity and commercialization
- Women's empowerment and equal access to decision making.

Equal Access to Land and Water Resources

Opening up the Indian agriculture to world market and subsequent withdrawal of subsidy from fertilizer and selected inputs has created impediments for many women in using and conserving scare resources in a sustainable manner. This is further aggravated with the mounting pressure to export to highly specialized markets and meet tough quality control standards. In addition, agricultural services are being commercialized and land tenure systems/arrangements are under threat, reducing the good land available to small farmers as the better land is allocated to cash crops. These small farmers, mostly the women, are increasingly pushed on the marginal plots that are dispersed, remote and usually less fertile. This land may be ill-situated to continuous cultivation and vulnerable to soil and water degradation, particularly through erosion and deforestation. Such insecure land tenure reduces the incentive to invest in land improvement and good crop husbandry particularly in permanent crops. This also leads to lack of collateral for credit for improved inputs and fertilizers.

Agrarian reform programmes of resettlement and land distribution have not tackled this problem and failed to recognize the land ownership rights of married women, women heads of household and women producers with partial or temporary land rights. In many cases they have simply awarded new tenancies to men despite the women's responsibility for growing all the family subsistence food supply.

Women also need to be actively involved in policies, programmes and projects in the water and sanitation sector. Rural women and men often have different roles and responsibilities in this area, with men usually concerned with water for irrigation or for cattle and playing a greater role than women in public decision making about water and sanitation use. Women, on the other hand, are usually the ones who collect, use and manage water in the household. Women are also traditionally responsible for disposing of household waste and maintaining sanitation facilities (World Bank, 2000). Despite the recognition of the central role of women in water and sanitation sector, the design of programme has not reflected this in the rural areas.

Access to Credit and Other Support Services

Even if the women have access to land, they have very little access to agricultural support services like credit, proper advice and training in agricultural and allied technologies. This limited access may arise from a range of factors such as legal restriction, lack of collateral, lack of information about credit availability and lack of small scale services like micro-credit scheme. To overcome this problem, efforts are being made to stimulate the economy at the local level. Such schemes are potentially of great usefulness in helping rural women create their own paths out of poverty. In the design and implementation of such schemes, a number of gender-related issues arise.

- ☆ Women must have access to loans for all types of profitable agricultural activities, not just those prescribed by gender stereotyping.
- ☆ Credit facilities should be accompanied by agricultural technical skills and human development training both for women and community leaders to enable them to utilize and receive full benefit from loans. Collective enterprises undertaken by women's groups for farming enterprises on leased land or joint ownership can be successful when the groups are provided with management training and support.
- ☆ The staffing pattern of micro-credit programmes needs to be carefully worked out. Field experience, local knowledge and knowledge of participatory approach are vital considerations alongwith adequate training.
- ☆ Equal access to credit means more that just micro-credit. The full range of economic policies including banking regulations and access to commercial credit as well as other financial services needs regular review to ensure that these are not explicit or implicit barriers to full gender equity and equality.

Gender Differences in Role and Activities

Analysis of the gender division of labour has revealed that women typically take on three types of role in terms of the paid and un-paid labour they undertake (Razavi and Miller, 1997). Thy are:

- ☆ The productive role
- ☆ The reproductive role and
- ☆ The community management role

Productive role refers to market production and home/subsistence production undertaken by women which generates an income (either financial or 'in kind').

Reproductive role refers to child bearing and child-rearing responsibilities borne by women (essential workforce) to the production of community management role refers to activities undertaken by women to ensure the provision of resources at the community level, as an extension of their reproductive role (Razavi and Miller, 1997). An understanding of these three roles can inform gender aware planning that takes into account the differential impact of programmes and projects on women and men because of women's triple role. Under-estimations and under-valuations of rural women's work, frequently observed, in the spheres of development planning may lead to serious consequences. Failure to acknowledge the importance of women's multiple contributions to rural survival means that practical policy interventions are frequently detrimental both to women themselves and to their communities.

Gender and Agricultural Extension and Research

In the changed agricultural scenario of India, the role of women becomes even more critical and there are immediate opportunities for community-led ventures to be successful in rapidly developing markets. The key problem faced is how to improve the interchange of information between the farmer, the researcher, and the extension agent. At present, women and their views are poorly represented in agricultural research, even though they possess first hand knowledge and insights in to such things as local weather patterns, crop varieties and planting methods. Inadequate link between research station staff and farmers has resulted in acute lack of take-up of research advances in the field. It is now generally recognized that extension based on group contact and using community-based organizations (CBOs) can be very effective. This allows the women who are at the forefront of community-based efforts not only to benefit from extension but also to influence the research agenda and the method of delivery. This is a major change and NGOs are increasingly being used as intermediaries between state and end-users, both in terms of delivering services on behalf of the state and in representing the interests of the poorer farmers. The challenge therefore is to ensure that community organizations truly have a gender dimension, that the NGOs are fully aware of the importance of women farmers, and that the research community increasingly recruits women and takes note of demands from the field which reflect the concerns of women.

Agricultural Biodiversity and the Commercialization of Agriculture

As predominantly small farmers, women have been largely responsible for activities such as selection, improvement and adaptation of plant varieties. This has both supported and increased agricultural biodiversity. Food and Agriculture

Organization has defined it as "the variety and variability of animals, plants and micro-organisms which are necessary to maintain the structure, processes and key functions of the agricultural ecosystem for, and in support of, food production and food security".

The growing commercialization of agriculture has led to efforts to optimize production and access to market. This trend will force change and provide opportunities to many. However, in the move to greater commercialization and privatization, there is also a danger that small-scale farming will suffer. Diverse food production systems are under threat and, with them, the accompanying local knowledge, culture and skills of the food producers. The scale of loss to agricultural biodiversity is extensive and with the disappearance of local breeds and cultivars particularly after green revolution era, poses greater threat in the eventuality of food crisis. As per study of Mulvani (1999), 75 per cent of agricultural crop varieties disappeared due to industrialization and green revaluation.

The poorest farmer living in the remote areas are ill served by market-driven changes. Amongst these are many women who have very little access to cash, since they are limited to household and food crops. The gulf between them and the government research and extension workers is much wider than the farmers in the more favoured area. There is a need for both government and NGOs sector to pay attention not only to production but also to marketing and the creation of business based on agriculture and natural resources. And this will undoubtedly need to involve women to be successful in production, processing and marketing.

Empowerment and Access to Decision Making

This challenge is perhaps the most difficult and long-term. Externally imposed programmes, globalization of trade, widening up of agriculture, penetration of multinational corporations are bound to create a negative impact on the lives of women in a developing country like India. Thus, there is a need for women to be empowered to the point where they can exert influence and participate in decision making on issues that affect their lives. In fact, gender equity is impossible without empowerment of women. In the agricultural sector, gender-awareness in decision making is needed at a number of levels in the field, in government departments like agriculture, animal husbandry, fishery, forestry and at the level at which macro-economic and other macro-level decisions are made.

At the micro-level changes can be brought through access to credit and other facilities. As a short-term process, more material resources can be made available for land, credit and technology.

At the macro-level, however, changes are coming slowly and will depend on a more favourable gender balance at all level of power structure, from ministry to researchers and field extension agencies. This will, in long term, require an essential change in the education and training women and the girl child, and the creation of opportunities to promote the advancement of women so that they can ensure that gender-equitable policies can be put into practice. The starting point may be set at preventing school drop-out rate of girl child, encouragement to the girls to go for higher technical education, promotion of agriculture as a career for women, so on and so forth.

Conclusion

The biggest challenge for women empowerment and gender mainstreaming is to structure a response to the fundamental changes taking place at global, national and local levels and to ensure that gender concerns are not lost in the flux of changing priorities. Any endeavour to meet this challenge will require careful examination of previous efforts to address gender concerns and learn from past mistakes. Lessons learnt from previous interventions can be summarized as follows:

Separate, small, women-specific agricultural and rural development programmes are not usually successful as financial support allocations are limited, women-specific agricultural projects are poorly designed and often staffed with persons less skilled in agriculture. However, demonstration of feasibility of particular types of programme targeting women followed by required skill training may enable them to effectively participate in mainstream agricultural programmes.

The most appropriate strategy to reach and assist greater numbers of rural women would be to integrate them in mainstream agriculture programmes with the objective of reaching both women and men with all types of agricultural services and resources. The integration of women in agricultural programmes may be achieved by specially including women as target in all major agricultural components like credit, technological skills and other training, delivery of extension and inputs, access to expanding markets, agricultural researches and education and price support of agricultural products.

The changes need to make in the existing policies, programmes and projects for close monitoring and evaluation of gender sensitive perspective. Gender management system may be developed for this purpose with an emphasis on holistic approach as well as forging linkage between all stakeholders, GOs and NGOs.

In order to make such an approach effective, a women's Desk or Gender Focal Point may be established at the Ministry Level staffed with Senior women with professional training and experience in agriculture and a sound knowledge in agricultural research and technical areas. The integration of women into mainstream programmes may appear to reduce the necessity of emphasizing women's special needs.

Hence, there is a need to build into mainstream programmes safeguard for women to ensure their involvement. This will lead to proper women empowerment and gender mainstreaming for the overall development of the farming community without being antagonists to each other *i.e.* men and women.

References

Food and Agriculture Organization, 1999. *Rome Declaration on World Food Security and World Food Summit Plan of Action*. FAO, Rome.

Gender Mainstreaming in Agriculture and Rural Development: A Reference Manual for Governments and other Stakeholders, 2006. Commonwealth Secretariat, London.

Mulvani, Patrick, 1999. *Agricultural Diversity and Food Security, in Paths to Prosperity: Science and Technology in the Commonwealth 1999/2000*. Commonwealth Secretariat, London.

Razavi, S. and Miller, C., 1997. *Conceptual Framework for Gender Analysis within the Development Context*. UNDP Gender in Development Programme, New York.

Chapter 54

Time Utilization of Women Labour in Dairy Activities

☆ ***Sonika Gupta[1], J.P. Singh[2] and R.P. Singh[3]***

Introduction

In India, livestock is an important segment of the agricultural economy and contributes substantially to food security, employment and poverty alleviation. Livestock ownership is more equally distributed.compared with the land and thus helps to improve income distribution, reduce poverty, and increase consumption of animal products, reducing the protein energy malnutrition among the young children and pregnant mothers. Women constitute 70 per cent of labour force in the livestock production looking after the management of livestock, starting from collection and chopping of fodder, feeding, milking, preparation of milk products,,cleaning cattle sheds, collection of cow dung for manure pits, for making fuel cake and their storage (Taneja, 2004).

Despite the fact that women in India do most of the work in animal husbandry, yet their contribution has largely been ignored and inadequately acknowledged. As a result our agricultural extension system is mostly men oriented and hardly' any

---

1 Research Scholar, Department of Agricultural Extension, R.B.S. College, Bichpuri, Agra.

2 Head, Department of Agricultural Extension, R.B.S. College, Bichpuri, Agra.

3 Associate Professor, Department of Agricultural Communication and Rural Development, SVBP University of Agriculture and Technology, Modipuram, Meerut.

thought is given to the transfer of new technologies to the farm women. Thus, the women get second hand information through a male partner. Further the production potential of women can only be increased if they get the necessary technical know how and support of all types. In view of all these points, it was considered relevant to quantify the women's work in the field of dairy farming (Kaur and Singla, 2001)

Methodology

The study was conducted in randomly selected ten villages, five from each panchayat samiti namely, Sewar and Kumher of Bharatpur district of Rajasthan state. From each selected village, thirty respondents under three categories namely landless, marginal and small farm families were selected randomly on the basis of their proportion in the village. Thus, total number of respondents selected for the study was three hundred farm women. The data were collected through personal interview with the help of interview schedule developed for the present study. The data thus collected were analysed in light of the objectives of the study.

Results and Discussion

Among the total respondents 44.67 per cent belonged to the middle age group of 31 to 45 years, majority of the respondents (99.67 per cent) were married, more than half of the landless respondents (52.78 per cent) were from scheduled caste and scheduled tribe while in the small and marginal farm families, most of the respondents 47.06 per cent and 44.44 per cent, respectively belonged to the other backward classes. Majority of the small (78.43 per cent) and marginal (76.98 per cent) farm families had farming as their main occupation, whereas, majority of the landless families (63.89 per cent) earned their major source of income by working as labourer, the percentage of the nuclear families (59.67 per cent) is more than the joint families (40.33 per cent) among the total respondents. Majority of the respondents (71.00 per cent) belonged to the medium family size (5 to 8 members). Majority of the respondents (66.00 per cent) were illiterate. Most of total respondent families (79.33 per cent) had annual income upto 0.5 lakh. Majority of the respondents had poor social participation with 68.33 per cent of farm families having no membership in any organization. The maximum of the respondents (51.33 per cent) rarely visited urban areas. It was found that nearly half of the small (49.02 per cent) and marginal (49.21 per cent) respondents had medium extension contacts and majority (84.72 per cent) of landless respondents had low extension contacts. Less than 50 per cent respondents of small (40.20 per cent) and marginal (46.83 per cent) and majority, (65.27 per cent) of landless had low mass media exposure.

Majority of the respondents had *kuchha* cattleshed (73.33 per cent). Majority of the respondnets in small (721.55 per cent), marginal (80.16 per cent) and landless (87.50 per cent) possessed 1-2 animals.

Time Utilization of Women in Dairy Activities

The average time utilization of women respondnets in dairy activities has been presented in Table 54.1.

Table 54.1: Average Time Utilization of Women Respondents in Dairy Activities (Man days/year)

Activities	*Small N-102*	*Marginal N-126*	*Landless N-72*	*Total N-300*
Milking of animals	9.6	8.1	6.7	8.3
Grazing	4.7	9.1	11.2	8.1
Bringing grass/fodder from field	21.7	18.4	17.1	19.2
Chaff cutting, preparation of concentrate mineral mixture	15.6	13.9	11.6	13.9
Feeding and watering animals	13.8	12.7	12.1	12.9
Cleaning animals/cattleshed and utensils	11.8	9.7	8.9	10.2
Making cow dung cakes	15.0	13.8	12.1	13.8
Preparation of milk products	17.1	15.5	11.9	15.2
Total	109.3	101.2	91.6	101.7

The figures in the parentheses indicate percentage to the total respondents in each category Average time utilization of respondents in dairy activities was found to be the highest in bringing grass/fodder from the field (19.2 MD/yr), followed by preparation of milk products (15.2 MD/yr). Next in the order are feed preparation (13.9 MD/yr), making dung cakes (13.8 MD/yr), feeding and watering animals (12.9 MD/yr), cleaning cattle shed (10.2 MD/yr), milking (8.3 MD/yr) and grazing (8.1 MD/yr). The total average time utilization was 101.7 MD/yr for all the respondents.

It was found that the average time utilization of small farm family respondent was higher than the other two categories, marginal and landless, in most of the activities *viz.* bringing grass/fodder from field (21.7 MD/yr), preparation of milk products (17.1 MD/yr), feed preparation (15.6 MD/yr), making dung cakes (15.0 MD/yr), feeding and watering animals (13.8 MD/yr), cleaning of cattle shed (11.8 MD/yr) and milking (9.6 MD/yr).

However, in grazing the average time utilization of respondents from small farm families is 4.7 MD/yr, which is lower than the landless (11.2 MD/yr) and marginal (8.1 MD/yr) farm family respondents. The average time utilization of respondents from marginal and landless category was 18.4 MD/yr and 17.1 MD/yr in bringing grass fodder from field, 15.5 MD/yr and 11.9 MD/yr in preparation of milk products, 13.8 MD/yr and 12.1 MD/yr in making dung cakes,

12.7 MD/yr and 12.1 MD/yr in feeding and watering animals, 13.9 MD/yr and 11.6 MD/yr in feed preparation, 9.7 MD/yr and 8.9 MD/yr in cleaning cattle shed, and 8.1 MD/yr and 6.7 MD/yr in milking, respectively.

The total average time utilization in all the selected dairy activities is 109.3 man days, 101.2 man days, and 91.6 man days per year in small, marginal and landless category, respectively. Similar findings have been reported by Joshi (1999) and Sasidhar *et al.* (2001).

Time utilization of women in dairy is quite high (101.7 man days per year). Better management and feeding of farm animals and processing and preparation of milk products form an important part of their training.

References

Joshi, M., 1999. An analytical study on participation of women in farming systems of semi-arid eastern plains of Rajasthan. *Ph.D. Thesis*, Department of Home Science Extension Education, Rajasthan Agricultural University, Bikaner.

Kaur, M. and Singla, S., 2001. Participation of rural women in dairy management practices. *J. Res. Punjab Agric. Univ.*, 38(3-4) : 294–300.

Sasidhar, P.V.K., Rao, B.S. and Reddy, M.S., 2001. Time utilization, suitability of month, duration and place of training programme for farm women. *J. of Res. ANGRAU*, 29(2–3): 82–86.

Taneja, V.K., 2004. Livestock biodiversity and food security. *Indian Farming*, 54(8): 9–11.

Chapter 55

Nutrition Garden: An Alternate Livelihood for Farm Women

☆ ***Kuldeep Singh and Pankaj Kumar[1]***

Introduction

Now days due to economic security, monoculture is prevalent in Punjab. Majority of the area is under paddy-wheat rotation. This crop rotation has directly or indirectly caused a lot of ecological, economical and health problems. Health problems arisen due to irrational consumption of food items. It is recommended by the Indian Council of Medical Research that a healthy person should consume 300g vegetables, 85g pulses and 50g fruits per day. However, in the rural areas there is average consumption of 180g vegetables, 40g pulses and negligible quantity of fruits per person per day. (Romana and Singh, 2007). A study cited by Pandher (2006) revealed that half of the number of children below the age of four year was malnourished and 60 percent women were anemic. This malnutrition can be due to less availability and costliness of pulses and vegetables. Majority of the farmers failed to fulfill their family requirements of pulses and vegetables from their own farm and they have to purchase these from the market. Due to high prices they compromise most of the times with daily consumption. Keeping all these points in view, Punjab Agricultural University, Ludhiana has developed a nutrition garden model. In this nutrition garden, 3 kanal (1500 m^2) land is required. Out of this area, one kanal area is utilized for vegetable production and 2 kanal for pulses. The fruit plants are grown around the boundary of the nutrition garden. This model fulfills the dietary requirements of 8 member family. There are 874000 operational land holdings in Punjab having more than 1 ha

---

1 KVK, Nurmahal, Jalandhar.

land (Anonymous, 2008) If all these families adopt this nutrition garden than 327750 acre land can be diversified. So to check the feasibility and wide spread of this model a project was formulated by the Krishi Vigyan Kendra, Nurmahal (Jalandhar) with the following objectives:

1. To educate the farmers to ensure nutritional security of their families through production of food ingredients at their own farms.
2. To demonstrate the profitability and viability of integrated nutritional garden.

Methodology

To implement the project, four villages were selected. From the villages a total of 50 farm women were selected who can spare 3 kanal area for nutritional garden. These selected women were provided free seed of different vegetables for one kanal area and pulses seed for 2 kanal area both in Rabi and Kharif season along with technical know-how. This one kanal area under vegetables was partitioned into 20 plots of one marla (25 m^2). Sowing of kharif vegetables was done in May-June, 2007. As these vegetables were sown for home consumption, so no pesticide spray or minimum spray was advised. In rest 2 kanal area moong and mash pulses were sown on one kanal each. All the practices recommended by PAU, Ludhiana were followed for growing vegetables and pulses. Similarly sowing of the *Rabi* vegetables and pulses (gram and lentil) was done in Oct-Nov. Yield data of the different vegetables and pulses were recorded regularly from all the selected farmwomen.

Results and Discussion

Yield of Vegetables

As shown in the Table 55.1 average yield of *kharif* vegetables was 45 kg/plot in summer squash, 65 kg/plot in bottle gourd, 22 kg/plot in bitter gourd, 20 kg/plot in round melon, 30 kg/plot in sponge gourd and there on. Similarly in Rabi vegetables the yield was recorded as 75 kg/plot in radish, 75 kg/plot in carrot, 60 kg/plot in turnip, 75 kg/plot in spinach etc. Radish and brinjal was grown in both of the season, but yield in *Rabi* season *i.e.* 75 kg/plot and 100 kg/plot respectively was higher than *kharif* season (65 kg/plot in both)

Yield of all mentioned vegetables was less than its potential yield because of the reason that these vegetables were grown for domestic purpose with no or limited use of chemical sprays to control different pests.

Yield of Pulses

Yield of the different pulses is depicted in the Table 55.2. In case of *kharif* pulses *i.e.* moong and mash, the yield was 40 kg/kanal and 42 kg/kanal respectively. In case of *Rabi* pulses *i.e.* gram and lentil the yield was 40.5 kg/kanal and 20 kg/kanal respectively. Again yield was far less than its potential yield in both *rabi* and *kharif* season and again the reason may be assigned to use of very less pest control measures.

Table 55.1: Average Yield of the *Rabi* and *Kharif* Vegetables

Sl.No.	*Kharif Vegetables*	*Average Yield kg/25 m^2*	*Rabi Vegetables*	*Average Yield kg/25 m^2*
1.	Summer squash	45	Raddish	150 (50 m^2)
2.	Bottle gourd	65	Carrot	150 (50 m^2)
3.	Bitter gourd	22	Turnip	60
4.	Round melon	20	Spinach	150 (50 m^2)
5.	Sponge gourd	30	Tomato	75
6.	Cucumber	70 (50 m^2)	Potato	67
7.	Long melon	40	Peas	15
8.	Brinjal	130 (50 m^2)	Cauliflower	60
9.	Okra	50(50 m^2)	Cabbage	60
10	Cowpea	25	Onion	75
11.	French bean	20	Garlic	50 (50 m^2)
12.	Raddish	130 (50 m^2)	Brinjal	100
13.	Pumpkin	70	Methi	10
14.	Wanga	20	-Methe	11
15.	Chillies	70 (50 m^2)	Coriander	11.5
16.	-	-	Chinese cabbage	32.5

Table 55.2: Average Yield of Pulses

Sl.No.	*Pulse*	*Av Yield kg/500sq. m*
	***Kharif* pulses**	
1.	Moong	55 kg
2.	Mash	51 kg
	***Rabi* pulses**	
1.	Gram	76.5 kg
2.	Lentil	51 kg

Economics of the Nutrition Garden

Economics of the nutrition garden was calculated by considering the prevailing prices of vegetables and pulses in the market and average yield. Details of the economics are given in Table 55.3. As it is clear from the table that total income incurred from 3 kanal (1500 m^2) nutrition garden was Rs 28668. We did not consider the cost of inputs as seed was provided free of cost by the Krishi Vigyan Kendra. Further spare time of the farm women was utilized for the management of nutrition garden, so we did not include the labour cost in the economics calculation.

Table 55.3: Economics of the Kitchen Garden

Sl No.	*Vegetables*	*Av. Yield kg/marla*	*Rate (Rs/kg)*	*Turn Over (Rs)*
		***Kharif* Vegetables**		
1.	Summer squash	45	15	675
2.	Bottle gourd	65	12	780
3.	Bitter gourd	22	20	440
4.	Round melon	20	18	360
5.	Sponge gourd	30	18	540
6.	Cucumber	70 (2 marla)	10	700
7.	Long melon	40	12	480
8.	Brinjal	130 (2 marla)	12	1560
9.	Okra	50 (2 marla)	20	1000
10	Cowpea	25	25	625
11.	French bean	20	20	400
12.	Raddish	130 (2 marla)	20	2600
13.	Pumpkin	70	15	1050
14.	Wanga	20	10	200
15.	Chillies	70 (2 marla)	10	700
			Total	**12110**
		***Rabi* Vegetables**		
1.	Raddish	150 (2 marla)	4	600
2.	Carrot	150 (2marla)	5	750
3.	Turnip	60	5	300
4.	Spinach	150 (2 marla)	10	1500
5.	Tomato	75	10	750
6.	Potato	67	5	335
7.	Peas	15	15	225
8.	Cauliflower	60	5	300
9.	Cabbage	60	4	240
10	Onion	75	15	1125
11.	Garlic	50 (2 marla)	25	1250
12.	Brinjal	100	4	400
13.	Methi	10	2	20
14.	Methe	11	2	22
15.	Coriander	11.5	6	69
16.	Chinese cabbage	32.5	3	97
			Total	**7983**

Contd...

Table 55.3–Contd...

Sl No.	*Pulse*	*Av. Yield kg/marla*	*Rate (Rs/kg)*	*Turn Over (Rs)*
		***Kharif* pulses**		
1.	Moong	55 kg (20 marla)	40	2200
2.	Mash	51 kg (20 marla)	40	2040
	Total	4240		
		***Rabi* pulses**		
1.	Gram	76.5 kg (20 marla)	30	2295
2.	Lentil	51 kg (20 marla)	40	2040
			Total	**4335**
			G. Total	**28668**

As the income from nutrition garden is Rs 28668 we can say this much amount can be saved per year from spending on vegetables and pulses for a family of 8 members. So in this context we can say that nutrition garden is economically viable.

Increase in Consumption of Vegetables and Pulses

All the 50 respondents were asked about their consumption pattern of vegetables and pulses before and after the adoption of nutrition garden. As given in Table 55.4 there was a significant increase in the consumption of vegetables and pulses. This increase was 3–5 g/person/day in case of different pulses and 15-35 g/person/day in case of different vegetables. Although there is more scope to achieve the recommended consumption but still the results are very promising. We can attribute all this achievement through the nutrition garden concept.

Table 55.4: Consumption of Vegetables and Pulses on Daily Basis

Sl.No.	*Crop*	*Earlier Consumption (g/person/day)*	*Present Consumption (g/person/day)*	*Increase (g/person/day)*
		Pulses		
1.	Moong	10	15	5
2.	Mash	10	15	5
3.	Gram	15	18	3
4.	Lentil	5	9	4
		Vegetables		
1.	Green	45	80	35
2.	Leafy	45	75	30
3.	Root	60	90	30
4.	Others	30	45	15

Summary

With the growing pressure of population and limited scope of increasing the income through crop production the diversification in agriculture is being considered necessary and imperative (Rifat *et al.,* 2002). Nutrition garden concept which is developed by PAU, Ludhiana which can meet the vegetable and pulse requirement of 8 member family can help in diversification also. If all the 874000 operational land holdings (having more than 1 ha land) adopt this model then 327750 acre land can be diversified. Farmwomen can get an average income to the tune of Rs 28668 by using their family labour only. In this way problem of unemployment/under employment among farmwomen can also be checked. As the easy availability and accessibility to vegetables and pulses increased the consumption is also increased. The consumption of vegetables and pulses which was 180 g/person/day and 40 g/person/day previously, had changed to 290 g/person/day and 57 g/person/day respectively after the adoption of nutrition garden. Overall we can say that nutrition garden is good for nutritional security of the farmers, profitable and viable. Hence, it can act as a alternate livelihood option for farm women.

References

Anonymous, 2008. *Punjab Agricultural Handbook*. Punjab Agricultural University, Ludhiana, India, pp. 7.

Pandher, Sarabjit, 2006. Kitchen gardens for every rural house hold recommended. Online edition of the *The Hindu*, February 2.

Romana, B.S. and Singh, Kuldeep, 2007. *Ghar dian loran poorian karan layee gherelu bagichi da ikk model.* Agricultural Technology Management Agency, Sangrur, India.

Rifat, Firdos, Arneja, C.S. and Khurana, G.S., 2002. Determinants of entrepreneurship among members of Punjab Kisan Club. In: Ranjit Singh, *Dairy Cooperatives in Punjab: Problems and Prospects*. Punjab Agricultural University, Ludhiana, pp. 7–28.

Chapter 56

Utilization Pattern of Leisure Time Activities of Rural Women

☆ *Anupam Kumari[1] and D.K. Sujan[2]*

Introduction

In Indian situations the leisure time refers such as the middle classes goes to the cinema or increasingly watches television. During school holidays families may visit relatives or go briefly to hill resorts where it is cooler. In rural areas, a slack time in the agricultural cycle allows families to go on pilgrimage or attend weddings, which include much feasting. India has many religious festivals, which provide occasions for even more feasting and conversation, perhaps accompanied by music or a dance or folk theater performance. Time is a highly valued resource and its use is of vital importance to rural women. To survive, every farm family has to manage its resource in such a way as to support its material needs and keep the group operating. The rural women must find ways of getting the household chores in fixed twenty four hours of the day. Thus, the study was planned and findout how much leisure time rural women normally have and how they spent the same.

Objectives

1. To study the leisure time activities of rural women.
2. To study the association of personal and socio-economic variables with leisure time of rural women.

---

1 Research Scholar.

2 Department of Extension Education, Institute of Agricultural Sciences, BHU, Varanasi, U.P.

Methodology

The study was conducted in purposively selected district of Jaunpur. Since the district is consists of 21 CD Blocks, hence for in-depth and comprehensive 10 per cent (two blocks namely Karanjakala and Jalalpur) of the blocks were selected randomly. There selected CD Blocks were Karanjakala and Jalalpur thereafter, 10 per cent villages from two selected CD Blocks were randomly selected. Firstly the lists of farm families of each selected village were prepared. Thereafter 10 per cent families from the list of each selected village were selected randomly. The required information was collected from only female head of the family of the selected family as respondent. Finally, the total number of respondents selected for this study was 282 rural women from all the selected 27 villages.

The data were collected in light of the objectives of study with the help of a structured schedule by conducting personal interview. The leisure time activity score was calculated for each individual on the basis of number of leisure time activities of the individual performance. The co-efficient of correlation test was employed to access the relationship of independent variables with the dependent variables.

Results and Discussion

The Table 56.1 indicates that cent percent respondents (100.00 per cent) were spends their leisure time in noncreative activity like taking rest and the average time spent daily in said activity was highest *i.e.* two hours and one and half hour. Majority of the respondents spent their leisure time in noncreative activities like gossiping (95.39 per cent), attend ceremonies (92.91 per cent) followed by visit friends and neighbours (83.69 per cent) and religious activities (76.59 per cent) respectively. These activities were followed by creative activities like preparation of preservative food (94.33 per cent) and radio listening (83.33 per cent). The other creative leisure time activities in descending order were watching television (79.43 per cent), prepare fancy articles (75.18 per cent), marketing (71.63 per cent), learn and under take sewing (67.20 per cent), decorate the house (63.83 per cent), preparation of special dishes (59.93 per cent), discussion of agricultural problems (50.71 per cent), reading newspaper (25.89 per cent), social gathering (23.40 per cent) and reading books and magazine (18.44 per cent). A sufficient number of respondents *i.e.* 23.05 per cent pass their leisure time in go to movies/cultural programmes and 2.84 per cent in playing cards. Similar types of findings were reported by Mahurkar (1986) and Jondhale, et.al (1992). The leisure time activities mentioned at I^{Ist} IInd IIIrd and IVth rank order *i.e.* resting, gossiping, preparation of preservative food and attend ceremonies were spent most of the leisure time activities of respondents.

Table 56.2 reveals that the time devoted for leisure time activities was found to be positive and significantly associated with age, caste, land possession, family, household material possession, socio-economic status and number of female adults. Thus it may be concluded that these variable had direct influence on time devoted for leisure time activities means the variables increases the leisure time of respondents also increases. Likewise, the household material possession provides the opportunity to complete the work within a short duration as compared to the manually operated work, therefore, the leisure time increases among the rural women and caste was related to performance of leisure time activities due to tendency of higher caste rural women not to labour on farm.

Table 56.1: Distribution of Utilization Pattern of Leisure Hours of the Respondents According to their Leisure Time Activities (N= 282)

Sl. No.	*Activities*	*Frequency of the Respondents*	*Rank*
	Creative activities		
1.	Preparation of preservative food	266(94.33)	III
2.	Radio listening	235(83.33)	VI
3.	Watching television	224(79.43)	VII
4.	Prepare fancy articles like baskets, dolls, fans, etc.	212(75.18)	IX
5.	Marketing (shopping etc)	202(71.63)	X
6.	Learn and undertake sewing etc	189(67.02)	XI
7.	Decorate the house	180(63.83)	XII
8.	Preparation of special dishes	169(59.93)	XIII
9.	Discussion of agricultural problems	143(50.71)	XIV
10.	Reading newspaper	73(25.89)	XV
11.	Attend social gathering- SHGs and other meetings	66(23.40)	XVI
12.	Reading books and magazines	52(18.44)	XVIII
	Noncreative activities		
1.	Resting	282(100.00)	I
2.	Gossiping	269 (95.39)	II
3.	Attend ceremonies-marriages, birth, death etc.	262(92.91)	IV
5.	Visit friends and neighbours	236(83.69)	V
6.	Religious activities	216(76.59)	VIII
7.	Go to movies/cultural programmes	65(23.05)	XVII
8.	Playing cards	8(2.84)	XIX

Figure in parentheses indicate percentage.

The table indicated that, the time devoted for leisure time activities was found positive and significantly associated with education, family occupation, social participation and farm material possession of the respondents and it may be concluded that the leisure time was not affected by their education, family occupation, social participation and farm material possession. The variables which were significant but negative correlation were animal possession. The negative trend clearly indicates that as the animal possession increases the leisure time decreases. The similar finding has also been stated by Jondhale, et.al. (1992). Contrary results were found by Pandey (2001) that education and farm power were found to be significant and positively associated with leisure hours of farmers. Therefore it may be concluded that the rural women has considerable leisure hours which may be utilized as per their interest in productive purposes.

Table 56.2: Association of the Personal and Socio-Economic Characteristics with Leisure Time of the Respondents

Sl.No.	*Independent Variable*	*Leisure Time(r)*
1.	Age	0.355**
2.	Education	0.026
3.	Caste	0.184**
4.	Family Occupation	0.095
5.	Social Participation	0.088
6.	Land Possession	0.121*
7.	Family	0.316**
8.	Farm Material Possession	0.095
9.	Household Material Possession	0.229**
10.	Animal Possession	-0.122**
11.	Socio-economic Status	0.217**
12.	Number of Female Adults	0.346 **

*: Significant at 0.05 probability level; **: Significant at 0.01 probability level.

On the basis of the study, it may be suggested that local leaders and voluntary agencies should engage in the leisure time of rural women in social and development activities. This will help to restrict the tendencies of the rural women for involving noncreative leisure time activities. Secondly developmental agencies may encourage rural women to start subsidiary occupations so as to be engaged in gainful activities.

References

Jondhale, S.G., Khandagale, G.B. and Kamble, R.M., 1992. Creative leisure time activities of farmers. *Maha. Jour. of Ext. Ed.*, 11: 34–36.

Mahurkar, D.S., 1986. Leisure time activities of the students of Marathwada Agricultural University, Parbhani. *Annual Report (Agresco)*, 1986-87.

Pandey, D.K., 2001. Leisure time activities of the farmers: review and critical analysis. *Ph.D. (Ag.) Thesis* (Unpublished). Inst. Agril. Sciences, BHU, Varanasi.

Chapter 57

Participation of Farm Women in Agricultural and Allied Activities

☆ ***Reeta Singh[1], Maya Kumari[2], Shivnath Das[3] and Madan Singh[4]***

Introduction

Concept of 'Bharat Nirman', 'Shining India', 'Feel good', 'Rainbow', 'Socio–economic development', 'Mainstream', 'National building process', Grassroot administration, and 'Good governance' is not possible without participation and empowerment of rural downtrodden women. UNDP report indicates that while 67 per cent of the world's work is done by women, only 10 per cent of the global income is earning by women and a mere 1 per cent of global property is owned by women(Singh and Kumari, 2007).

More than 80 per cent of the economically active women are engaged in agriculture and allied activities. In India, as high as 76 per cent of rural population is engaged in agricultural production and rural women comprise about 50 per cent of

---

1 SMS (Home Science) KVK., Bikramganj, Rohtas, RAU, Pusa , Bihar.

2 SMS (Home Science) KVK, Pakur, Maheshpur, Jharkhand.

3 Junior Scientist-cum-Assistant Professor (Agronomy), IRS., Bikramganj , Rohtas, RAU, Pusa, Bihar.

4 Director Extension Education, RAU, Pusa, Bihar.

agricultural force. There are more than 30 million women working in rural sector, 20 million as agricultural labourers and 10 million employed in animal husbandry, handicraft and related activities (Rath et al, 2007). In family farm labour, women play an important role in all stages of crop raising, starting from land preparation to harvest and post-harvest processing. In view of the tremendous contribution that women are making and can yet make, it is not a matter of little concern, rather time has come when it has been realized that for speedy and overall development of agriculture, their involvement needs to be properly assessed and known. In the production-oriented programmes for rural development, there has been very little or practically no effort to evaluate the role played by the rural women.

Today, 44 per cent of the world's food is produced by women which indicate how important their role is in farming. Rural women in general and farm women in particular are engaged in different activities. Despite their substantial contribution, women continue to be marginalized, undervalued and unrecognized. Recent studies reveal that 50-90 per cent of agricultural operations are carried out by women only (Rath et al, 2007). It can therefore, be forcefully established that their contribution to agriculture and allied activities in rural India is immense. In reality rural farm women are invisible in statistics, they remain unreached and unattended in planned development efforts and are with no access and control on the resources. Rural Indian women are extensively involved in agricultural activities. Keeping in view the role of women and their importance in agricultural and allied activities, present study was conducted with the following objectives to find out the involvement of farm women in agricultural and allied activities according to their a contribution of activity in agriculture and allied sector and involvement of farmwomen in agriculture and allied activities according to their economic status of the families and to study the constraints/problems faced by them, while performing these activities.

Methodology

This study was conducted in four village (Laxamanpur, Shivpur/Varuna, Dhangain and Tenduni) of Bikramganj in Rohtas district of Bihar. Twenty Seven (27) farm women from each village were selected randomly for making the total number of respondents to 108. A pre-tested interview schedule was used to collect the data from selected respondents. The data were collected from women included in the sample in respect of their involvement in different agricultural and allied activities according to their contribution in related fields. The data collection were also made on the basis of involvement of farm women in agriculture and allied activities as per their economic status of the families and various problems faced by them during performing these activities through interviewing. The data were processed and analyzed.

Results and Discussion

Participation of Farm Women in Agricultural and Allied Activities as per their Contribution

Table 57.1 depicts the usual activities performed by farm women under different heads: (A) Crop Production–1. Land preparation 2.Nursery raising 3. Management practices 4. Harvesting and processing 5. Farm mechanization.

Table 57.1: Contribution of Farm Women in Agricultural and Allied Activities

Activity		*f*	*Per cent*	*Rank*
A. Crop production				
1. Land preparation	Removal of weeds/stubbles	93	86.11	II
	Application of manure	80	74.07	VI
	Pit/line making	67	62.04	X
	Land levelling	51	47.22	XIV
	Ploughing/puddling	38	35.19	XIX
2. Nursery raising	Seed collection (crops)	75	69.44	VIII
	Seed collection (forestry)	75	69.44	VIII
	Sowing	61	56.48	XII
	Seed treatment	71	65.74	IX
	Nursery management	34	31.48	XXI
3. Management practices	Weeding	83	76.85	IV
	Sowing	77	71.30	VII
	Transplanting	71	65.74	IX
	Fertilizers application	60	55.55	XII
	Irrigation	56	51.85	XIII
	Top dressing of fertilizers	50	46.30	XV
	Plant protection measures	36	33.33	XX
	Monitoring and supervision	32	29.62	XXII
4. Harvesting and processing	Manual harvesting	98	90.74	I
	Winnowing, drying	89	82.41	III
	Transportation to threshing floor	81	75	V
	Sorting/grading/packaging/bagging	77	71.30	VII
	Storage (for seed/consumption)	68	62.96	X
	Value addition	65	60.18	XI
	Marketing	48	44.44	XVII
5. Farm mechanization	Storage	64	59.26	XII
	Drying	49	45.37	XVI
	Cleaning and grading	47	43.52	XVIII
	Tillage	10	9.26	XXVII
	Sowing	24	22.22	XXV
	Weeding and interculturing	38	35.19	XIX
	Plant protection	15	13.89	XXVI
	Harvesting	30	27.78	XXIII
	Threshing	27	25	XXIV
	Handling and transportation	7	6.48	XXVIII

Contd...

Table 57.1–Contd...

Activity		f	Per cent	Rank
B. Animal Resource Management	Feeding, Watering	102	94.45	I
	Grazing	94	87.04	II
	Milking	90	83.33	III
	Cleaning of shed	87	80.56	IV
	Bathing of cattle	85	78.70	V
	Breeding and castration	32	29.63	XII
	Purchase of ingredients	47	43.52	X
	Selling of milk and milk products	54	50.00	IX
	Preparation of indigenous milk products	40	37.04	XI
	Collection of green fodder	77	71.30	VI
	Bringing to and from the shed	65	60.19	VIII
	Cutting of paddy straw for feeding	71	65.74	VII
	Selling of cattle in market/health care including deworming/vaccination	28	25.93	XIII
C. Household Activities	Home cleaning	96	88.89	II
	Washing of clothes and utensils	88	81.48	III
	Fetching water	81	75.00	IV
	Care of children, sick and old	72	66.67	VI
	Cooking	98	90.74	I
	Kitchen gardening	43	39.81	IX
	Nurturing children	78	72.22	V
	Budgeting	34	31.48	X
	Home decoration	54	50.0	VII
	Preparation of cow dung cakes	50	46.30	VIII

Animal Resource Management and Household Activities

Under the first category of activities *i.e.* land preparation, removal of weeds/ stubbles (86.11 per cent) and application of manures (74.07 per cent) were the major activities performed by farmwomen. They helped in pit/line making (62.04 per cent) and land levelling (47.22 per cent).There was a least contribution of farm women in nursery raising, only 69.44 per cent of them engaged in seed collection. Similarly, seed treatment operations involved 65.74 per cent of the farmwomen and actual sowing operations were done by 56.48 per cent of farmwomen. In management practices, they were found to share 76.85 per cent,71.30 per cent, 65.74 per cent, 55.56 per cent and 51.85 per cent respectively for weeding, sowing, transplanting, fertilizer application and irrigation. This finding is in conformity with the Bhagat (1980), Mahapatra (1987) and Chaudhari and Gonorkar (1992).

Regarding harvesting and post harvesting activities (processing), they were found to play very important role as compared to male counterparts. The major role performed by them was seen in manual harvesting *i.e.* 90.74 per cent whereas, in other operations grouped under this category, farmwomen were also found to supercede the males. In winnowing and drying, 82.41 per cent respondents showed their involvement while in the operation of transportation to threshing floor, their involvement were found 75.0 per cent. The lowest contribution of women in this category were seen in sorting/ grading/packaging/bagging (71.30 per cent),storage for seed/consumption(62.96), value addition, (60.18 per cent), Transportation of produced (55.56 per cent) and Marketing (44.44 per cent). These findings were supported by the results obtained by Bhagat (1980), Singh and Bhattacharya (1990) and Sen (2002) in his study about rural women.

Their least participation were seen in farm mechanization operation. In farm mechanization operation, farm women were found to share 43.52 per cent, 45.37 per cent and 59.26 per cent respectively for cleaning and grading, drying and storage operation. Farmwomen had least participation in farm mechanization operation due to women illiteracy, low productive work, their meager exposure to the outside world and also women have limited access to modern technology *viz.* credit, training and to other facilities. To overcome the least participation of farmwomen in 'farm mechanization operation', they should have gender based technology, training and extension services. This findings are in line with the Sharma (1992), Singh and Sharma (1999) and Ravishankar and Pargunan (2002).

Besides contributing and taking part in main occupation of agriculture, they were also found to be major sharer in allied agricultural activities. In animal husbandry, there were substantial contribution of farmwomen in feeding and watering (94.45 per cent), Grazing (87.04 per cent), Milking (83.33 per cent) and cleaning of shed (80.56 per cent). Farmwomen were found to share 78.70 per cent, 71.30 per cent, 65.74 per cent, 60.19 per cent and 50.0 per cent, respectively for Bathing of cattle, Collection of green fodder, Cutting of paddy straw for feeding, Bringing to and from the shed and selling of milk and milk products. These findings are in conformity with the findings of Bhagat (1980), Chaudhari and Gonorkar (1992) and Sen (2002).

In animal husbandry, women have a multiple role. An estimated 75 million of women as against 5 million men are engaged in the Indian dairy sector. Their activities vary widely ranging from care of animals, grazing, fodder collection, cleaning of animal shed to processing milk and livestock products. Though women play a significant role in livestock management and production, but their control over livestock and its products is negligible(Singh and Kumari, March 2007).

Women are an integral component of the society and play a pivotal role in home management. Table 57.1 reveals that cooking, home cleaning, washing of clothes and utensils, fetching water, nurturing children and care of children, sick and old were performed mainly by farm women *i.e.* 90.74 per cent, 88.89 per cent, 81.48 per cent, 75.00 per cent, 72.22 per cent and 66.67 per cent respectively. They were also found to be major sharer in Home decoration (50.0 per cent), Preparation of cowdung cakes (46.30 per cent) and kitchen gardening (39.81 per cent). This findings are in conformity with the finding of Bhagat (1980), Chaudhari and Gonorkar (1992), Singh and Kumari (2007) and Rath *et al.* (2007).

Participation of Farm Women in Agricultural and Allied Activities as per their Economic Status

Women are as equal as men involved in agriculture related activity, United Nation (2000) reported on the status of women that National averages of female workers in the agricultural labour force vary, but globally women have a principal role in agribusiness, food processing and consumer related activity. Women, who work in the field as producers, help in feeding their families as well. But it goes beyond farming alone, to encompass marketing and value added food processing. There is common belief that women play a marginal role in agriculture. Official statistics recognize that women now make up about 40 percent of the agricultural labour force world wide, and about 67 percent in developing countries (Chaudhari and Gonorkar, 1992).

There is well recognition of the role played by women in agriculture. Women are involved in various activities from land preparation to post harvest operations in crop production, livestock production and management, fish farming etc. in addition to house keeping activities. The women belonging to rural areas are different from those of the urban areas in various socio-economic aspects as well as nature and extent of their contributions to the family income. They are mostly engaged in farm and home related activities.

Generally, there were number of sources in the families belonging to poor well-being groups, viz wage labour, backyard poultry, horticulture, goatery and other small enterprises for food and income generation. But the families of moderate well-being groups besides agriculture have different enterprises like orchard, fishery, poultry and dairy for sustainability in food and income generation in the family. Families of good well-being groups have major enterprises in agriculture, horticulture and dairy for sustenance of their livelihoods.

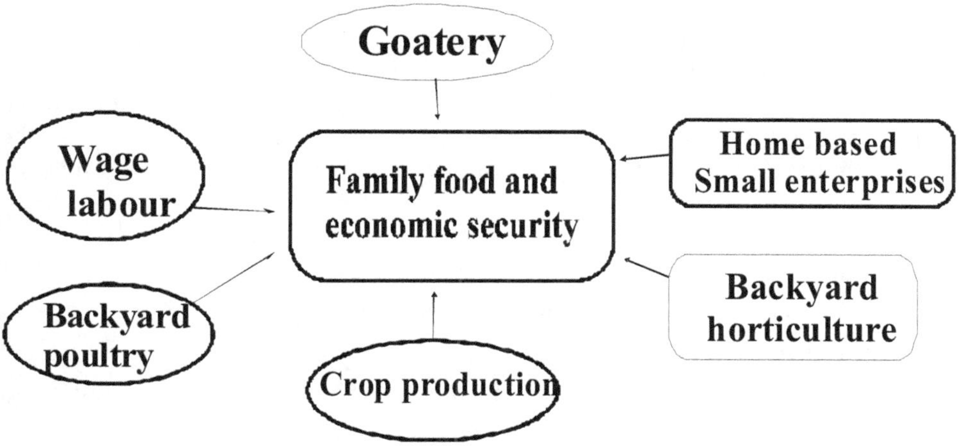

Figure 57.1: Main Sources of Family Food and Economic Security Among Poor Well-Being Families

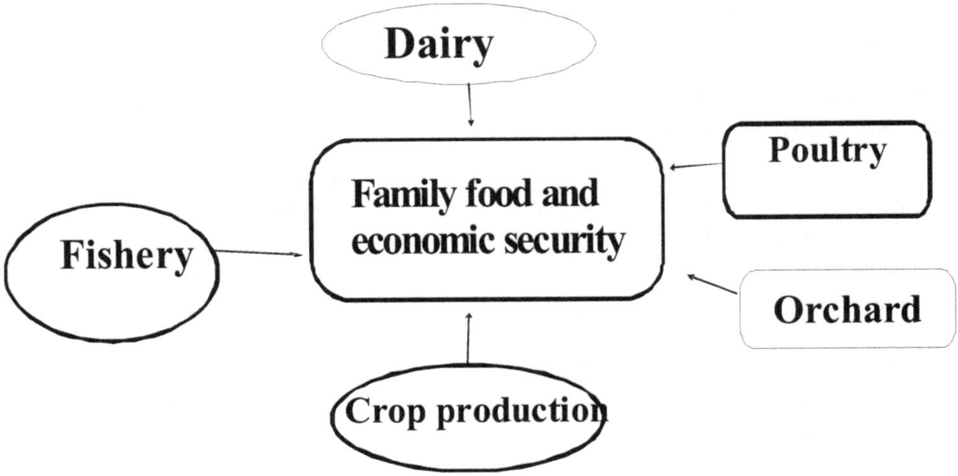

Figure 57.2: Main Sources of Family Food and Economic Security Among Moderate Well-Being Families

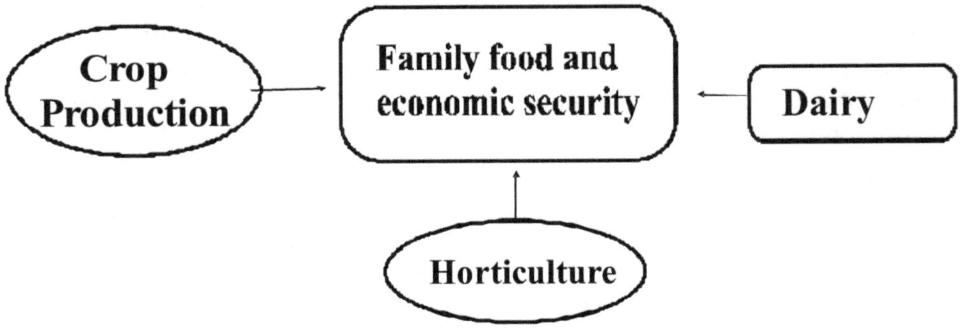

Figure 57.3: Main Sources of Family Food and Economic Security Among Well-Being Families

Involvement of farm women in agriculture and allied activities vary with economic status of the families. Farmwomen of poor economic status participate in all the agricultural activities including animal rearing which is in contradiction to women of good and moderate well-being categories. The salient findings on the major activities done by the women according to well-being categories are as follows:

Poor Well-Being Categories

Crop Production

- Land preparation, Particularly for removal of stubbles.
- Nursery raising
- Crop establishment and management, particularly for transplanting and weeding in paddy and other crops.
- Crop harvesting, winnowing, drying, cleaning of grains, bagging and storage.
- Value addition to agricultural produce.

Animal Resource Management

- ☆ Cleaning of animal shed, feeding,grazing, watering, milking and collection of green fodder.
- ☆ Milking.
- ☆ Preparation of milk products.

Household Activities

- ☆ Preparation of cow dung cakes.
- ☆ Preparation of food items.
- ☆ Care of children and sick family member.
- ☆ Kitchen gardening.
- ☆ Home decoration.

Moderate Well-Being Categories

Crop Production

- ☆ Transplanting and weeding in paddy and other crops.
- ☆ Winnowing, drying, cleaning of grains, bagging and storage.
- ☆ Post–harvest value addition to agricultural produce.

Animal Resource Management

- ☆ Cleaning of animal shed, feeding and collection of green fodder, watering and milking.
- ☆ Preparation milk products.

Household Activities

- ☆ Preparation of cow dung cake.
- ☆ Preparation of food items.
- ☆ Care of children and sick family member.
- ☆ Kitchen gardening.
- ☆ Home decoration.

Good Well-Being Categories

Crop Production

- ☆ Post-harvest value addition to agricultural produce.

Animal Resource Management

- ☆ Feeding and watering.
- ☆ Preparation milk products.

Household Activities

- ☆ House cleaning.
- ☆ Care of children and sick family member.
- ☆ Preparation of food items.
- ☆ Kitchen gardening.
- ☆ Home decoration.

It was observed from the daily activity profile that women of poor well-being group do not find time for leisure, which is an impediment for initiating interventions for entrepreneurship development. In contrast, women of moderate and good well-being groups have time that can be utilized in such activities. However, avenues for family based income for poor farm women can be made open through land based activities such as backyard poultry, duckery and beekeeping.

Major Constraints/Problems for Participation of Women in Agricultural and Allied Activities

Table 57.2: Major Constraints/Problems for Participation of Farm Women in Agricultural and Allied Activities

Sl.No.	*Constraints/Problems*	*f*	*Per cent*	*Rank*
1.	Discrimination in wages, Low wages for women	95	87.96	III
2.	Insufficient loan facilities	82	75.92	V
3.	Input problems			
	(a) Procurement of agri-inputs at higher rates	77	71.30	VI
	(b) Non-availability of inputs at time	74	68.52	VII
4.	Gender based technology, training and extension services	98	90.74	II
5.	Inadequate irrigation facility	65	60.19	VIII
6.	Women have limited access to modern technology *viz.* credit, training and to other facilities	103	95.37	I
7.	Irregular contact with extension agent	85	78.70	IV
8.	Due to women illiteracy their exposure to outside world is less and productivity is less	98	90.74	II
9.	Job security and availability is less due to seasonal nature of agricultural production	50	46.30	XI
10.	Labour problems			
	(a) Unavailability of Labourers	54	50	X
	(b) Lack of skilled labourers	60	55.56	IX

The respondents were also asked about day-to-day problems faced by them in performing agricultural and allied activities. The major constraints for participation of women in agriculture were having limited access to modern technology *viz.* credit, training and to other facilities (95.37 per cent), due to women illiteracy, their meager exposure to outside world and less productivity (90.74 per cent), Gender based technology, training and extension services (90.74 per cent), discrimination in wages, low wages for women (87.96 per cent), and irregular contact with extension agent (78.70 per cent). These findings are in conformity with the result obtained by Sharma (1992), Ravishankar and Pargunan (2002) and Sen (2002).

The financial difficulties regarding inadequacy of loans were recorded by 75.92 per cent.While dealing with this aspect, they said that due to more cost of inputs during main season, the cost of production was high (71.30 per cent). According to them, supply of agricultural inputs became less (68.52 per cent) during the peak period of crop season. Similarly, 60.19 per cent farmwomen has been facing the problem of irrigation. Labour problem is important among the other problems existing in rural areas. In the present study, 50.00 per cent and 55.56 per cent respondents were found facing the problem of unavailability of labours and lacking of skilled labour especially during crop season. Such findings were similar to the reports of Mahapatra (1987), Chaudhari and Gonorkar (1992), Singh and Sharma (1999) and Sen (2002).

Conclusion

Women are an integral part of any society and play a crucial role in home management. Women play a major role in farm operations besides their important responsibility at home. In Indian agriculture, women are the epicentres around which, all our agricultural works rotate. They are producers of food as well as cooks, they look after agriculture as well as their homes. Women's work day is several hours longer than a man's and often, more arduous.The present study was conducted in four villages of Bikramganj of Rohtas district of Bihar. The major contribution of farm women in agriculture were removal of weeds/stubbles, weeding, transplanting, application of fertilizers. In agriculture, harvesting and processing were mainly performed by farm women. In Harvesting and processing, winnowing and drying, sorting/grading/packaging/bagging, storage and value addition were mainly done by farm women. In animal husbandry, there were substantial contribution of farm women in feeding and watering, milking, grazing, cleaning of shed, bathing of cattle, collection of green fodder, cutting of paddy straw for feeding. In household activities, farm women were found to share major role in cooking, home cleaning, washing of cloths and utensils, nurturing children etc. Involvement of farmwomen in agriculture and allied activities vary with economic status of the families. Farm women of poor economic status participate in all the agricultural activities including animal rearing which is in contradiction to women of good and moderate well-being categories. There were number of sources in the families belonging to poor well-being groups, viz wage labour, backyard poultry, horticulture, goatery and other small enterprises for food and income generation. But the families of moderate well-being groups besides agriculture have different enterprises like orchard, fishery, poultry and dairy for sustainability in food and income generation in the family. Families of good well-being groups have major enterprises in agriculture, horticulture and dairy for sustenance of their livelihoods.The major constraints/problems for participation of women in agriculture and allied activities were they have limited access to modern technology *viz.* credit, training, and to other facilities, Due to women illiteracy their exposure to outside world is less and productivity is less, gender based technology, training and extension services etc.

References

Bhagat, R., 1980. Family life of working rural women: A few cases. *IJEE*, 16: 70.

Chaudhari, M.D. and Gonorkar, P.I., 1992. Involving farm women in agricultural activities. *Kurukshetra*, pp. 25–27.

Sharma, Kumud, 1992. Grassroot organization and women's empowerment: Some issues in the contemporary debate. In: *Samya Shakti, Journal of Women Studies*, Vol. 6.

Mahapatra, K., 1987. *India: The Role of Women in Agricultural Extensions*. Draft consultant report prepared for World Bank, New Delhi.

Ravishankar, P. and Pargunan, M., 2002. Women's development policy and gender equality: Myth and reality. *Kurukshetra*, 50(1).

Sen, D., 2002. Mainstreaming women in agriculture: Needs and constraints. In: Pa*per Presented in the Winter School on "Recent Advances in Participatory Extension Methodology in Agricultural and Rural Development*, held at B.C.K.V.V., W.B. during December 2–31, pp. 27–37.

Singh, B. and Sharma, Punam, 1999. Empowerment of rural women: Issues and implications. In: *Paper Presented in the Winter School on "PRA Techniques for Sensitization and Empowerment of Women*, held at IARI, New Delhi during October 5–14, pp. 64–71.

Singh, K.N. and Bhattacharya, H.C., 1990. Crop production technologies for farm women. *Women in Agriculture: Technological Perspective*, pp. 59–60.

Singh, Indrabhusan and Kumari, Usha, 2007. Rural development and women empowerment. *Kurukshetra*, 55(5): 3–8.

Rath, N.C., Das, Lipi, Mishra, S.K. and Lenka, S., 2007. Social and institutional framework for mainstreaming women in agriculture. *Kurukshetra*, 55(5): 21–27.

Chapter 58

Adoption of Improved Apiculture Practices by the Rural Women

☆ *Anuradha Ranjan Kumari[1], Meera Singh[2] and Sudhir Das[3]*

Apiculture is an ideal, absorbing, instructive and economically profitable hobby. It is especially suitable for women, because it does not involve heavy physical work, allows time flexibility, provides gainful employment near to their house and ultimately provides financial security. It solves problem of unemployment if adopted on commercial scale or as cottage industry.

Any number of improvement in our science and technology will not fulfill unless until this will be adopted by our farm women who are the real user of our modern technology or who are the real beneficiaries. Suppose, we do invent any thing for the development of our society and if it is not being transferred upto to the gross root level. Then that invention is meaningless, worthless. In the modern scenario when extent of agriculture is the main source for 65 per cent of our population and 52 per cent of our women who are illiterate, we have to search alternate sources of employment to strengthen our rural economy and at this juncture of global economy when India is standing with WTO sanctions, apiculture has increased as a new venture were our

---

1 Research Associate (Home Science), KVK, Sheohar, R.A.U., Pusa.

2 Dean, College of Home Science, R.A.U., Pusa.

3 Programme Coordinator, KVK, Sheohar, R.A.U., Pusa.

rural mass can get self employment and they can empower themselves economically. Apiculture is new enterprise, which does not require very big investment, nor does it require a complex infrastructure with high-energy investment to start with. In this direction only to transfer our technology a center of excellence has been established by RAU (Apiculture Research Training Center) to disseminate the technology among rural masses of surrounding areas.

Keeping this view in the mind that how much adoption has taken place the study was conducted focussing on these two major objectives.

1. To study the socio-economic status of trained and untrained women
2. To study the adoption rate of improved apiculture practices

Methodology

The study was carried out in Pusa block of Samastipur district selected purposively from different block of the district, seven villages were chosen for this study. A list of rural women practicing apiculture was obtained from Apicultural Research Training Centre (ARTC), Pusa. Thus, the sample for the study comprised a total of 100 respondents. For the measurement of socio-economic characteristics and other aspect of apiculture, a comprehensive schedule was developed, personal interview, direct observation, meeting and discussion were the methods used for data collection.

A regression equation was computed where an estimate of knowledge was made by appropriate values of selected variables from X_1, X_2, X_3 ....X_n.

In General Formula

$Y = a+b_1+b_2x_2+b_3x_3$

Dependent Variable

Y = Extent of adoption

Independent Variables

X_1: Age, X_2: Caste, X_3: Personal education, X_4: Family education, X_5: Occupation, X_6: Family size, X_7: Family type, X_8: Size of land holding, X_9: Annual income, X_{10}: House type, X_{11}: Household material possession,, X_{12}: Social participation, X_{13}: Cosmpoliteness and X_{14}: Economic motivation.

The extent of adoption of improved apiculture practices of apiculturists was studied with reference to five recommended improved apiculture practices.

Findings and Discussion

Socio-Economic Profile of the Beekeepers (Trained and Untrained)

In order to know the background and socio-economic aspects of the respondents, it is important to analyze these characterizations in depth; in all fourteen socio-economic variables were structured using appropriate tools. The relevant data, collected were analysed and the results are presented as follows:

In case of total respondents (*i.e.* trained and untrained), majority (55 per cent) of the respondents practicing apiculture were from middle age group (31-50 years). The maximum proportion of the respondents *i.e.* 53 per cent belongs to backward caste and 34 per cent were educated upto 8th standard where as 9 per cent of them were illiterate. Majority of the respondents (35.00 per cent) were engaged in labourer and belonged to medium sized family (upto 6-10 members). Maximum (74 per cent) of respondents belonged to nuclear family. About half of the respondents do not possess any land in both the groups. Hence they have opted for beekeeping occupation for their livelihood and (40.00 per cent) of the respondents had got pucca house. Nearly two third (66.0 per cent) of the respondents had available traditional materials like radio, chair, cycle and watch. Majority (91.00 per cent) of the respondents were having no membership of any organization. Majority (45.00) of the respondents had low exposure of mass media and 44 per cent of respondents had under the medium economic motivation category.

Effect of Apiculture Training on SES of Rural Women

Table 58.1 indicated that the mean value of trained women in relation to age, caste, personal education, family education, occupation, family size, family type, size of land holding, annual income, type of house, household material porsession, social participation, cosmopoliteness and economic motivation were 26.68, 2.14, 1.92, 1.66, 2.52, 1.52, 1.38, 1.96, 2.04, 2.34, 2.18, 1.12, 12.18 and 5.4 respectively.

Table 58.1: Comparison of Socio-Economic Characterizes of Trained and Untrained Rural Women

Sl.No.	*Socio-economic Indicators*	*Mean Value of Independent Variables*		*'t' value*
		Trained	*Untrained*	
1.	Age	26.68	26.78	–7.2371*
2.	Caste	2.14	1.88	2.2647*
3.	Personal Education	1.92	1.56	1.8833
4.	Family Education	1.66	1.88	2.8300*
5.	Occupation	2.52	2.02	2.5198*
6.	Size of the Family	1.52	1.32	1.8708
7.	Type of the Family	1.38	1.22	0.9417
8.	Size of land holding	1.96	1.64	1.4607
9.	Annual income	2.04	1.68	2.1685*
10.	Type of house	2.34	2.02	2.2655*
11.	Household material possession	2.18	1.86	2.9473*
12.	Social participation	1.12	1.08	0.6286
13.	Compositeness	12.18	4.84	10.9851**
14.	Economic Motivation	5.4	5.60	0.3960

In case of untrained women the mean value of independent variables were 26.78, 1.88, 1.56, 1.18, 2.02, 1.32, 1.22, 1.64, 1.68, 2.02, 1.86, 1.08, 4.84 and 5.6 respectively.

The Table 58.1 further indicated that the computed t-value 10.9851 is highly significant at 1 percent level of significance in respect of cosmopoliteness. Thus it could be concluded that both the trained and untrained beekeeper differed significantly for the cosmopoliteness. This also indicated that training/programme on beekeeping might have, some how on other played crucial role in developing cosmopoliteness in trained beekeepers. The variables *i.e.* caste, family education, occupation, annual income, type of house, household material possession was significant at 5 per cent. Thus, it could be inferred that the trained and untrained women differ significantly to each other.

Correlates of Adoption

With the assumption that adoption is influenced by various socio-economic characteristics of the respondents, the relationship of various socio-economic variables with adoption of apiculture practices among trained and untrained respondents, interms of coefficient of correlation between adoption and 14 selected independent variables were computed and compared. The results are presented in Table 58.2.

Table 58.2: Correlates of Adoption of Improved Apiculture Practices

Sl.No.	*Variable*	*Correlation Value*	
		Trained	*Untrained*
1.	Age	-0.3063**	-0.3299*
2.	Caste	0.6686**	0.2849*
3.	Personal Education	0.4087**	0.3408*
4.	Family Education	0.4987**	0.3209*
5.	Occupation	0.0287	0.0367
2.	Size of the Family	0.0484	-0.0860
3.	Type of the Family	0.2399	-0.0223
4.	Size of land holding	-0.0393	0.0991
5.	Annual income	0.3713**	0.3493*
6.	Type of house	0.1794	0.0714
7.	Household material possession	-0.0363	0.1024
8.	Social participation	-0.1446	-0.0796
9.	Compositeness	0.5729**	0.3340*
10.	Economic Motivation	0.3745**	0.3199*

*: Significant at 1 per cent level of probability; **: Significant at 5 per cent level of probability.

It is observed from Table 58.2 that caste, education, family education, annual income, cosmopolitenes and economic motivation of trained as well as untrained women were positive and significant relationship with the adoption of improved apiculture practices of 1 and 5 per cent level of significance. However, age was negatively but significantly correlated with level of adoption of trained and untrained women. The remaining variables were non-significant.

It is clear from the above findings that there was significant correlation between majority of characteristics of respondents and extent of adoption of improved apiculture practices. The variables showing positive and significant relationship need greater attention on the part of extension agency to enhance adoption of apiculture technology in the area.

The Regression Analysis (Multiple Regressions)

The multiple regression analysis was performed to find out the effect and extent of influence of each variable (fourteen independent variables) towards the level of adoption of improved apiculture practices. The results of analysis involving 14 independent variables are presented in Table 58.3.

Table 58.3: Relative Contribution of Independent Variables Towards Level of Adoption (Dependent Variables) of Trained Rural Women

Independent Variables	*Regression Co-efficient*	*Standard Error (SE)*	*Standard Regression Co-efficient (SRC)*	*T-value*	*R^2*	*f-Value*
Age (X_1)	0.0699	0.08882	0.09157	–0.79		
Caste (X_2)	2.0172	0.78570	0.39357	2.57*		
Personal education (X_3)	–0.6051	0.57313	–0.16078	–1.06		
Family education status (X_4)	1.5093	0.64927	0.30976	2.32*		
Occupation (X_5)	0.1395	0.69412	0.3598	0.26		
Size of the family (X_6)	–0.3582	0.79152	–0.05482	–0.45		
Type of the family (X_7)	1.5691	0.68871	0.24848	2.30*	0.83213	5.63
Size of land holding (X_8)	–0.3232	0.55740	–0.04527	–0.58		
Annual income (X_9)	0.3765	0.78663	0.09471	0.48		
Type of house (X_{10})	0.8661	0.54813	0.16090	1.58		
Household material possession (X_{11})	–2.1003	1.24673	–0.25228	–1.68		
Social participation (X_{12})	0.4476	1.08044	0.04300	0.41		
Cosmpoliteness (X_{13})	0.2913	0.14697	0.28015	1.98		
Economic motivation (X_{14})	0.3958	0.22332	0.21838	1.77		

*: Significant at 5 per cent level of probability; **: Significant at 1 per cent level of probability.

It may be seen from Table 58.3 that out of fourteen selected independent variables used in multiple regression analysis, three contributed positively and significantly in the prediction of the level of adoption of recommended technologies. There were caste, family education and family type. The contributions of remaining eleven variables were found non-significant.

With regard the magnitude of the other in dependent and of dependent variable (level of adoption) partial regression coefficient bi (I = 1, 2, 3, 4..........14) were worked out and tabulated in Table 58.3. The coefficient of determination (R^2 value) was 0.8321, which indicated that 83.21 per cent variation in the adoption gain on apiculture was explained by the fourteen independent variables taken together. The regression coefficient of Y upon caste bi_2–2.017, family education bi_4 = 1.509 and family type bi_7 = 1.569 was significant. If the caste of rural women is improved by one unit, their level of adoption score will be improved at the rate of 2.57. Similarly, if the family education is improved by one unit their level of adoption score will be improved at the rate of 2.32.

Table 58.4 indicated that out of fourteen variables under study none of the variables contribution towards dependent variables *viz.* age, caste, education, family education, occupation, family size, family type, size of land holding, annual income, house type, household material possession, social participation, cosmopoliteness and economic motivation.

Table 58.4: Relative Contribution of Independent Variables Towards Level of Adoption (Dependent Variables) of Untrained Rural Women

Independent Variables	*Regression Co-efficient*	*Standard Error (SE)*	*Standard Regression Co-efficient (SRC)*	*T-value*	*R^2*	*f-Value*
Age (X_1)	–0.0513	0.03248	–0.26193	–1.58		
Caste (X_2)	0.0513	0.37456	0.02703	0.34		
Personal education (X_3)	–0.0258	0.33840	–0.01548	–0.08	0.62027	1.56
Family education status (X_4)	0.4486	0.40575	0.22632	1.11		
Occupation (X_5)	–0.2154	0.28221	–0.16779	–0.76		
Size of the family (X_6)	–0.1461	0.54763	–0.05258	–0.27		
Type of the family (X_7)	0.1089	0.55941	0.03479	0.19		
Size of land holding (X_8)	–0.2104	0.42213	–0.09623	–0.50		
Annual income (X_9)	0.3046	0.39771	0.34199	1.52		
Type of house (X_{10})	0.0237	0.30213	0.01341	0.08		
Household material possession (X_{11})	–0.2256	0.45928	–0.11015	–0.49		
Social participation (X_{12})	–0.5136	0.75170	–0.10750	0.68		
Cosmpoliteness (X_{13})	0.0658	0.09971	0.13676	0.66		
Economic motivation (X_{14})	0.1299	0.08907	0.21565	1.46		

*: Significant at 5 per cent level of probability; **: Significant at 1 per cent level of probability.

Data shown through Figure 58.1 indicated that the two categories of rural women differed significantly from each other. Computed t-value obtained in respect of five components of apiculture technology *viz.*, use of high yield stock from the existing colonies (4.07), artificial management practices and seasonal management (17.53), use of different recommended apiculture equipment (18.57), preventive measures taken against pests and diseases (9.63) and scientific packing of honey (5.44) were significant.

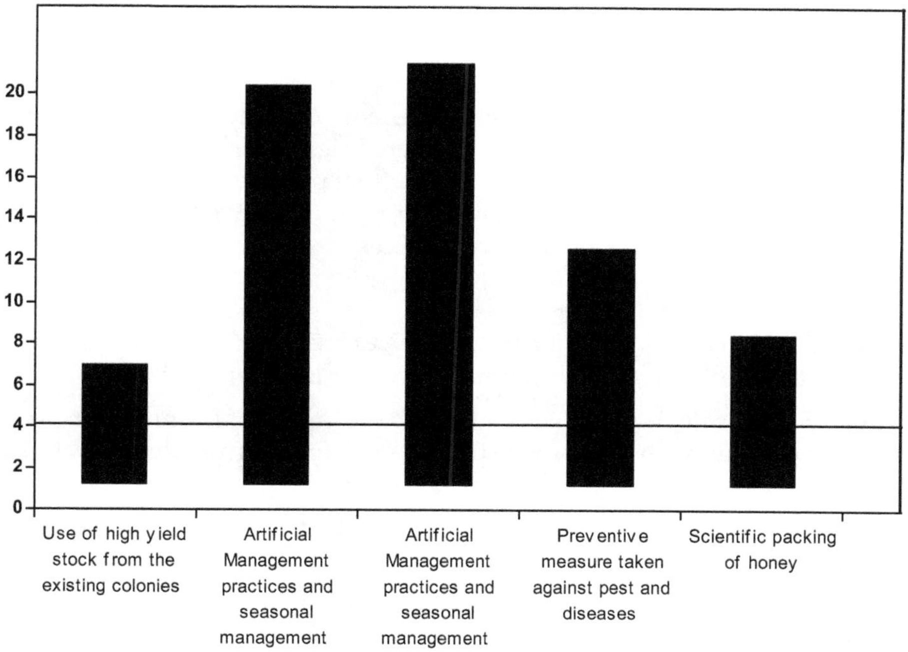

Figure 58.1: Components Wise Distribution of Respondents According to Extent of Adoption

Table 58.5 indicated that majority (88.0 per cent) of trained women had higher adoption level followed by 12.00 per cent of women fell under medium adoption level and not even a single women having low adoption level, whereas in case of untrained women majority (76.00 per cent) had medium adoption followed by 20.0 per cent having low adoption and only 4.0 per cent having high adoption.

Table 58.5: Distribution of Respondents Among Different Categories of Women with Respect to Level of Adoption

Level of Adoption	*Women's Category*					
	Trained (n=50)		*Untrained (n=50)*		*Total (n-100)*	
	Frequency	*Per cent*	*Frequency*	*Per cent*	*Frequency*	*Per cent*
Low	0	0.0	10	20.0	10	10.0
Medium	6	12.0	38	76.0	44	44.0
High	44	88.0	2	4.0	46	46.0

Conclusion

After going through data mentioned in all above table and presented in Figure 58.1, it can be very well concluded that six, out of 14 variables were found significant at 0.01 and 0.05 level of probability of trained and untrained with adoption of improved apiculture practices among selected respondents.

Three variables were found to be contributed towards the beekeepers adoption of improved apiculture practices namely caste, family education and family type. It could be concluded that trained women differed significantly over the untrained women in respect of adoption regarding apiculture practices, the findings that the proportion of trained women was higher than the untrained women appearing high level of adoption of improved apiculture practices.

References

Bhatkar, S.V., Bhinde, P.S. and Bhople, R.S., 1997. Correlates of adoption of sugarcane production technology by the farmer. *Maha. J. of Ext. Edn.*, 16: 18–22.

Jondhale, S.G., Ingale, L.A. and Fatak, U.N., 2000. Impact of krishi vigyan kendra training on adoption of improved practices of summer groundnut. *Maha. J. Ext. Edn.*, 19: 109–111.

Keshev, M.A. and Hossain, B., 1992. Adoption behaviour of sugarcane growers. *Indian J. Ext. Edn.*, 28(1 and 2): 92–96.

Motamed, M.K. and Singh, B., 2003. Correlates of adoption of improved sericulture practices. *Indian J. Ext. Edn.*, 39(1 and 2): 51–57.

Senthanarai, G., Mandharan, M. and Paul, Man Singh, J.T., 1997. Socio-personal and psychological characteristics of farm women. *Journal Ext. Edn.*, 8(5): 1607–1608.

Sinha, Par and Prasad, R., 1971. Sources of information related to adoption process of some improved farm practices. *Indian J. Ext. Edn.*, 2: 1 and 2.

Srinivasa, G., Dolli, S.S., Narendra, M. and Iyengar, M.V.S., 1996. Socio-economic factors and their relation to adoption of improved sericulture practices. *Indian J. of Sericulture*, 35(1): 43–35.

Author Index